Information Processing in Medical Imaging

Computational Imaging and Vision

Volume 3

Information Processing in Medical Imaging

14th International Conference, Ile de Berder, France, June 1995

Edited by

Yves Bizais

Biophysique,
Faculté de Médecine,
Université de Nantes,
Nantes, France
and
Laboratoire d'Automatique de Nantes,
Ecole Centrale de Nantes,
Nantes, France

Christian Barillot

Laboratoire Signaux et Images en Médecine,
Université de Rennes,
Rennes, France

and

Robert Di Paola

Unité INSERM 66,
CHU Pitié-Salpêtrière, Paris,
Paris, France

KLUWER ACADEMIC PUBLISHERS
DORDRECHT / BOSTON / LONDON

A C.I.P. Catalogue record for this book is available from the Library of Congress

ISBN 0-7923-3593-7

Published by Kluwer Academic Publishers,
P.O. Box 17, 3300 AA Dordrecht, The Netherlands.

Kluwer Academic Publishers incorporates
the publishing programmes of
D. Reidel, Martinus Nijhoff, Dr W. Junk and MTP Press.

Sold and distributed in the U.S.A. and Canada
by Kluwer Academic Publishers,
101 Philip Drive, Norwell, MA 02061, U.S.A.

In all other countries, sold and distributed
by Kluwer Academic Publishers Group,
P.O. Box 322, 3300 AH Dordrecht, The Netherlands.

Printed on acid-free paper

TABLE OF CONTENTS

PART 1: LONG PAPERS

Image formation

PART 2: SHORT PAPERS

Image formation, tomographic reconstruction and image quality

Shape description

PREFACE

This book includes all the papers and the posters presented at the XIVth International Conference on Information Processing in Medical Imaging (IPMI 95). Following the highly successful meeting in Flagstaff, Arizona, USA, in 1993, IPMI 95 was held in June 1995, in Ile de Berder, a small island in the Gulf of Morbihan, France, where attendees were "confined" for a week. This turned out to be an efficient means to stimulate discussions and exchange ideas, as well as to enjoy French food and Celtic scenery.

IPMI meetings are unique in several ways. They focus on methodological aspects in Medical Imaging whilst other meetings are more clinically oriented or highly theoretical. In addition, clinical relevance and effective clinical validation are highly regarded IPMI topics. Consequently new approaches to acquiring, processing, analyzing an displaying medical images have often been presented for the first time at IPMI meetings. IPMI 95 was no exception as this book demonstrates. Only plenary sessions are organized during which ample time is provided for presenting (30 min) and discussing papers (30 min in the programme, virtually unlimited in practice). This allows in depth analysis of the 28 presented works corresponding to high quality, journal-like papers in the Proceedings. The negative counterpart is that many excellent submissions could not be accepted. For this reason 30 additional papers were presented as short communications and posters. They are not of inferior quality when compared to the regular papers but they concern more specific topics or need further development and validation. They are included in the Proceedings as short papers, and will certainly appear as long papers in medical imaging journals. Finally, IPMI meetings were created by François Erbsmann and IPMI board members to promote young scientists. Most first authors of IPMI 95 papers are under 30, and are working in leading institutes all over the world. This also explains why IPMI meetings and their Proceedings are so important in the history of medical imaging advances.

IPMI 95 covers most fields of medical imaging: image formation and reconstruction, image quality, image processing and analysis using external knowledge and *a priori* information, segmentation applied to recognition and characterization of anatomical and physiological entities, registration of multimodality images, clinically relevant applications. Several papers present significant advances in topics already discussed at previous meetings while the others concern new topics and methodologies, opening new horizons in medical imaging. Short papers are an efficient way to alert readers to such new approaches, even though additional work is needed to firmly establish them.

The scientific quality of the meeting and of the proceedings have been made possible by the hard work provided by the authors and the scientific committee which had to review more than 100 full-length papers in less than one month. The organizers consider the transfer of knowledge from basic research to medically useful products as extremely important, and would like to thank the companies that participated in and sponsored the meeting. In conclusion, it is hoped that the reader finds these Proceedings both informative and enjoyable.

Sponsoring companies and institutions

INSERM, Région Bretagne, Département du Morbihan, District de Vannes
Université de Nantes, Université de Rennes
General Electric Medical Systems Europe, Philips Medical Systems
Siemens Medical Systems
Control Data, IBM, Sun Microsystems Inc.
Canon France, Dome
CIS-BIO, Mallinkrodt

Scientific committee

DESIGN OF A ROBUST STRATEGY TO MEASURE INTRAVASCULAR ELECTRICAL IMPEDANCE

MAURITS K. KONINGS, WILLEM.P.TH.M. MALI

AND

MAX A. VIERGEVER
Department of Radiology and Nuclear Medicine,
University Hospital Utrecht, Heidelberglaan 100,
3584 CX Utrecht, The Netherlands

Abstract. A new Intravascular Impedance Catheter (IIC) system has been developed allowing a reliable assessment of the longitudinal intravascular electrical impedance distribution in order to discriminate between fatty and non-fatty atheroslerotic lesions. Existing impedance catheter systems cannot be used for this purpose because artefacts owing to deviating impedances from the extravascular surroundings have a major and irretraceable effect on the measurement. Standard algorithms used in attempt to compensate for these artefacts suffer from severe instability problems. We have developed a mathematical model that describes intravascular impedance measurement, as well as an algorithm that reconstructs from the data a two-layered impedance distribution. A number of conditions are derived which have to be satisfied by catheter design, measurement strategy and current field patterns in order to achieve the necessary stability of the algorithm. Following these guidelines, we have designed an IIC system in which two new elements are added to the standard concept: 1st) a differential measurement using three measuring electrodes and a fast switch, and 2nd) an extra current source producing a current field opposite in direction to that of the standard current field, thus creating a resulting current field with properties that differ strongly from the conventional dipole field. An experimental IIC system has been built according to this design. With this catheter, a $30mm^3$ plastic model of an arterial fatty lesion can be detected reliably.

Y. Bizais et al. (eds.), Information Processing in Medical Imaging, 1–12.

1. INTRODUCTION

Intravascular therapy has progressed considerably over the last 25 years, thanks to the development of a number of increasingly sophisticated therapeutic devices. The quality of diagnostic methods has not kept pace with this advance; the determination of localisation, size and nature of atherosclerotic plaques is poor in comparison with the therapeutical possibilities. For example, X-ray angiography only reveals the location of a stenosis, not its material composition. The angioscope does not reveal the structure inside a plaque and needs substantial injection of saline. Intravascular Ultrasound (IU) offers a transaxial view of the vessel wall. With respect to the composition of plaques it only reliably discriminates calcium.

Recent studies [3] show that the presence of a lipid pool in the atherosclerotic vessel wall is a risk factor for acute occlusion of bloodvessels as a result of plaque disruption. Detection and localisation of lipid pools would, therefore, be of relevance for a proper diagnosis; at present, however, there is no reliable in vivo method available to show the presence of these pools. Because fat and calcium have a high specific electrical resistivity compared to other possible plaque compounds [4], the assessment of the electrical impedance would be a suitable indicator for the presence of a lipid pool or calcium cristals in a plaque. Once a plaque of high electrical impedance would be detected, IU can been used to determine whether or not calcium was responsible for the high impedance value. The IU catheter will play a crucial role in the in-vivo application of the Intravascular Impedance Catheter (IIC) system that we have developed.

Up to now, electrical impedance catheters (conductance catheters) have been developed to measure e.g. ventricular volume during the heart-cycle[1]. One of the main premisses to make this technique work is that the catheter remains on a *fixed position* during the measurement, thereby leaving the global geometrical order of all surrounding impedances the same. These catheters can, however, not be used to perform so local a measurement as needed to determine the impedance of plaque material in vivo: The electrical fields generated by these catheters are not at all limited to the volume of a blood vessel, which causes deviating impedances in a wide environment of the catheter to have a major and irretraceable effect on the measurement. Because in the case of an intravascular impedance measurement the catheter *is moving* through the artery, the environment changes continuously. *The purpose of this paper is to design a new impedance catheter system capable of detecting fatty intravascular lesions reliably without being disturbed by deviating impedances from the environment of the blood vessel.*

A description of the inverse problem related to impedance measurements shows that the major problem is stability (section 2). The design of a new catheter is given a solid basis by formalizing the principle of impedance measurements into a matrix model, and formulating an algorithm that reconstructs the intravascular conductivity distribution from the measured data (section 3). From this algorithm, we derive design guidelines to be satisfied by the electrode configuration and the current field pattern of the new impedance catheter in order to achieve a sufficiently robust method of measurement and a reconstruction algorithm of sufficient stability. In section 4 we describe the design of a new impedance catheter system according to these guidelines.

2. MATHEMATICAL BACKGROUND

We consider the general case of an isotropic, Ohmic, conductive environment with given conductivity distribution $\sigma(x)$, in which x is a three-dimensional spatial coordinate. The continuity equation for the current density, which relates the potential distribution $\phi(x)$ to $\sigma(x)$, reads:

$$\nabla \cdot (\sigma(\overline{x}) \nabla \phi(\overline{x})) = 0 \tag{1}$$

From this equation, given a certain conductivity distribution and a set of appropiate boundary conditions, the potential distribution $\phi(x)$ may be calculated. This calculation is called the *forward potential problem*. In the case of the Intravascular Impedance Catheter system however - and, more generally, in Electrical Impedance Tomography (EIT) - one has to deal with the *inverse problem*: given - a limited number of - measured potential differences, calculate the conductivity distribution. Inverse problems are generally ill-posed [2], i.e. they suffer from inconsistency, non-uniqueness, instability, or a combination.

In the case of the IIC system, we have to deal with non-uniqueness and instability.

Non-uniqueness problem

Consider the general case of an impedance catheter lying in a stretched position in a straight artery.

We define cilindrical coordinates (x,r,Θ) around the catheter (fig.1a), with the x axis being the length-axis of the catheter, while r denotes the distance to this axis. The third coordinate, the angle Θ, will be irrelevant in our discussion because electrical measurements from catheters with cilindrical electrodes cannot discriminate between points with equal x and r but distinct Θ. During the catheterization, the catheter is moving from $x_{\mathbf{start}}$ to $x_{\mathbf{end}}$, and measurements of potential differences are performed at N equidistant positions along the trajectory, separated by a distance $\delta x = (x_{\mathbf{end}}-x_{\mathbf{start}})/N$.

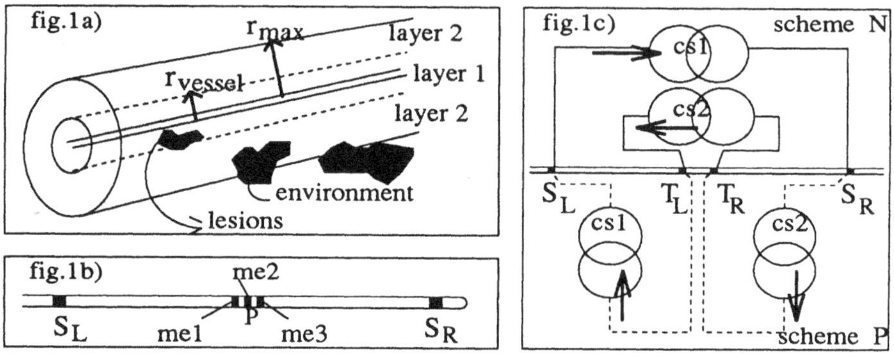

Figure 1. a) Two-layered cylinder-symmetrical conductivity distribution around the catheter. b) Positioning of three measuring electrodes very near to each other in behalf of the differential measurement. me2 coincides with the reference point P. S_L and S_R are driven electrodes creating a dipole current field. c) Nested current connection scheme N (solid lines), and parallel scheme P (dotted lines). cs1 and cs2 are AC current sources, π rad out of phase with each other.

Furthermore, we define two layers (fig. 1a):

First, a layer ('layer 1') in the immediate environment of the catheter which includes the lumen of the artery and the inner part of the vessel wall (intima and media). In this layer local spots of high impedance are to be classified as abnormalities within the blood-vessel, i.e. an atherosclerotic lesion. The boundary of this layer is denotated as r_{vessel}.

Secondly, the layer ('layer 2') for which $r_{vessel} < r < r_{max}$: within this layer local spots of high impedance still have a reasonable effect on the measured $\delta\phi$, but are not to be regarded as an abnormality of the blood-vessel itself (no atherosclerotic lesion exceeds beyond the media). For the detection of fatty intravascular lesions, it is essential to be able to decide whether observed abnormal effects are due to deviating impedances in layer 1 or in layer 2.

We divide both layers into cylindersymmetrical, annular volume elements (voxels) of width δx. The task that we assign to the IIC, is to assess the average impedance within each of the 2N voxels. A requirement for uniqueness is that on each of the N positions two distinct measurements of potential differences be performed, say ϕ^A(x) and ϕ^B(x), using two current field patterns A and B that are not linearly dependent. This solves the non-uniqueness problem completely.

Instability problem

The instability problem is more delicate than the non-uniqueness problem and less easy to deal with. In order to attain the desired stability and thus solve the inverse problem, we have designed a new impedance catheter and a measuring strategy. The design involves the development of a matrix model of the intravascular impedance catheterization problem. On basis of this model, we derive design guidelines that have to be satisfied by the IIC system to be developed. This is the subject of the next section.

3. RECONSTRUCTION ALGORITHM AND STABILITY

Let ϕ^{A}_{meas} be an N-dimensional vector of potential differences, measured with current field pattern A, with elements $\phi^{A}_{meas}(x_{start} + k\delta x)$ (with k running from 0 to N-1); ϕ^{B}_{meas} is defined analogously for current pattern B. Let furthermore σ_1 be the N-dimensional conductivity vector of layer 1 with elements $\sigma_1(x_{start} + k\delta x)$, i.e. $\sigma_1(x_{start}+k\delta x)$ is the average conductivity of the annular region at $x_{start}+k\delta x$ in layer 1; σ_2 is defined analogously for layer 2.

For the above defined two-layered problem of impedance catheterization, we are interested in the detection of boundaries between areas of different impedances within each layer, rather than the absolute values of the impedances. We therefore introduce the two vectors $\mathbf{s}_1$ and $\mathbf{s}_2$ as the set of variables of which the values are to be assessed by the IIC system, these variables being defined as the 'discrete axial derivatives' of the σ_1-values of layer 1 and the σ_2-values of layer 2 respectively:

$$\begin{aligned} s_1(x) &= \sigma_1(x) - \sigma_1(x - \delta x) \\ s_2(x) &= \sigma_2(x) - \sigma_2(x - \delta x) \end{aligned} \tag{2}$$

The unknowns $\mathbf{s}_1$ and $\mathbf{s}_2$ are reconstructed from the measured ϕ^{A}_{meas} and ϕ^{B}_{meas} by means of a Newton-Raphson iteration algorithm. As initialization we set: $s_1^{(0)}(x)\equiv 0$ and $s_2^{(0)}(x)\equiv 0$ for all x.

Let $\phi^{A(n)}(s_1^{(n)},s_2^{(n)})$ and $\phi^{B(n)}(s_1^{(n)},s_2^{(n)})$ be two model vector functions solving the forward potential problem for mode A and mode B respectively, given the vectors $s_1^{(n)}$ and $s_2^{(n)}$ at some point in the iteration process; the superscript (n) denotes the iteration number. At each iteration step, the approximation of $s_1^{(n)}$ and $s_2^{(n)}$ is updated.

An important criterion will be the stability of the algorithm, which depends on certain characteristics of the compound matrix

$$\begin{bmatrix} J_1^{A(n)} \, J_2^{A(n)} \\ J_1^{B(n)} \, J_2^{B(n)} \end{bmatrix} \tag{3}$$

in which $\mathbf{J}_q^{A(n)} = \{ \, \partial(\phi^{A(n)})_i/\partial(s_q^{(n)})_j \, \}$ is the Jacobian matrix of model vector function $\phi^{A(n)}$ in layer q with q∈{1,2}, with $(\phi^{A(n)})_i$ denoting the i-th element of $\phi^{A(n)}$, and $(s_q^{(n)})_j$ the j-th element of $s_q^{(n)}$ (i,j $\in 0, ..., N-1$).

For model vector function $\phi^{B(n)}$, the Jacobians $\mathbf{J}_1^{B(n)}$ and $\mathbf{J}_2^{B(n)}$ are defined analogously. For n>0, the model vector functions $\phi^{A(n)}$ and $\phi^{B(n)}$ depend on $\mathbf{s}_1^{(n)}$ and $\mathbf{s}_2^{(n)}$. Complete stability guidelines would involve all possible values for (n). Derivation of such guidelines would not only have to take account of intrinsic catheter features, but of all possible impedance distributions around the catheter as well. For reasons of simplicity however, we start with design guidelines based on step n=0 only. In section 4, we design the IIC system based on these guidelines for n=0. In section 5 we will show that the resulting IIC has favorable characteristics for the stability for $n > 0$ as well.

Stability guidelines

We derive these stability guidelines in three steps. To start with, consider the four Jacobians $\mathbf{J}_q^{M(0)}$, with q∈ {1,2} and M(0)∈ {A(0),B(0)} denoting the measurement mode. The Jacobians $\mathbf{J}_q^{M(0)}$ are all Toeplitz matrices (because of shift invariance for n=0), i.e. matrices that have constant elements along each diagonal. Hence we can specify $\mathbf{J}_q^{M(0)}$ completely when we can find some function $\psi_q^{M(0)}$ in which the elements of the first row and the first column can be expressed:

First row: $[\ \mathbf{J}_q^{M(0)}\ ]_{1,k} = \psi_q^{M(0)}((k\text{-}1)\delta x)$

First column: $[\ \mathbf{J}_q^{M(0)}\ ]_{k,1} = \psi_q^{M(0)}((1\text{-}k)\delta x)$

The characteristic shape of each of the functions $\psi_1^{A(0)}(\xi)$, $\psi_2^{A(0)}(\xi)$, $\psi_1^{B(0)}(\xi)$ and $\psi_2^{B(0)}(\xi)$, in which ξ is the relative position along the x-axis with respect to a fixed point P on the catheter ($\xi = 0$ at P), follows directly from the physics of the chosen measurement strategies for the modes A and B. Since $\psi_q^{M(0)}$ represents the sensitivity of that specific measurement mode M(0) to some nonzero $s(\xi + x_P)$ in layer q at position $\xi + x_P$, we will refer to the graphs of $\psi_q^{M(0)}$ as 'sensitivity curves'. The design guidelines will be translated into a set of constraints on the possible shapes of these four $\psi(\xi)$ curves.

As second step we write down explicit expressions for $\mathbf{s}_1$ and $\mathbf{s}_2$ (omitting the superscript (0) for brevity):

$$\mathbf{J}_1^{A}\mathbf{s}_1 + \mathbf{J}_2^{A}\mathbf{s}_2 = \phi^{A}_{meas} \quad and \quad \mathbf{J}_1^{B}\mathbf{s}_1 + \mathbf{J}_2^{B}\mathbf{s}_2 = \phi^{B}_{meas} \tag{4}$$

Provided $det(\mathbf{J}_1^A\mathbf{J}_2^B - \mathbf{J}_2^A\mathbf{J}_1^B) \neq 0$ and $[\mathbf{J}_2^B]^{-1}$ and $[\mathbf{J}_1^A]^{-1}$ exist, we have:

$$\begin{aligned} \mathbf{s}_1 &= \left[\mathbf{J}_1^A - \mathbf{J}_2^A[\mathbf{J}_2^B]^{-1}\mathbf{J}_1^B\right]^{-1}\left(\phi^A_{meas} - \mathbf{J}_2^A[\mathbf{J}_2^B]^{-1}\phi^B_{meas}\right) \\ \mathbf{s}_2 &= \left[\mathbf{J}_2^B - \mathbf{J}_1^B[\mathbf{J}_1^A]^{-1}\mathbf{J}_2^A\right]^{-1}\left(\phi^B_{meas} - \mathbf{J}_1^B[\mathbf{J}_1^A]^{-1}\phi^A_{meas}\right) \end{aligned} \tag{5}$$

in which we used that $\phi^A(s_1,s_2)=0$ and $\phi^B(s_1,s_2)=0$. We remark that the superscripts A and B can be interchanged; their nomenclature was chosen arbitrarily.

As third and last step we will now formulate the guidelines for the design of the Intravascular Impedance Catheter, based on the stability of the solutions for $\mathbf{s}_1$ and $\mathbf{s}_2$. In the case of eq.(5), stability is guaranteed when none of the matrices that are inverted is ill-conditioned; so the guideline to optimize the stability of the algorithm is to minimize the condition numbers of the following four matrices:

$$\mathbf{J}_1^A, \mathbf{J}_2^B, \left[\mathbf{J}_1^A - \mathbf{J}_2^A[\mathbf{J}_2^B]^{-1}\mathbf{J}_1^B\right], \; and \left[\mathbf{J}_2^B - \mathbf{J}_1^B[\mathbf{J}_1^A]^{-1}\mathbf{J}_2^A\right]$$

Minimization of the condition numbers of the two matrices $\mathbf{J}_1^A$ and $\mathbf{J}_2^B$ is sufficiently accounted for by replacing this demand with the more specific guideline that $\mathbf{J}_1^A$ and $\mathbf{J}_2^B$ be approximately band-diagonal matrices. In the terminology of the $\psi(\xi)$-functions, approximate band-diagonality of a $\mathbf{J}_q^M$ translates into $\psi_q^M(\xi)$ being almost zero almost everywhere, except for a sharp peak around $\xi=0$.

Minimization of the condition numbers of $[\mathbf{J}_1^A - \mathbf{J}_2^A[\mathbf{J}_2^B]^{-1}\mathbf{J}_1^B]$ and $[\mathbf{J}_2^B - \mathbf{J}_1^B[\mathbf{J}_1^A]^{-1}\mathbf{J}_2^A]$ can both be replaced by the single guideline that expression Z_{AB} is to be minimized:

$$Z_{AB} = \frac{|\lambda_{[J_2^A]}|_{max} |\lambda_{[J_1^B]}|_{max}}{|\lambda_{[J_1^A]}|_{min} |\lambda_{[J_2^B]}|_{min}}$$

in which $|\lambda_{[Q]}|_{max}$ and $|\lambda_{[Q]}|_{min}$ are the maximum and minimum eigenvalue of some matrix Q respectively. In the terminology of $\psi(\xi)$-functions, this boils down to the guideline that there should be one measurement mode A in which the impedance vector $\mathbf{s}_1$ of layer 1 dominates the measurement, and one mode B in which the $\mathbf{s}_2$ of layer 2 is dominant.

In summary, the guidelines for the $\psi_q^M(\xi)$ read:

- Find a measurement mode A featuring $\psi_1^A(\xi)$ and $\psi_2^A(\xi)$, such that $\psi_1^A(\xi)$ is almost zero almost everywhere, except for a sharp peak at $\xi=0$, and $\psi_2^A(\xi)$ is dominated by $\psi_1^A(\xi)$.
- Find a measurement mode B featuring $\psi_1^B(\xi)$ and $\psi_2^B(\xi)$, such that $\psi_2^B(\xi)$ is almost zero almost everywhere, except for a sharp peak at $\xi=0$, and $\psi_1^B(\xi)$ is dominated by $\psi_2^B(\xi)$.

4. THE INTRAVASCULAR IMPEDANCE CATHETER

In order to comply with the above formulated design guidelines, we add two new features to the existing concept (see introduction) of the impedance catheter, viz. the *differential measurement with three measuring electrodes* and the *CounterCurrent field*.

Differential measurement

The first feature, which is referred to as differential measurement, is designed to make the $\psi_1^{A(0)}(\xi)$ and $\psi_2^{B(0)}(\xi)$ sharply peaked at ξ=0 and almost zero elsewhere. To realize this, three measuring electrodes are placed equidistantly and in close vicinity of each other (at a distance μ_{me}), the middle one coinciding with the point of symmetry, P. (fig.1b).

During the catheterization, at each catheter position x (i.e. at each position x of point P, omitting the subscript C for brevity) out of the series of N catheter positions { $x_{start}+\delta x$, $x_{start}+2\delta x$,..., x_{end} }, and for each of the measurement modes A and B (to be specified later on), two potential differences are measured: ϕ_{12} between the first and the second measuring electrode, and ϕ_{23} between the second and the third measuring electrode. The differential measurement procedure consists of the subtraction of these potential differences to obtain the measurement signals:

$$\begin{aligned} \phi^A_{meas}(x) &= \phi^A_{23}(x) - \phi^A_{12}(x) \\ \phi^B_{meas}(x) &= \phi^B_{23}(x) - \phi^B_{12}(x) \end{aligned} \qquad (6)$$

These ϕ_{meas}(x) form the input for the reconstruction algorithm. As a rule, in subtractions like eq.(6), the signal-to-noise ratio of the result of the subtraction is worse than that of the original signals. How this will work out in the specific case of eq.(6) will be discussed in section 5.

In order to illustrate the effect of the differential measurement procedure, we first consider a simple dipole field generated from driving electrodes S_L and S_R, and use only me1 and me3 to perform a conventional impedance measurement, i.e. to measure a potential difference ϕ_{13}^{Dipole} at each position x. Furthermore, we choose $r_{cath}/\mu_{me} = 1.2$, $r_{vessel}/\mu_{me} = 3.5$

The resulting $\psi_1^{Dipole,me13}(\xi)$ (fig.2a) does not meet the guidelines at all, owing to major peaks around the driven electrodes ($\xi=\xi_{SL}$ and $\xi=\xi_{SR}$), which is a natural consequence of the conventional impedance catheter concept.

However, the differential measurement
$\phi^{Dipole,diff}(x) = \phi^{Dipole,me23}(x) - \phi^{Dipole,me12}(x)$ provides a sensitivity curve
$\psi_1^{Dipole,diff}(\xi) = \psi^{Dipole,me23}(\xi) - \psi^{Dipole,me12}(\xi)$, which has the desired properties (see fig.2b).

Because the measuring electrodes are so near each other, the differences between $\psi^{Dipole,me12}(\xi)$ and $\psi^{Dipole,me23}(\xi)$ are small for the regions surrounding the *driven* electrodes, whereas around the *measuring* electrodes

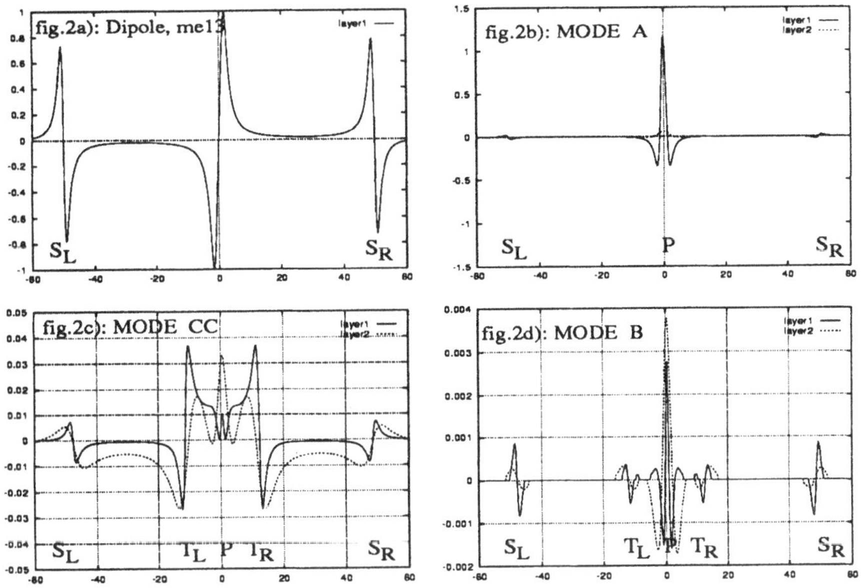

Figure 2. Graphs of $\psi(\xi)$ as function of ξ for various MODES. All $\psi(\xi)$ have been normalised with the same reference constant K, which was chosen such that the maximum value present in graph a equals 1. Solid lines represent $\psi_1^{MODE}(\xi)$, dotted lines represent $\psi_2^{MODE}(\xi)$. a) using only me13 in dipole field, b) using differential measurement in dipole field, c) MODE=CC, d) MODE=B. See text.

the two curves are very distinct. The curves $\psi_1^{\text{Dipole,diff}}(\xi)$ and $\psi_2^{\text{Dipole,diff}}(\xi)$ both meet the design guideline for measurement mode A.

Therefore we define measurement mode A as follows:

$$\phi^A_{meas}(x) = \phi^{Dipole,diff}(x)$$

with $\psi_1^A(\xi) = \psi_1^{Dipole,diff}(\xi)$ and $\psi_2^A(\xi) = \psi_2^{Dipole,diff}(\xi)$

The CounterCurrent Field

The second feature consists of the addition of a second current source (cs2) to the system, whose output wires are connencted to the catheter in such a way that the current is ejected in the opposite direction with respect to the first current source (cs1). See fig.1c. Since impedance measurements in biological materials use AC current sources to avoid artefacts due to polarization effects, the AC currents produced by cs1 and cs2 are π rad out of phase with respect to each other. We use two ways of connecting the two current sources to the driven electrodes: the 'nested' connection (connection scheme N) and the parallel connection (connection scheme P). With this 'CounterCurrent' set-up, current field patterns can be created that differ entirely from the Dipole current pattern:

We will now introduce measurement mode B step by step, starting with the general idea of the CounterCurrent field (measurement mode CC), which as such does not yet meet the guidelines. We will extend mode CC with a first refinement, which leads to measurement mode E, and after that make a second refinement, which yields the final recipe for mode B.

To start with, we define CC to be the current field in connection scheme N in which cs2 and cs1 are used simultaneously, and in which the intensity of cs2 is tuned in such a way that its current field annihilates the current field produced by cs1 at point P. For step n=0 with $\mathbf{s}_1^{(0)}(x)\equiv 0$ and $\mathbf{s}_2^{(0)}(x)\equiv 0$, this situation is reached according to Gauss' law when the current I_2 of cs2 is approximately equal to:

$$I_2 \approx -I_1\beta \;\; with \;\; \beta = \left(\frac{d(T_L,P)}{d(S_L,P)}\right)^2$$

with $d(S_L,P)$ being the distance between electrode S_L and P. When at some location ξ on the catheter the current density approximates zero, so does $\psi(\xi)$ on that location. Therefore the sensitivity of the CounterCurrent field for the region near $\xi=0$ is low. On the other hand it is obvious that whenever the two current fields annihilate each other at point P, this equilibrium cannot hold for any point Q in layer 2 lying in the plane $\xi=0$, separated from the catheter by some distance r ($>r_{vessel}$), since the ratio $d(S_L,Q)^2/d(T_L,Q)^2$ strongly depends on r. (We placed the T-electrodes near the measuring electrodes in order to enhance this dependency on r). Thus, defining the region $\mathcal{M}$ as the region between the measuring electrodes me1 and me3 ($\mathcal{M}$: $\xi_{\mathcal{M}} \in [\xi_{me1}, \xi_{me3}]$), we create a situation in which $\psi_2^{CC}(\xi_{\mathcal{M}})$ dominates $\psi_1^{CC}(\xi_{\mathcal{M}})$ in the $\mathcal{M}$-region (see fig.2c) However, outside $\mathcal{M}$, $\psi_2^{CC}(\xi_T)$ and $\psi_1^{CC}(\xi_T)$ ($\xi_T \notin \mathcal{M}$) still have peaks near the T-electrodes that dominate $\psi_1^{CC}(\xi_{\mathcal{M}})$, which is in conflict with the design guidelines for stability.

In order to overcome these shortcomings of CC, we define a measurement mode E, in which β is assessed by a feedback procedure for each catheter position. The recipe for measurement mode E is the following:

Perform the following steps for each catheter position x:

- Measure $\phi_{13}^S(x)$, $\phi_{13}^T(x)$, $\phi_{12}^S(x)$, $\phi_{23}^S(x)$, $\phi_{12}^T(x)$ and $\phi_{23}^T(x)$, in which the superscript S denotes that the S electrodes eject a current of unit strength and the T electrodes are mute. Superscript T is defined analogously.
- Calculate $\phi_{diff}^S(x)=\phi_{23}^S(x)-\phi_{12}^S(x)$ and $\phi_{diff}^T(x)=\phi_{23}^T(x)-\phi_{12}^T(x)$ for all x, and calculate $\beta(x)$ from:

$$\phi_{13}^S(x) + \beta(x)\phi_{13}^T(x) = 0$$

Finally, $\phi_{diff}^E(x)$, being the output of the procedure, is calculated according to:

$$\phi_{diff}^E(x) = \phi_{diff}^S(x) + \beta(x)\phi_{diff}^T(x)$$

In this way we use $\phi^{E}_{13}(x)=0$ as the criterion for *'current density = 0 at P'*, thus restoring the balance between the current field from cs1 and the current field from cs2.

The rationale behind this procedure is that the differences between the $\beta(x)$ for the various catheter positions x mainly arise from the shift of the sensitive areas around the T-electrodes along the x-axis during the movement of the catheter. According to the method of images, the effect that the emergence of a local boundary $s(x_{bd})\neq 0$ on some position x_{bd} has on the measured potential differences is in first order approximation the same as that of a small virtual current source I^{virt} on that position x_{bd}. When this position x_{bd} is approximately equal to the position of some current electrode, as is the case with the relevant s(x) in relation to $\phi(x)$, this I^{virt} simply adds to the current ejected from that electrode. However, the difference between $d(T_L,\xi_{\mathcal{M}})$ and $d(T_R,\xi_{\mathcal{M}})$ for the various $\xi_{\mathcal{M}} \in \mathcal{M}$ is not negligible, which makes a second refinement of the recipe necessary: having restored balance between S (S_L and S_R) and T (T_L and T_R) electrodes with the procedure described above, it is now desired to restore the balance between the left (S_L and T_L) and the right (S_R and T_R) electrodes as well. This is done by replacing ϕ^{T}_{diff} by ϕ^{TLR}_{diff}, in which ϕ^{TLR}_{diff} is assessed according to a feedback procedure analogous to the one described above, using the parallel scheme P instead of the nested scheme N. Thus we obtain the final result $\phi^{B}_{meas}(x)$:

$$\phi^{B}_{meas}(x) = \phi^{S}_{diff}(x) + \beta(x)\phi^{TLR}_{diff}(x)$$

We calculated the graphs $\psi^{B}_{1}(\xi)$ and $\psi^{B}_{2}(\xi)$ by computer simulation of the measurement process, and applied on both $\phi^{B}_{q}(x)$ and $\psi^{B}_{q}(x)$ (q=1,2) a filter that removes low spatial frequencies. (This can be done because very small s(x)-values are clinically irrelevant). The result is plotted in fig 1d.

The peak in $\psi^{B}_{2}(\xi)$ for $\xi=0$ dominates $\psi^{B}_{2}(\xi_T)$ for $\xi_T \notin \mathcal{M}$, and dominates $\psi^{B}_{1}(\xi)$ for $\xi=0$ as well; thus mode B complies with the design guidelines.

5. CONCLUSIONS

We have designed a new intravascular impedance catheter system on the basis of a robust strategy of measurement, capable of discerning a two-layered impedance distribution around the catheter. The differential measurement procedure does not give rise to an intolerable signal-to-noise ratio: even in the case of a relatively small lesion (30 mm^3) the measurement error in the peak values of ϕ^{A}_{meas} was only 9% in our experimental set-up. Since the non-zero values of the ψ^{A}_{1}-curve are sharply concentrated around $\xi=0$, we expect the Jacobians $\mathbf{J}^{A(n)}_{1}$ and $\mathbf{J}^{A(n)}_{2}$ for n>0 to largely resemble $\mathbf{J}^{A(0)}_{1}$ and

$\mathbf{J}_2^{A(0)}$ in shape. We therefore conclude that, as far as MODE A is concerned, this IIC system has the desired properties for stability for $n > 0$ as well. Furthermore, we found by computer simulation that when using measurement mode A only, the sensitivity in layer 1 is about 9 times that of layer 2. We therefore conclude that severe fatty lesions can already be detected using measurement mode A only, provided that some a priori information is available about extreme variations in conductivity in layer 2. This information may probably be sufficiently provided by the IU-catheter; we expect that the use of the complete IIC with both modes A and B at least will be helpful in obtainig this information about layer 2. Whether or not the IIC will be capable of assessing the two-layered $\mathbf{s}_1$,$\mathbf{s}_2$ distribution in vivo on its own, is a question to be answered by future in-vitro experiments, as well as calculations of $\mathbf{J}_1^{B(n)}$ and $\mathbf{J}_2^{B(n)}$ for n>0.

In order to be able to assess interior impedance within a lesion (stenosis) from the mean impedance values of the annular sections of layer 1 as obtained with the IIC, it is essential to know how the blood/lesion boundary was positioned with respect to the catheter axis during the measurement. To this end, the IIC and the existing Intravascular Ultrasound Catheter should be combined into one device.

ACKNOWLEDGMENTS

This research is supported by the Technology Foundation of the Netherlands (STW). The authors are indebted to prof. J. Baan (University Hospital Leiden) for his advise concerning impedance measurements.

REFERENCES

1) J. Baan, E.T. Van Der Velde, H.G. De Bruin, G.J. Smeenk, J. Koops, A.D. van Dijk, D. Temmerman, J. Senden, B. Buis, "Continuous measurement of left ventricular volume in animals and humans by conductance catheter", Circulation 70: pp.812-823, 1984.

2) W.R. Breckon, M.K. Pidcock, "Ill-posedness and non-linearity in Electrical Impedance Tomography", In: C.N. De Graaf, M.A. Viergever (eds): Information Processing in Medical Imaging, Plenum, New York, pp. 235-244. 1988.

3) V. Fuster, L. Badimon, J.J. Badimon, J.H. Chesebro, "The pathogenesis of coronary artery disease and the acute coronary syndromes", New Engl. J. Med., vol.326, No.4 pp.242-250 & No.5 pp.310-318, 1992.

4) C.J. Slager, A.C. Phaff, C.E. Essed, N. Bom, J.C.H. Schuurbiers, P.W. Serruys, "Electrical impedance of layered atherosclerotic plaques on human aortas", IEEE Trans. Biomed. Eng. 39, pp.411-419, 1992.

STRATEGIES FOR MOTION CORRECTION IN DIFFUSION-WEIGHTED MRI

Arthur GMITRO, Theodore TROUARD, and Yashvinder SABHARWAL
Department of Radiology, University of Arizona, Tucson AZ

1. INTRODUCTION

Magnetic Resonance Imaging (MRI) is an extraordinary imaging modality that, under most circumstances, produces exquisite and diagnostically meaningful images of the human body. However, patient motion is one of the problem areas in MRI; it leads to artifacts that degrade image quality and, in some cases, even negates the diagnostic utility of the exam. Several practical methods have been devised to partially combat the problem of patient motion including active patient restraint, cardiac and/or respiratory gating, and faster imaging sequences. While these have greatly increased the number of successful MRI studies, they have not completely solved all problems associated with motion. This is especially true for diffusion-weighted imaging techniques, that employ strong gradient fields to produce image contrast dependent on the degree of random diffusional motion. This paper is concerned with techniques for motion correction that are applicable to diffusion-weighted imaging (DWI).

The mechanisms that produce motion artifacts in MRI are discussed in the next section, followed by a comparison of three motion-correction techniques. The first method is an iterative reconstruction algorithm that attempts to correct the data errors using known constraints in the image and data domains. This algorithm is a variant of the Gerchberg Saxton technique[1] and falls into the category of a generalized projection onto convex sets (POCS) algorithm[2]. The second method, known as the navigator echo technique[3], incorporates an additional data acquisition step, which is used to estimate and correct the errors in the data. The final technique uses an alternative data-acquisition strategy, whereby data are collected along radial lines in Fourier space rather than in a conventional rectilinear format. This combined with a modified projection reconstruction algorithm is capable of correcting for motion[4].

2. DESCRIPTION OF MOTION ERRORS

A general description of the effects of three-dimensional translational and rotational motions on magnetic resonance images has been derived. For the sake of brevity, a simplified two-dimensional (2D) model, which demonstrates most

Y. Bizais et al. (eds.), Information Processing in Medical Imaging, 13–26.

of the significant features of the theory, is presented here. Gradient magnetic fields are used in MRI to encode spatial information as well as to produce specific image weighting such as in DWI. We assume that a thin slice volume in the z direction is excited by a frequency-selective RF pulse combined with a gradient $G_Z(t)$ in the z direction, and that imaging gradients $G_X(t)$, and $G_y(t)$ are applied. For a stationary object the time-varying MRI signal $S(t_a)$, during the measurement time window t_a, is given by

$$S(t_a)=\iint f(x,y)e^{-ixK_x(t_a)}e^{-iyK_y(t_a)}dxdy \quad , \qquad \text{(Eq. 1)}$$

where

$$K_x(t_a)=\gamma\int_0^{t_a} G_x(t)dt \quad \text{and} \quad K_y(t_a)=\gamma\int_0^{t_a} G_y(t)dt \qquad \text{(Eq. 2)}$$

represent the spatial-frequency variables conjugate to x and y; γ is the gyromagnetic ratio relating the resonance frequency ω of the nuclear spins to the strength of the magnetic field, i.e., $\omega=\gamma B$. Equation 1 states that the MR signal is a trajectory through the Fourier space of the object. In conventional 2D Fourier imaging, with the so-called read direction in x and the phase direction in y, a series of measurements is made with different phase gradients G_{y_n} so that a single line of Fourier data, $S_n(t_a)=F(K_X(t_a), K_{y_n})$, is obtained each time the spins in the slice volume are excited. Repeated measurements (n=1,2, ... N) fill in the Fourier space in a rectilinear fashion, and a simple inverse Fourier transform from $F(K_X, K_y)$ to $f(x,y)$ reconstructs the image. Figure 1 shows this basic concept of the data acquisition and image reconstruction.

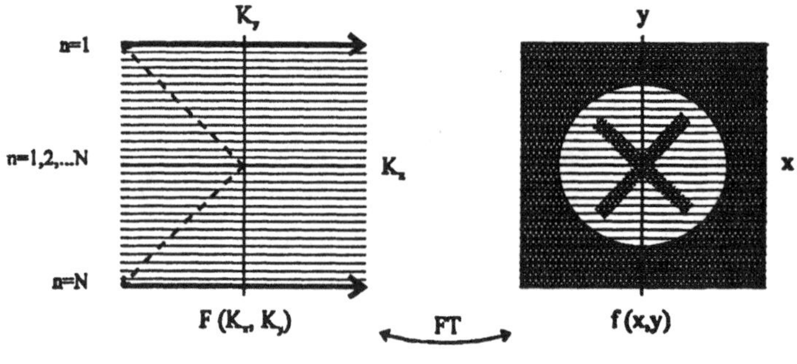

Fig. 1: Data acquisition in MRI.

When there is object motion during the scan, the measured data are corrupted. Consider first the case of translational motion of the object. A coordinate system x' and y' fixed to the moving object is related to the fixed coordinate system x and y by

$$x'=x+\Delta x_n+\delta x_n(t) \quad \text{and} \quad y'=y+\Delta y_n+\delta y_n(t). \qquad \text{(Eq. 3)}$$

where Δx_n and Δy_n represent the shift of the object at the start of the *n*th acquisition step (*n*th line of Fourier data) and $\delta x_n(t)$ and $\delta y_n(t)$ are motions that occur during the *n*th step. In this case the MR signal becomes

$$S_n(t_a) = \iint f(x',y') e^{-i\int_0^{t_a}\left(xG_x(t)+yG_y(t)\right)dt} dx'dy'$$
$$= e^{i\Delta x_n K_x(t_a)} e^{i\Delta y_n K_{y_n}} e^{i\varphi_n(t_a)} \iint f(x',y') e^{-ix'K_x(t_a)} e^{-iy'K_{y_n}} dx'dy' \quad \text{(Eq. 4)}$$

where

$$\varphi_n(t_a) = \int_0^{t_a} \delta x_n(t) G_x(t) dt + \int_0^{t_a} \delta y_n(t) G_y(t) dt \quad \text{(Eq. 5)}$$

If $\delta x_n(t)$ and $\delta y_n(t)$ are zero, then φ_n is zero, and the effect of the object shift on the *n*th measurement is a linear phase term in K_x and a constant phase offset in K_y, which multiply the *n*th line of Fourier data. In general, the values of Δx_n and Δy_n vary with *n*, and therefore, the phase terms are different on each line of Fourier data. This modulation of the phase, as *n* varies, leads to the standard MRI motion artifact - blurring or ghosting of the image in the direction of the phase encoding gradient. If $\delta x_n(t)$ and $\delta y_n(t)$ are not zero, then there is an additional phase error φ_n that varies from line to line, which also causes motion artifact in the reconstructed image.

In DWI a special balanced gradient waveform is utilized in addition to the standard imaging gradients. This gradient has a temporal integral of zero. Spins that are stationary acquire no phase shift from this gradient; however, spins that are moving do. If the motion of the spins within a pixel is random, as in diffusion, then there is a random distribution of phase shifts and a net reduction in the signal intensity by the factor e^{-bD}, where D is the diffusion coefficient and b is a factor that depends on the strength and timing of the diffusion gradients. If the motion is uniform for all spins, as it would be if the whole object moves, then the phase shifts of all the individual spins are equal (i.e. that given by Eq. 5)

In general, φ_n is time varying during the data measurement. However, in DWI, the majority of the phase error is accumulated before the measurement of the signal $S_n(t_a)$; φ_n is more-or-less a constant phase offset on each line of data. The phase offsets do, however, vary from line to line in the Fourier data because the motion varies in time. Again, the result is motion artifact. DWI sequences are extremely sensitive to motion because of the large gradients used to produce the diffusion weighting. Even very small sub-pixel shifts can lead to objectionable errors. Thus, for DWI it is the constant phase offset errors that dominate.

Next, we consider the effects of rotational motion of the object about an axis perpendicular to the image plane. In this case the coordinate system x' and y' attached to the object is related to the fixed coordinate system x and y by

$$x' = x\cos\{\Delta\alpha_n + \delta\alpha_n(t)\} + y\sin\{\Delta\alpha_n + \delta\alpha_n(t)\}$$
$$y' = -x\sin\{\Delta\alpha_n + \delta\alpha_n(t)\} + y\cos\{\Delta\alpha_n + \delta\alpha_n(t)\} \quad , \qquad \text{(Eq. 6)}$$

where $\Delta\alpha_n$ represents the rotational position of the object at the start of the nth acquisition step and $\delta\alpha_n(t)$ represents the rotation during the nth step. In this case the MR signal becomes

$$S_n(t_a) = \iint f(x',y') e^{-ix'K_x'(t_a)} e^{-iy'K_y'(t_a)} dx'dy' \quad , \qquad \text{(Eq. 7)}$$

where

$$K_x'(t_a) = \gamma \int_0^{t_a} \left[\cos\{\Delta\alpha_n + \delta\alpha_n(t)\} G_x(t) + \sin\{\Delta\alpha_n + \delta\alpha_n(t)\} G_y(t)\right] dt$$
$$K_y'(t_a) = \gamma \int_0^{t_a} \left[-\sin\{\Delta\alpha_n + \delta\alpha_n(t)\} G_x(t) + \cos\{\Delta\alpha_n + \delta\alpha_n(t)\} G_y(t)\right] dt \qquad \text{(Eq. 8)}$$

The signal $S_n(t_a)$ is still a measurement of the Fourier transform of the object $f(x',y')$, but the trajectories through Fourier space are altered according to Eq. 8. If $\delta\alpha_n(t)$ is zero, then

$$K_x'(t_a) = \cos\Delta\alpha_n K_x(t_a) + \sin\Delta\alpha_n K_{y_n}$$
$$K_y'(t_a) = -\sin\Delta\alpha_n K_x(t_a) + \cos\Delta\alpha_n K_{y_n} \quad , \qquad \text{(Eq. 9)}$$

which is just a rotation of the Fourier domain. In other words, the nth line of Fourier data is rotated by $\Delta\alpha_n$. If $\delta\alpha_n(t)$ is not zero, then the path in Fourier space is altered and will not necessarily be a straight line. In either case, the variation of motion from line to line leads to variations in the measured data whose consequence is motion artifact.

For small rotations the deviation of the trajectory is typically small and does not significantly affect the image quality. However, for DWI, the strong gradients during a portion of the sequence can produce significant errors even when the rotations are small. In this case the integrals in Eq. 8 can be evaluated over the periods of the diffusion gradients and yield constants ΔK_{x_n} and ΔK_{y_n}. The MR signal becomes

$$S_n(t_a) = \iint f(x',y') e^{-ix'\Delta K_{x_n}} e^{-iy'\Delta K_{y_n}} e^{-ix'K_x'(t_a)} e^{-iy'K_{y_n}'} dx'dy' \quad , \qquad \text{(Eq. 10)}$$

where it has been assumed that the rotations are small. The measured signal is the Fourier transform of the object f(x',y') times linear phase factors in x' and y'. This implies that the measurement $S_n(t_a)$ is equal to a line of data in a shifted Fourier domain (i.e. $S_n(t_a)=F(K_x'(t_a)+\Delta K_{x_n}, K_{y_n}'+\Delta K_{y_n})$). Once again, the variations of the object motion from line to line lead to differences in the shifts in Fourier data from line to line and result in motion artifact. It is common to use diffusion gradients in only one direction. If a G_x diffusion gradient is used, then the dominant shift is ΔK_{y_n}. If a G_y gradient is used, then the dominant shift is ΔK_{x_n}. This is explained by Eq. 8, where the cosine terms are approximately equal to one for small rotations and the balanced nature of the diffusion gradient waveform leads to an integral close to zero. The sine terms, on the other hand, give a finite contribution. The net effect of using a G_y diffusion gradient is a line of data in Fourier space that is at the correct K_y location, but that is shifted along K_x.

The first step in reconstructing an MR image via the 2D discrete Fourier transform is to perform a 1D discrete Fourier transform on each row of data. This is typically done utilizing the fast Fourier transform (FFT) algorithm. We shall refer to these data (i.e. where each row of data has been Fourier transformed) as the hybrid space. The next step in the reconstruction is to perform 1D discrete Fourier transforms on each column of the hybrid space data. This results in a complex reconstruction whose magnitude is displayed as the final image. If a G_y diffusion gradient is used, then the shift along K_x due to rotation of the object becomes a linear phase factor $\exp(-i\Delta K_{x_n}x')$ in the hybrid space. This linear phase factor, of course, varies from line to line.

To summarize, translational motion leads to constant as well as linear phase errors on each line of data in Fourier space. Rotations of the object cause rotations, alterations of trajectories, and shifts of data in Fourier space. Both types of motion lead to errors in the measured data and result in artifacts in the images. In DWI sequences, motion during the sequence can lead to severe artifacts even when the net motion is small. In this case, translations lead to constant phase errors on individual lines of Fourier data and rotations lead to shifts. In the hybrid space the constant phase errors remain constant phase errors and the shifts become linear phase errors. These constant and linear phase errors vary from line to line and it is these errors we seek to correct with the techniques described below.

3. TECHNIQUES FOR MOTION CORRECTION

3.1. Generalized POCS Algorithm

Three techniques were investigated for correcting the phase errors encountered in DWI. The first technique is an iterative post-processing algorithm that utilizes constraints in the image and data domains to correct the errors. The algorithm described here is similar to the Gerchberg-Saxton algorithm[1] for phase recovery that was modified by Hedley et. al.[5-6] to correct for motion artifacts in MRI. It differs from that algorithm in that it is able to deal with rotations as well as translations. The computational flow of the algorithm is shown in Fig. 2.

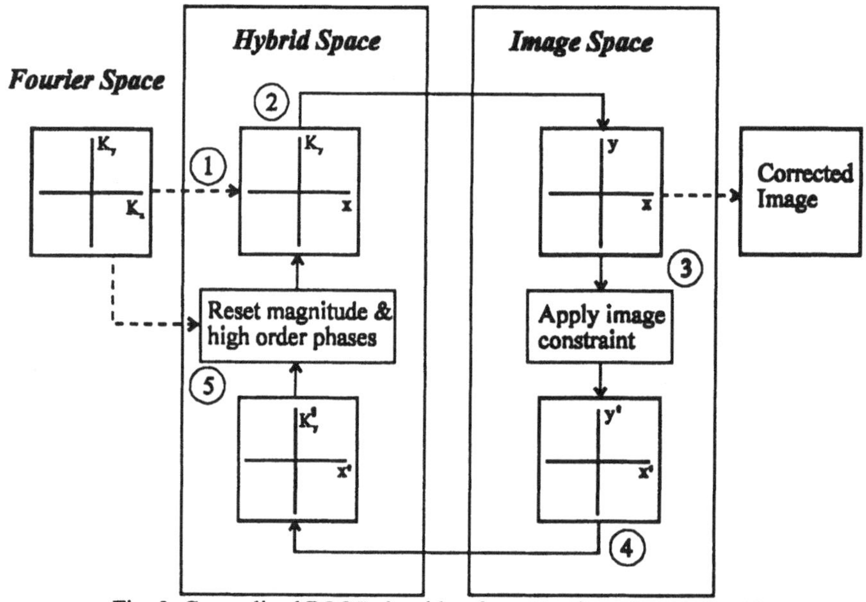

Fig. 2. Generalized POCS algorithm for correction of motion artifacts.

We assume that the Fourier data are measured with diffusion gradients in the y direction so that the effect of bulk motion (translation and rotation) is constant phase offsets and lateral shifts in each line of Fourier data. The first step of the algorithm is to perform a 1D Fourier transform on each line of data to transform to the hybrid space. In the hybrid space the lateral shift becomes a linear phase and the constant phase remains a constant phase. The task is to correct the zero order (constant phase term) and first order (linear phase) phase errors caused by the motion. The assumption is made that the magnitude and higher-order phase terms of each line of data in the hybrid space are correct. This is the constraint (or projection operator) in the hybrid space. On the first iteration the hybrid-space data are transformed to the image domain by taking 1D Fourier transforms of columns; this yields the motion-corrupted image. In our approach, the compact support of the object is used as an image domain constraint. The object boundary is determined via an automated algorithm operating on an image that is obtained without diffusion gradients, and hence no motion artifact. A binary mask function is derived that is equal to zero outside the boundary and one within the boundary. The support constraint can be applied to the corrupted image by multiplying the object by the mask function, which has the effect of zeroing out the image intensity in the region outside the object. This new object is then transformed back to the hybrid space (by transforming columns of the image - step 4 in the diagram) yielding new hybrid-space data. In general, both the magnitude and phase of the new data are different from the original hybrid-space data. Application of the hybrid-space constraint results in modified hybrid-space data that have the magnitude and higher-order phase of the original data on each line, but zero and first-order phase terms that match the new incoming hybrid-space data. The implementation of the hybrid-domain constraint is complicated by the cyclic nature of the phase function. A search procedure is used to find the zero and first-order phase terms that, when added to the phase of the original hybrid-space data, produce the closest match (in an RMS sense) to the phase of the iterate. This is done independently for each line of data. The result is a modified hybrid-space data set that satisfies the constraint and is as close to the hybrid-space data of the support-constrained object as possible.

The algorithm iterates back and forth between the image space and the hybrid space applying the appropriate constraint at each step, and it continues until it converges or is stopped by the operator. Convergence is defined as a consistent solution that satisfies the constraints in both domains. There are convergence proofs for POCS algorithms when the constraints are convex sets. Unfortunately, the hybrid-space constraint is not a convex set, and there is no proof of convergence for this algorithm; yet in many applications the algorithm works extremely well.

3.2. Navigator Echo

Another approach, known as the navigator echo method, for correcting motion artifacts in MRI was introduced by Ehman and Felmlee[2] and later adapted to correct for motions in DWI.[7-9] In this technique the pulse sequence and gradient waveforms are modified so that an additional signal $S_n^{Nav}(t)$ is measured on the K_y =0 line in Fourier space right after the signal $S_n(t_a)$ is acquired. Thus, for every line of Fourier data, a "navigator" line of data is acquired. If the errors are caused primarily by the motion during the application of the diffusion gradients, which occurs before the data line and the navigator line are acquired, then the errors should be the same in both. Because the same line of data at $K_y = 0$ is obtained every time, the relative error can be estimated by comparing the differences in the navigator data from one line to the next. The data measurement for the navigator echo method is shown in Fig. 3.

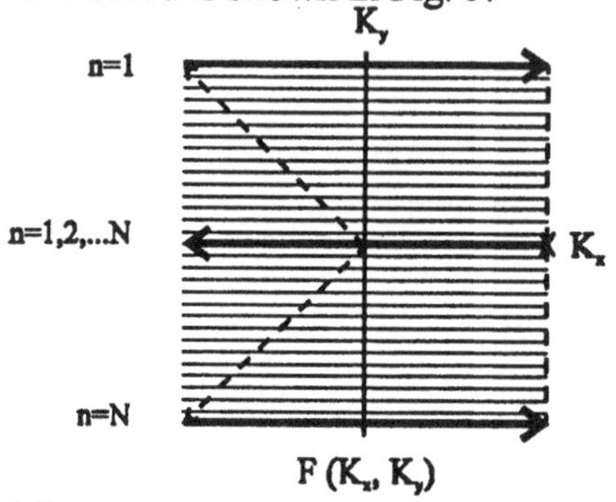

Fig. 3 Data measurement with the navigator-echo method.

To correct the zero-order phase errors due to translational motion and the shifts due to rotation, the hybrid space is used, which converts the errors to zero and first-order phase errors on each line of the hybrid-space data. Both the Fourier data and the navigator data are transformed to the hybrid space. The first line of navigator data $S_1^{Nav}(t)$ is used as the reference to which subsequent lines of navigator data are compared. The idea is to compute the change in the zero and first-order phase on a line of navigator data relative to the reference line of navigator data and then to modify the phase of the corresponding line of actual data by this amount to correct for the error due to motion.

3.3. Radial-Scan Method

The final method for correcting the error due to motion is to alter the data-acquisition strategy. Instead of collecting data in a rectilinear fashion in Fourier space, data are collected in a polar format. Figure 4 indicates the geometry. This modification is straightforward in terms of the gradient waveforms required to collect the data. Translations in the presence of diffusion gradients produce constant phase errors, j_n, which now occur along each radial line of data in Fourier space. Rotations in the presence of a fixed gradient, say in the *y* direction, produce shifts of the Fourier space in the K_x direction. For a radial line along the K_x direction, this is a shift along the line of data, just as it was for rectilinear data acquisition. For a radial line along the K_y direction the shift is perpendicular to the line so that the data no longer lie along a radial line in the Fourier space of the object. However, it is possible to alter the direction of the diffusion gradient in such a manner that it is always perpendicular to the radial line of data being acquired. In this case the effect of object rotation is a shift of the data along the radial line.

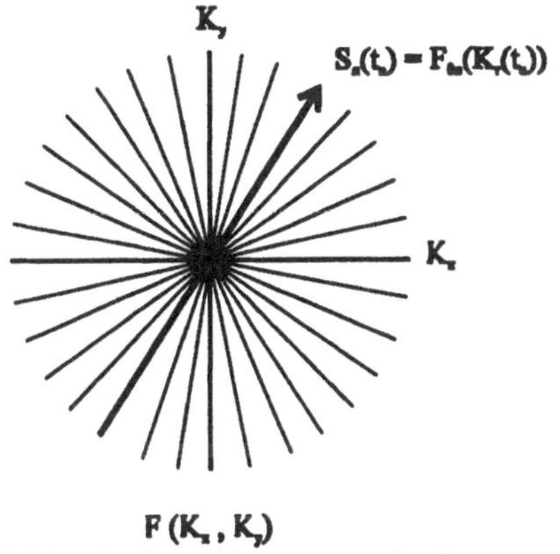

Fig. 4 Data acquisition in the radial-scan method.

Reconstruction of the image from data acquired in a polar format requires either interpolation of the data onto a rectilinear grid followed by a standard 2D Fourier transform, or a direct implementation of a 2D Fourier transform in polar coordinates. The latter is related to the filtered-backprojection algorithm because, by the central slice theorem, each radial line of Fourier data is equal to the 1D Fourier transform of a projection of the object. The filtering step of the filtered-backprojection algorithm can be accomplished by multiplying each radial line of Fourier data by an appropriate filter function and taking 1D Fourier transforms. This yields filtered projections that are backprojected to reconstruct the image. A slight difference for MRI reconstruction, however, is that the filtered projections are complex and so a backprojection of complex data must be performed to yield a complex reconstructed image, whose magnitude is usually displayed. This approach works well when there is no motion. The nature of the motion artifacts is different, however, when there is motion. One can view the error as a modulation of the data as one moves azimuthally around the polar Fourier space. Although the artifacts are different, they still significantly degrade the image quality in DWI.

An alternative reconstruction algorithm can, however, be used to compensate the errors. By taking the 1D Fourier transform of each radial line of Fourier space, the data are transformed to the projection domain, which is similar to the hybrid space for the rectilinear data. In the projection domain the error due to translations remains a constant phase error on each projection. With diffusion gradients that follow the projection direction, the radial shift error in the Fourier data becomes a linear phase across the projection. Now, in general, the object being imaged in MRI is a positive real quantity, but due to inhomogeneities and other effects, a phase distribution, which is typically slowly varying, is introduced across the object. However, if we ignore this phase and consider the object as positive real, then the projections of the object should be positive real. The motion adds constant and linear phase to this positive real quantity. However, by taking the magnitude of the projection data, the desired positive real projection is obtained and the phase errors due to motion are eliminated. Then the object can be reconstructed by a standard filtered-backprojection algorithm. Figure 5 shows a data-flow diagram of this modified filtered-backprojection reconstruction algorithm. We have previously described the use of this data acquisition and reconstruction strategy for motion correction and have demonstrated its ability to correct for errors due to translational motions[4].

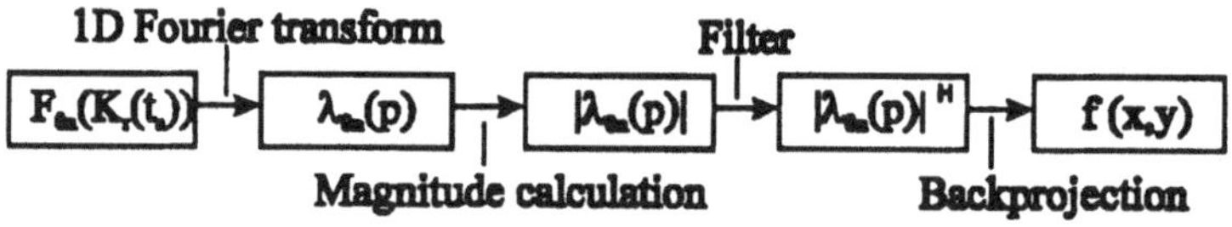

Fig. 5 Modified filtered-backprojection algorithm for motion correction.

4. RESULTS

Experiments were performed to demonstrate and evaluate the effectiveness of the three approaches to motion correction in DWI. A 4.7 T Bruker Biospec MRI instrument with a 13.9 Gauss/cm gradient insert was used for the experiments, and a stimulated-echo diffusion-weighted (STED) sequence[10], which we use regularly for DWI, was modified to incorporate a navigator echo or to acquire data in a polar format. The STED sequence allows control of the strengths of the diffusion gradients to alter the degree of diffusion weighting and, therefore, also the sensitivity to motion.

For the first series of experiments a phantom consisting of a cylindrical syringe filled with baby oil was used. A plastic insert in the shape of a cross (part of the syringe plunger) was placed inside the syringe. An MR image of the phantom with and without translational motion is shown in Fig. 6. The motion in this case is a small random motion along the axis of the cylinder (z axis). The diffusion gradients are along the z direction. This results in a signal error that is strictly a constant phase offset on each line of Fourier data. Even with very subtle (sub slice-thickness) motions, the resulting artifacts Fig. 6b are severe.The results using the three approaches for motion correction are shown in Fig. 7. Figure 7a shows the result obtained with the generalized POCS method. In this case the algorithm corrects only for phase offsets on each line of data.

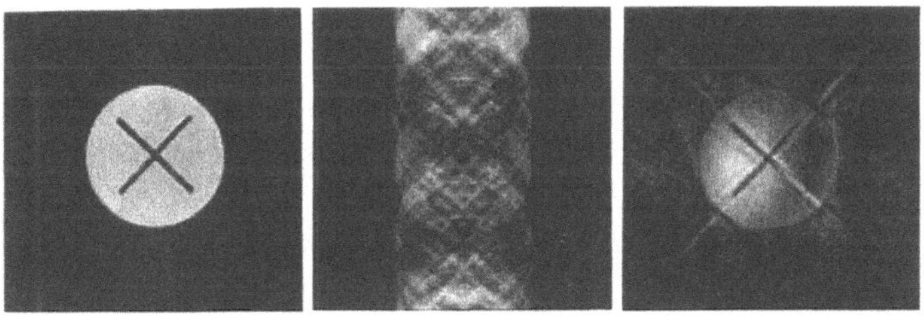

(a) (b) (c)

Fig. 6 MR images of phantom a) without motion, b) with motion - standard 2D Fourier method, and c) with motion - radial-scan acquisition and complex filtered backprojection reconstruction.

The support constraint derived from the image in Fig 6a is shown in Fig. 7b. Figure 7c shows the result with the navigator echo. Again, the correction is only for the constant phase offsets on each line of data. Fig. 7d shows the result with the radial-scan approach. As can be seen all three techniques work very well in this case.

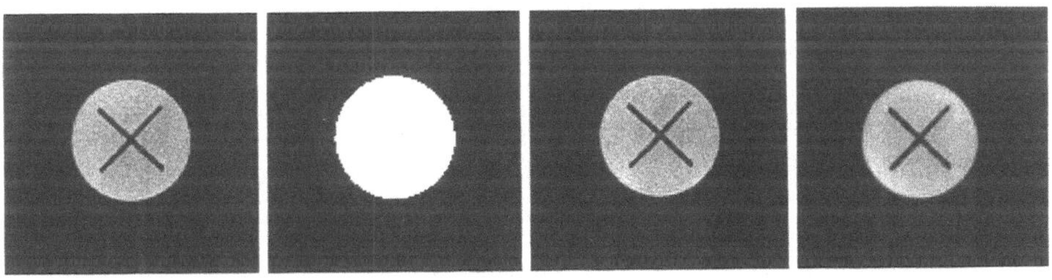

(a) (b) (c) (d)

Fig. 7. Corrected images using a) generalized POCS, c) the navigator-echo approach, and d) the radial-scan approach. Image (b) is the support constraint used in the POCS algorithm.

Experiments were also carried out with a more general in-plane motion, which was a combination of in-plane translation and rotation about the image isocenter. The diffusion gradients were oriented in the y direction. The motion was small in these experiments, but still leads to visible artifact as seen in Fig. 8a. Figures 8b through 8d show the results of the three motion-correction techniques. As can be seen, the POCS algorithm and the navigator-echo technique do improve the image quality, but do not completely correct for the motion-induced error. The radial-scan approach appears to give a better image with less motion artifact, although there is still some residual error. The final set of experiments was done in an anaesthetized rat. The imaging was performed in the abdominal region where significant breathing motion occurs. In this case the motion is spatially varying and the model of bulk motion (i.e. pure translation and rotation of the object) is no longer valid.

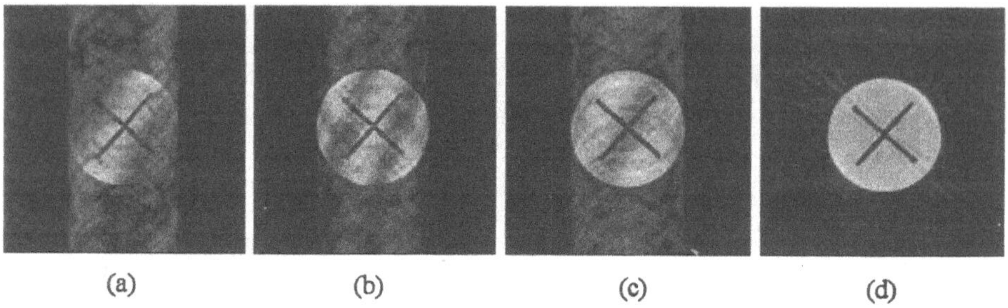

Fig. 8. Imaging results with translational and rotational motion. a) no correction, b) POCS algorithm, c) navigator echo technique, and d) radial-scan approach.

The results of these experiments are shown in Fig. 9. Figure 9a shows the results with standard 2D Fourier acquisition and no motion correction. Figures 9b, 9c, and 9d show the results of the POCS, navigator-echo, and radial-scan approaches, respectively. The POCS and navigator-echo techniques show essentially no improvement and, qualitatively, even degrade the image quality. The radial-scan technique, on the other hand, shows marked improvement in reducing motion artifacts.

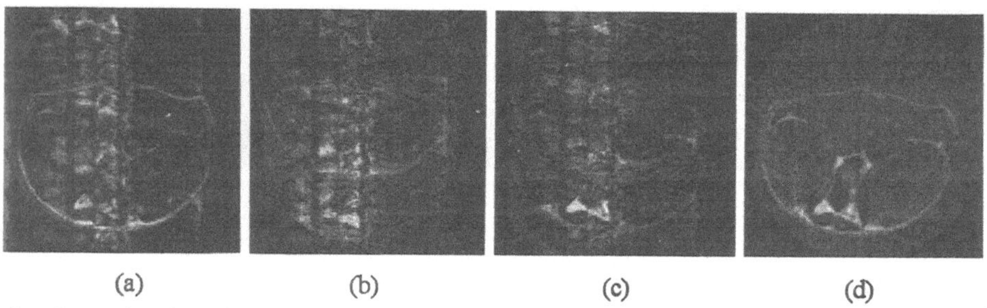

Fig. 9 In-vivo imaging of a rat. a) no correction, b) POCS algorithm, c) navigator-echo technique, and d) radial-scan approach.

5. DISCUSSION

The experiment with translational motion and diffusion gradients along the z axis was specifically designed to produce a constant random phase error on each line of Fourier data. The results shown in Fig. 7 indicate that all three techniques work very well when the assumption of a constant random phase error on each line of Fourier data is valid. In the more complex case of in-plane translation and rotation, motion during the diffusion gradients produces constant phase errors and shifts of the Fourier data. The algorithms are designed to correct for these errors. However, there are additional data errors associated with the motion that are not corrected. These include a linear phase term in K_x due to the translation, x_n, of the object (Eq. 6), rotations of the Fourier data (Eq. 10), and alterations of the trajectory in Fourier space due to the $\Delta\alpha_n(t)$ rotations during the imaging gradients. The results in Fig. 8 indicate that the correction algorithms do reduce the motion artifact, but do not completely eliminate it. This may be due to these residual (uncorrected) errors in the data.

When the motion varies from point-to-point in the object, then the assumptions of simple constant phase offsets and shifts of the Fourier data are no longer valid. In fact, there is no longer a true Fourier space of the object because the object varies in time. Figure 9 shows that, except for the radial-scan approach, the correction techniques are not very good in this situation.

The radial-scan technique with the modified filtered-backprojection reconstruction algorithm does appear to produce better results than the other two techniques. We attribute this to two factors. First, there is an inherent advantage of the radial-scan approach as seen in Fig. 6 (compare Fig. 6c to 6b), which shows that the nature of motion artifacts is very different for the radial-scan approach compared to the standard 2D Fourier method. In 2D Fourier data-acquisition, the motion produces modulation of the data in the K_y direction, which results in an image point-spread-function (psf) that is blurred along the y direction. In radial-scan data acquisition, the motion produces an azimuthal modulation of the data, which results in an image psf that blurs out in many directions. Although the total error is the same, it appears to be less objectionable in radial scanning, because the error is spread out over the whole 2D image plane rather than along a single direction. The second factor that may account for the superiority of the radial-scan approach is that the magnitude calculation of the modified filtered-backprojection reconstruction algorithm corrects for higher order phase errors than just the zero and first-order terms. The navigator-echo technique, however, did not show significant improvement even when the entire phase function was corrected via the navigator data, and therefore, we believe this second factor is of less importance.

It should be noted that the conclusions drawn from this work are based on qualitative assessment of image quality. A more careful comparison of the three correction techniques will require evaluation of observer performance in the context of a specific diagnostic task. Nevertheless, under certain circumstances these techniques do appear to provide a means of improving the performance and applicability of diffusion-weighted MR imaging.

6. REFERENCES

[1] R.W. Gerchberg and W.O. Saxton, "A Practical Algorithm for the Determination of Phase from Image and Diffraction Plane Pictures," *Optik*, 35, p237 1972

[2] H. Stark, ed. *Image Recovery: Theroy and Application*, Accademic Press, Orlando FL 1987.

[3] R.L. Ehman and J.P. Felmlee, "Adaptive Technique for High-Definition MR Imaging of Moving Structures," *Radiology* 173, p255 1989.

[4] A.F. Gmitro and A.L. Alexander, "Use of a Projection Reconstruction Imaging Method to Decrease Motion Sensitivity in Diffusion-Weighted MRI," *Magn. Reson. Med.* 29, p835 1992.

[5] M. Hedley, H. Yan, and D. Rosenfeld, "Motion Artifact Correction in MRI Using Generalized Projections," *IEEE Trans. Med. Imaging* 10, p40 1991.

[6] M. Hedley, H. Yan, and D. Rosenfeld, "An Improved Algorithm for 2-D Translational Motion Artifact Correction," *IEEE Trans. Med. Imaging* 10, p548 1991.

[7] R.J. Ordidge, J.A. Helpern, Z.X. Qing, R.A. Knight, and V. Nagesh, "Correction of Motional Artifacts in Diffusion-Weighted MR Images Using Navigator Echoes," *Magn. Reson. Imaging*, 12, p455 1994.

[8] V.J. Wedeen, R.M. Weisskoff, and B.P. Poncelet, "MRI Signal Void to In-Plane Motion is All-or-None," *Magn. Reson. Med.* 32, p116 1994.

[9] A.W. Anderson and J.C. Gore, "Analysis and Correction of Motion Artifacts in Diffusion Weighted Imaging," *Magn Reson. Med.* 32, p379 1994.

[10] M.I. Altbach, M.A. Mattingly, M.F. Brown, and A.F. Gmitro, "Magnetic Resonance Imaging of Lipid Deposits in Human Atheroma Via a Stimulated-Echo Diffusion-Weighted Technique," *Magn. Reson. Med.* 20, p19 1991.

USING GROUND-TRUTH DATA TO DESIGN PRIORS IN BAYESIAN SPECT RECONSTRUCTION

S.J. LEE AND G.R. GINDI
Departments of Electrical Engineering and Radiology
SUNY at Stony Brook, Stony Brook NY 11794 USA

AND

I.G. ZUBAL AND A. RANGARAJAN
Departments of Diagnostic Radiology and Computer Science
Yale University, New Haven CT 06510 USA

Abstract. We propose a method for estimating hyperparameters for prior models used in SPECT reconstruction. Our method uses a maximum-likelihood technique to fit hyperparameters to a training set that is itself representative of ground-truth objects to be reconstructed. We argue that autoradiography may be a good source for such exemplars. For a simple smoothing prior, we derive a closed-form expression for the hyperparameter, and for a more complex prior that promotes piecewise smoothness, we evaluate the likelihood expression numerically. Numerical experiments demonstrate that parameters derived from training sets are fairly close to parameter values optimal in the sense of various measures of image quality.

1. Introduction

Bayesian methods in SPECT are attractive in that they support accurate models of the imaging system, and allow incorporation of prior information into the solution. Many studies demonstrate that Bayesian reconstruction methods can result in improved, relative to conventional methods, image quality according to various metrics. The transition to practical use is, however, hampered by computational barriers. The optimizations resulting from Bayesian methods are numerically intensive and may be impractical to run if implemented on conventional hardware. Evolution in hardware and

Y. Bizais et al. (eds.), Information Processing in Medical Imaging, 27–38.

the prospects of utilizing special-purpose parallel computers have dampened these objections; but, a second computational obstacle remains.

Algorithms based on Bayesian approaches are usually iterative and contain free parameters, and the performance of these algorithms usually depends crucially on the values of these parameters. Some parameters, such as step size in a descent procedure, or number of iterations, are associated with the iteration procedure itself, and others, termed hyperparameters, are associated with the mathematical model for the prior probability. Empirical studies show a tradeoff in reconstruction bias and variance as these parameters are varied, but, depending on a particular task, there is often an optimal choice for a parameter setting. Many studies have assessed bias/variance tradeoffs as a function of iteration number for ML-EM and other, Bayesian, algorithms, but much less work has been directed towards determining the values of hyperparameters.

The choice of a prior model is tantamount to a belief in the spatial structure of the underlying object whose noisy and corrupted projection image we observe. One class of well-studied "bootstrap" methods [1, 2] seeks to determine the optimal values from the noisy and corrupted projection data. A second strategy [3] involves determining these values from a training set of noiseless objects, chosen judiciously to reflect the kinds of objects we expect to encounter in a clinical setting. The latter method is much less computationally demanding, but its utility ultimately relies on its ability to generalize from training data.

2. Autoradiography

Ideally, an exemplar should be a record of the radionuclide spatial density in a human patient. Since there is no way to record ground-truth functional patterns in humans, one often resorts to physical or software phantoms. These are typically idealized versions of the expected functional pattern [4]. An opportunity to record ground-truth functional patterns, does exist, however, for animal models in the form of autoradiography. Successful reconstruction of ground truth biological data is appealing, but the validity of the extension to humans from animal models, even advanced primates, remains open. Results in our work, however, may be applied to *any* functional phantom, biological or synthetic, that is deemed useful.

While autoradiography has become a standard research tool in much of biological science, it is less familiar in radiological imaging. In autoradiography [5], an animal is injected with a radiopharmaceutical, sacrificed, and thin (20μ) tissue sections are apposed to film in order to record high-resolution images of the radionuclide density. Though the image formation process for autoradiographs is characterized by its own quantities of noise and blur, these are so small, that from the standpoint of radiology, the autoradiograph can be considered as a ground-truth phantom. In our work, we have used digitized (64×64) versions of autoradiographs ("autorads") such as those in Fig. 1 (a)(b)(c). These are scaled geometrically to approximately match adult human dimensions so as to more realistically simulate the physical effects. For the studies reported here, we simply compute deterministic line integrals of the phantom and add an amount of independent Poisson noise appropriate for the count level desired. Attenuation is simulated by assuming a constant attenuation coefficient (0.12 cm^{-1}) equal to that of water. We note that these 2-D phantoms are crude; a better autorad phantom would be 3-D and contain spatial detail only up to a degree of resolution consistent with the imaging system under study.

3. Mechanical Models as Priors

Though the methods here apply to any form of prior, we have been exploring a class of "mechanical" models [6] inspired by observations of the spatial structure of autorads. The *weak membrane* (WM) prior encourages the formation of piecewise flat regions while the *weak plate* (WP) prior encourages piecewise linear regions. In addition, we consider a third prior, the *membrane*(MM), that encourages overall smoothness with no discontinuities.

We use the familiar Gibbs distribution to model prior distributions:

$$\Pr(\mathbf{F} = \mathbf{f}, \mathbf{L} = \mathbf{l}) = \frac{1}{Z}\exp(-E(\mathbf{f}, \mathbf{l})),$$

where $\mathbf{f}$ is a 2-D random field realization for the source intensity, with $\mathbf{F}$ the associated random field, and $\mathbf{L}$,$\mathbf{l}$ are the random field and 2-D realization for binary line-process variables [7] used to signal the suspension of continuity. Also, Z is a normalization factor (partition function) and $E(\mathbf{f}, \mathbf{l})$ the prior energy: $E_{WP}(\mathbf{f}, \mathbf{l})$ for WP, $E_{WM}(\mathbf{f}, \mathbf{l})$ for WM, and $E_{MM}(\mathbf{f})$ for simple membrane (MM). Our definitions of the priors are given by the energy functions:

$$E_{WM}(\mathbf{f}, \mathbf{l}) = \lambda \sum_{i,j} \left\{ V^v_{i,j}(1 - l^h_{i,j}) + V^h_{i,j}(1 - l^v_{i,j}) \right\} + \alpha \sum_{i,j} \left(l^h_{i,j} + l^v_{i,j} \right) \quad (1)$$

$$E_{WP}(\mathbf{f}, \mathbf{l}) = \lambda \sum_{i,j} \{ V_{i,j}(1 - l_{i,j}) \} + \alpha \sum_{i,j} l_{i,j}, \quad (2)$$

$$E_{MM}(\mathbf{f}, \mathbf{l}) = \lambda \sum_{i,j} \left\{ V^v_{i,j} + V^h_{i,j} \right\} \quad (3)$$

where $V^v_{i,j} \stackrel{\text{def}}{=} f^2_v(i,j)$ and $V^h_{i,j} \stackrel{\text{def}}{=} f^2_h(i,j)$, and $V_{i,j} \stackrel{\text{def}}{=} f^2_{vv}(i,j) + 2f^2_{hv}(i,j) + f^2_{hh}(i,j)$. The binary variable l is line process (for WM, l^v and l^h are vertical and horizontal line processes, respectively), and λ and α are our positive hyperparameters. In (1), f_v and f_h are the partial first derivatives along the vertical and the horizontal directions, respectively. The term $V_{i,j}$ in (2) is quadratic variation for a thin plate [6]. Here f_{vv} and f_{hh} are partial second derivatives along the vertical and horizontal directions, respectively, and f_{hv} is a partial cross second derivative. The terms involving $V_{i,j}^v$ and $V_{i,j}^h$ in (3) and $V_{i,j}$ in (2) encourage smoothness except where discontinuities occur (i.e. $l = 1$) and include a penalty α for the creation of discontinuities.

It turns out [6] that the above energies can be expressed in terms of $\mathbf{f}$ only, with the line processes accounted for implicitly by the form of a "broken parabola" function ϕ defined as

$$\phi_{\lambda,\alpha}(\xi) \stackrel{\text{def}}{=} \begin{cases} \xi^2 & \text{if } \xi^2 \le \alpha/\lambda \\ \alpha & \text{if } \xi^2 > \alpha/\lambda \end{cases} \tag{4}$$

In this case, the WM and WP energies may be expressed as:

$$\begin{aligned} E_{WM}(\mathbf{f}) &= \sum_{i,j} \left[\phi_{\lambda,\alpha}(\sqrt{V_{i,j}^v}) + \phi_{\lambda,\alpha}(\sqrt{V_{i,j}^h})\right] \\ E_{WP}(\mathbf{f}) &= \sum_{i,j} \left[\phi_{\lambda,\alpha}(\sqrt{V_{i,j}})\right] \end{aligned}$$

Figure 2 [8] shows regional bias and variance results on reconstructions using these priors with an autorad phantom. These optimistic results were obtained with a laborious manual search for hyperparameters.

4. Calculating Hyperparameters via Maximum Pseudolikelihood

Given ground truth data in the form of an autoradiograph, and a prior model, we may use a maximum-likelihood (ML) approach to estimate the hyperparameters. With $\mathbf{f}$ now denoting the (known) ground truth data, we formulate a maximum-likelihood criterion for optimal parameters $\hat{\alpha}$ and $\hat{\lambda}$:

$$(\hat{\lambda}, \hat{\alpha}) = \arg\min_{\alpha,\lambda}(-\log \Pr[\mathbf{F} = \mathbf{f}; \lambda, \alpha])$$

Note that, typically, we are estimating only one or two hyperparameters from a large number (64×64) of data points. For Gibbs priors, we may express the likelihood term as

$$\Pr[\mathbf{F} = \mathbf{f}; \lambda, \alpha] = \frac{1}{Z(\lambda, \alpha)} \exp(-E(\mathbf{f}; \lambda, \alpha)),$$

where the partition function $Z(\lambda, \alpha)$ is given by

$$Z(\lambda, \alpha) = \sum_{\{\mathbf{w}\}} \exp[-E(\mathbf{w}; \lambda, \alpha)].$$

Summation over $\{\mathbf{w}\}$ denotes a sum over all possible object vectors. For example, for a discretization of one pixel to 256 values and a 64x64 image, this sum has $256^{64\times 64}$ terms, clearly intractable! To sidestep this intractability, we utilize a "pseudolikelihood" approximation to ML [9] given, in our context, by

$$\begin{aligned} PL(\mathbf{f};\lambda,\alpha) &= \prod_{i,j} \Pr[F_{i,j} = f_{i,j} \mid F_{k,l} = f_{k,l}, (k,l)\epsilon\mathcal{N}_{i,j}; \lambda, \alpha] \\ &= \prod_{i,j} \frac{\exp\{-E(f_{i,j} \mid F_{k,l} = f_{k,l}, (k,l)\epsilon\mathcal{N}_{i,j}; \lambda, \alpha)\}}{\sum_w \exp\{-E(w \mid F_{k,l} = f_{k,l}, (k,l)\epsilon\mathcal{N}_{i,j}; \lambda, \alpha)\}}, \end{aligned}$$

where indices k, l refer to the pixels in the neighborhood $\mathcal{N}_{i,j}$ at the site i, j and w spans all possible (scalar) values that may occur at a site. Neighborhood $\mathcal{N}_{i,j}$ consists of all pixels that interact with site (i, j) through E. We again convert to a minimization problem by taking a negative log:

$$\begin{aligned} -\log PL(\lambda,\alpha) &= \sum_{i,j} E(f_{i,j} \mid F_{k,l} = f_{k,l}, (k,l)\epsilon\mathcal{N}_{i,j}; \lambda, \alpha) \\ &+ \sum_{i,j} \left[\log \sum_{w} \exp(-E(w \mid F_{k,l} = f_{k,l}, k, l\epsilon\mathcal{N}_{i,j}; \lambda, \alpha)) \right] \quad (5) \end{aligned}$$

For the WM prior, the above evaluates to:

$$\begin{aligned} -\log PL(\lambda,\alpha) &= \\ &\sum_{i,j} \left\{ \phi_{\lambda,\alpha}(\sqrt{V^h_{i-1,j}}) + \phi_{\lambda,\alpha}(\sqrt{V^h_{i,j-1}}) + \phi_{\lambda,\alpha}(\sqrt{V^h_{i,j+1}}) + \phi_{\lambda,\alpha}(\sqrt{V^h_{i+1,j}}) \right\} \\ &+ \sum_{i,j} \Big[\log \sum_{w} \exp\{ -\phi_{\lambda,\alpha}(w - f_{i-1,j}) - \phi_{\lambda,\alpha}(w - f_{i,j-1}) \\ &\qquad -\phi_{\lambda,\alpha}(w - f_{i,j+1}) - \phi_{\lambda,\alpha}(w - f_{i+1,j})\}\Big]. \quad (6) \end{aligned}$$

For the simple membrane prior, we derive a closed-form solution for the single parameter $\hat{\lambda}$. To streamline the derivation, we denote four nearest neighbor pixels to $f_{i,j}$ as N_{ij}, S_{ij}, E_{ij} and W_{ij} (north,south,east,west). The pseudolikelihood solution is then:

$$\begin{aligned} \hat{\lambda} &= \arg\min_{\lambda} \sum_{i,j} \Big\{ \lambda \left\{ (f_{ij} - N_{ij})^2 + (f_{ij} - S_{ij})^2 + (f_{ij} - E_{ij})^2 + (f_{ij} - W_{ij})^2 \right\} \\ &+ \sum_{i,j} \log \Big[\sum_{w} \exp\{ -\lambda[(w - N_{ij})^2 + (w - S_{ij})^2 + (w - E_{ij})^2 + (w - W_{ij})^2]\}\Big]\Big\}. \end{aligned}$$

(7)

We may complete the square in the sum in [small brackets] in (7) and reexpress it as $\left(\frac{w-\mu}{\sigma}\right)^2 + A$, where $\mu(N,S,E,W) = \frac{N+S+E+W}{4}$, $\sigma = \frac{1}{2}$, and $A(N,S,E,W) \stackrel{\text{def}}{=} N^2 + S^2 + E^2 + W^2 - \frac{(N+S+E+W)^2}{4}$. (In the definitions for A and μ we drop subscripts i and j on N,S,E,W for convenience.)

We replace the sum over w with an integral: $\sum_w \to \int_{-\infty}^{\infty} dw$, and the integration over w becomes:

$$\exp\{-\lambda A(N,S,E,W)\} \int_{-\infty}^{\infty} \exp\left\{-\lambda\left(\frac{w-\mu}{\sigma}\right)^2\right\} dw$$
$$= \frac{1}{2}\sqrt{\frac{\pi}{\lambda}} \exp\{-\lambda A(N,S,E,W)\}. \tag{8}$$

At this point we have evaluated the term in $\Big[$ large brackets $\Big]$ in (7). Application of $\sum_{i,j}\log(\cdot)$ to the right side of (8) results, after dropping terms independent of λ, in the expression

$$-\frac{N}{2}\log\lambda - \lambda\sum_{i,j} A(N,S,E,W). \tag{9}$$

Here, N, the number of pixels in the autoradiograph, does not include pixels outside the region of support of the image as these lead to an estimate of λ based partly and incorrectly on the "flat" parts of the image field.

The insertion of (9) into (7) results in the following restatement of the minimization problem:

$$\hat{\lambda} = \arg\min_{\lambda}\{\lambda\sum_{i,j}[(f_{ij}-N_{ij})^2 + (f_{ij}-S_{ij})^2 + (f_{ij}-E_{ij})^2 + (f_{ij}-W_{ij})^2]$$
$$-\frac{N}{2}\log\lambda - \lambda\sum_{i,j} A(N_{ij},S_{ij},E_{ij},W_{ij})\}$$
$$= \arg\min_{\lambda}\{-\frac{N}{2}\log\lambda + \lambda\sum_{i,j} G_{ij}\}, \tag{10}$$

where the term G_{ij} may, after some algebra, be identified as

$$G_{ij} = \frac{1}{4}\left[(2f_{ij} - N_{ij} - S_{ij}) + (2f_{ij} - E_{ij} - W_{ij})\right]^2. \tag{11}$$

Note that in (11), the two terms in (parentheses) may be identified as discrete second derivatives in the horizontal and vertical directions, respectively. We identify the latter term in (10) as the discrete version of the number:

$$G \equiv \sum_{i,j} G_{ij} \to \frac{1}{4}\iint \left(\nabla^2 f(x,y)\right)^2 dxdy. \tag{12}$$

The minimization problem has been further reduced to

$$\hat{\lambda} = \arg\min_{\lambda}\{-\frac{N}{2}\log\lambda + \lambda G\} = \arg\min_{\lambda}\{PL(\lambda)\}$$

which may be solved via $\frac{dPL(\lambda)}{d\lambda} = 0$ to get the solution

$$\hat{\lambda} = N/2G. \qquad (13)$$

Equation (13) together with the definition in (12) is our solution. Note that since

$$\frac{d^2PL}{d\lambda^2} = \frac{N}{2\lambda^2} > 0$$

our solution for $\hat{\lambda}$ is indeed a minimum, and that since N and G are clearly nonnegative, $\hat{\lambda}$ is also nonnegative as desired.

Intuitively, $\hat{\lambda}$ should be dependent upon the amount of noise (total bin counts) in the projection data. Yet the value of $\hat{\lambda}$ was derived from noise*less* training data. The noise level does enter, however, through the scaling of the autoradiograph values **f**, which are the means of Poisson densities for γ-emission at each location. With proper scaling of **f** to account for the typical number of bin counts, λ depends on noise after all. It is easy to show that **f** should be scaled according to

$$f_{i,j} = \frac{K}{\sum_{t,\theta} H_{t,\theta,i,j}} f_{i,j}^{\text{norm}}, \qquad (14)$$

where K equals the desired total bin counts, $H_{t,\theta,i,j}$ is the probability of a count at (t,θ) from pixel (i,j), and f^{norm} is a version of the autorad normalized so that $\sum_{i,j} f_{i,j}^{\text{norm}} = 1$. In (13), $\hat{\lambda}$ scales with the inverse square of the count level, i.e. more counts means less smoothing as expected. The value of $\hat{\lambda}$ also depends on geometrical scaling; if **f** is scaled by factor Δ (in 2-D), then $\hat{\lambda} \rightarrow \hat{\lambda}\Delta^4$. In summary, one first inserts noise level and geometrical information via (14) to obtain a scaled autoradiograph $f_{i,j}$ and inserts this into (13) to obtain $\hat{\lambda}$.

5. Numerical Studies

In order to test the performance of our optimal $\hat{\lambda}$ for the membrane prior, we created 40 noisy projection data sets of the phantom in Fig. 1(a). The count level, approx. 730,000 counts, was equivalent to that of a typical SPECT brain scan on a per-pixel basis. For each projection data set, reconstructions using the membrane prior were performed for nine values $\lambda = \lambda_0 \times 2^n, n = 0, \pm 1, \pm 2 \pm 3 \pm 4$, thus yielding 360 reconstructions ($\lambda_0 = 0.02$).

Figure 3(a) shows a plot of root mean-squared error (RMSE) vs. λ for one projection data set, and indicates the position of the value for $\hat{\lambda}$ obtained by training on the same phantom. As seen, the derived value is not far from the one that minimizes RMSE. Figures 3 (b) and (c) show the reconstructions using the λ that minimizes RMSE and the $\hat{\lambda}$ obtained from maximum pseudolikelihood (MPL), respectively.

Figure 4 shows bias-variance variation with λ. The definitions for regional bias and standard deviation are: $b_R = \sqrt{\sum_{i,j\in \mathrm{ROI}} b_{i,j}^2}$, where $b_{i,j}$ is the pointwise bias at (i,j), and $s_R = \sqrt{\sum_{i,j\in \mathrm{ROI}} \sigma_{i,j}^2}$, where $\sigma_{i,j}^2$ is the pointwise variance at (i,j). Figure 5 shows bias variance results for membrane reconstructions. Here, bias and variance were calculated for a random variable equal to the number of counts in an ROI.

Table 1. summarizes a few trials used to anecdotally explore issues of generalization. Each row displays $\hat{\lambda}$ and also λ_{RMSE}, the value that minimizes, for the phantoms in Fig. 1. Phantom A is a sagittal autorad primate phantom. Since phantoms B and C are transverse slice autorads, from a single primate, spaced at 5 mm, we expect the parameter values to be similar for each, and they are. (Training and reconstruction were done on only central regions of these phantoms.) By contrast, we expect the Hoffman brain phantom D, which consists of flat regions with sharp edges, to have fairly different values. This is indeed the case. Phantom E consists of pixels sampled independently from a uniform [0, 255] distribution. It has correspondingly different λ values. Note that there are general trends: "Rougher" (more edge content) phantoms have lower values of $\hat{\lambda}$ and λ_{RMSE}, as expected, and $\hat{\lambda}$ and λ_{RMSE} are correlated for a given phantom.

TABLE 1.

Phantom	$\hat{\lambda}$ obtained from MPL	λ for minimum RMSE
A	0.033	0.020
B	0.039	0.022
C	0.031	0.025
D	0.00167	0.00175
E	0.00228	0.00228

We evaluated RMSE's for WM and WP reconstructions of phantom A as a function of α and λ. These are shown in Fig. 6(a)(b). For these plots, iterations were terminated at minimum RMSE relative to the phantom. We also evaluated (6) numerically to explore the variation of WM pseudolikelihood with λ and α. The contour plot in Fig. 6 (c) shows a smooth unimodal surface over the range of α and λ used. This result implies that a descent procedure used to calculate $\hat{\lambda}$ and $\hat{\alpha}$ may work well.

6. Summary and Discussion

The essence of our approach has been to consider "ground-truth" exemplar phantoms as samples from an underlying Gibbs prior, and to fit hyperparameters of this prior model to the exemplars. Two unproven presumptions are that the optimal values thus derived will lead to reconstructions with an image quality not far from optimal, and that the training will generalize to objects that are similar to the training set. Our initial experiments reported here give optimistic anecdotal evidence for these presumptions. We note that the prospects for generalization is made somewhat more practical in that we show how to scale optimal parameters for count level and geometrical scaling.

References

1. V. E. Johnson, W. H. Wong, X. Hu, and C. T. Chen, "Image Restoration Using Gibbs Priors: Boundary Modeling, Treatment of Blurring, and Selection of Hyperparameter", *IEEE Trans. on Pattern Analysis and Machine Intelligence*, PAMI-13(5), pp. 413–425, May 1991.
2. Z. Zhou, R. M. Leahy, and E. U. Mumcuoglu, "A Comparative Study of the Effects of Using Anatomical Priors in PET Reconstruction", In *Proc. IEEE Nuclear Science Symposium and Medical Imaging Conference*, volume 3, pp. 1749–1753, November 1993.
3. S. Geman and C. Graffigne, "Markov random fields image models and their application to computer vision", In A. M. Gleason, editor, *Proc. of the Intl. Congress of Mathematicians 1986*, pp. 1496–1517, American Mathematical Society, Providence, 1987.
4. E. J. Hoffman, P. D. Cutler, W. M. Digby, and J. C. Mazziotta, "3-D Phantom To Simulate Cerebral Blood Flow and Metabolic Images for PET", *IEEE Trans. Nuclear Sci*, 37, pp. 616–620, 1990.
5. J. L. Lear, "Principles of Single and Multiple Radionuclide Autoradiography", In M. E. Phelps, J. C. Mazziotta, and H. R. Schelbert, editors, *Positron Emission Tomography and Autoradiography*, chapter 5, Raven Press, New York, NY, 1986.
6. A. Blake and A. Zisserman, *"Visual Reconstruction"*, Artificial Intelligence, MIT Press, Cambridge, MA, 1987.
7. S. Geman and D. Geman, "Stochastic Relaxation, Gibbs Distributions and the Bayesian Restoration of Images", *IEEE Trans. on Pattern Analysis and Machine Intelligence*, PAMI-6(6), pp. 721–741, November 1984.
8. S. J. Lee, A. Rangarajan, and G. Gindi, "A Comparative Study of the Effects of Using Higher Order Mechanical Priors in SPECT Reconstruction", In *IEEE Nuclear Science Symposium and Medical Imaging Conference (in press)*, November 1994.
9. J. Besag, "Efficiency of Pseudo-likelihood estimation for simple Gaussian Fields", *Biometrika*, 64, pp. 616–618, 1977.

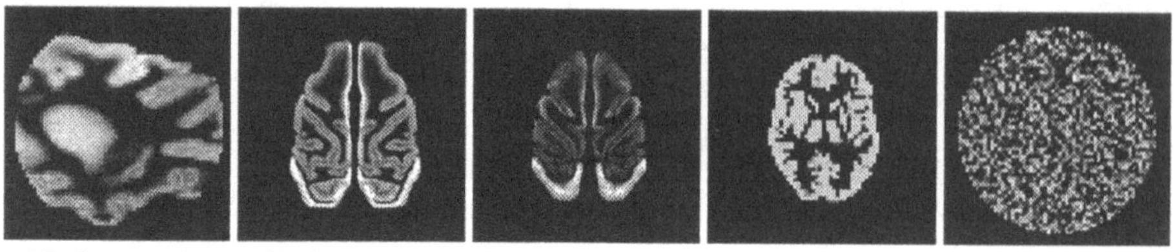

(a) phantom A (b) phantom B (c) phantom C (d) phantom D (e) phantom E

Figure 1. Phantoms used in the simulations. (a) Primate autoradiograph phantom obtained with the blood flow agent (^{99m}Tc-ECD). (b)(c) Primate autoradiograph phantoms obtained with the benzodiazepine neuroreceptor agent Iomazenil (^{123}I). (d) Hoffman brain phantom. (e) Circle phantom filled with uniform noise.

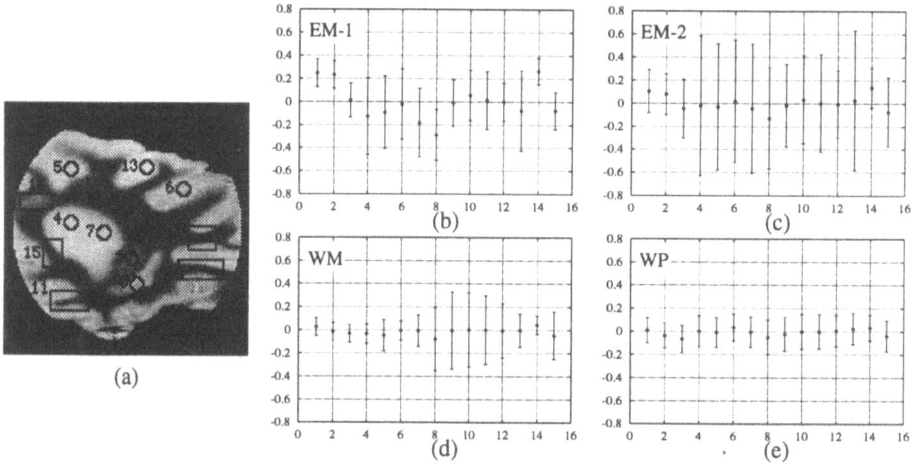

Figure 2. Regional bias-standard deviation plots for phantom A. (a) The phantom with region borders superposed. In (b)-(e), the abscissa indexes the region number, and plotted points with error bars indicate bias ± one standard deviation.

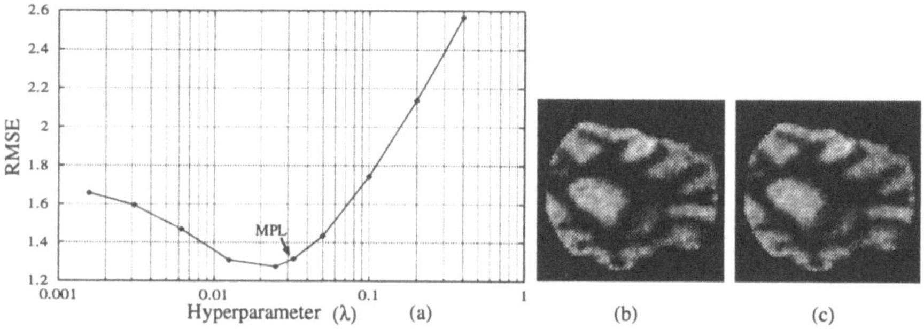

Figure 3. (a) Plot of RMSE vs. hyperparameter (λ) for the membrane reconstruction of phantom A. The arrow denotes λ obtained from MPL. (b) Reconstruction using λ for minimum RMSE. (c) Reconstruction using $\hat{\lambda}$ obtained from MPL.

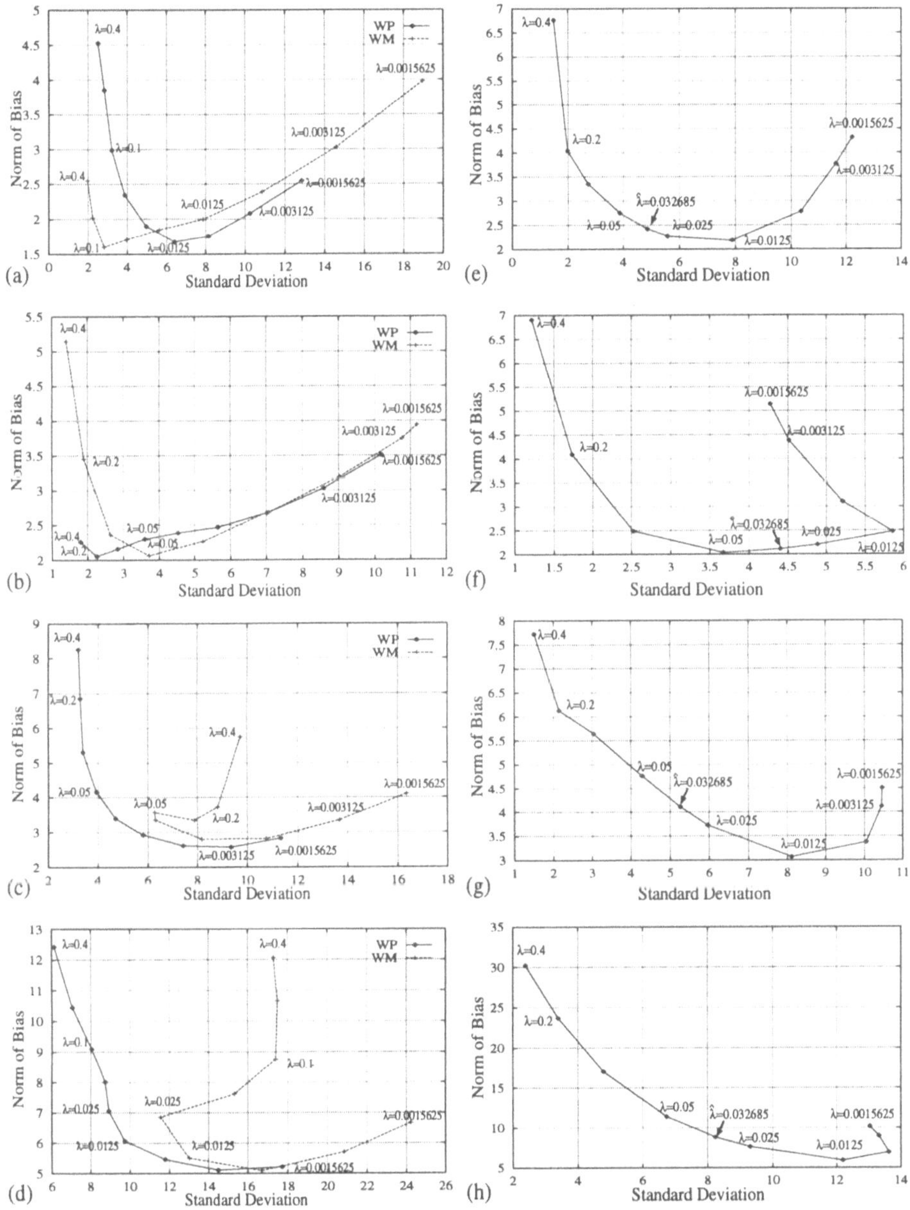

Figure 4. The effects of several different values for λ in WM, WP and membrane reconstructions. (a)-(d): Regional bias (b_R) standard deviation (s_R) plots for WM and WP reconstructions. (e)-(h): b_R and s_R plots for membrane reconstructions. The four representative regions in phantom A – (a)(e) high activity region (grey matter), (b)(f) low activity region (white matter), (c)(g) low contrast (ramplike) region, and (d)(h) high contrast region (anatomical boundary) – were measured. The arrows in (e)-(f) indicate the $\hat{\lambda}$ obtained from MPL.

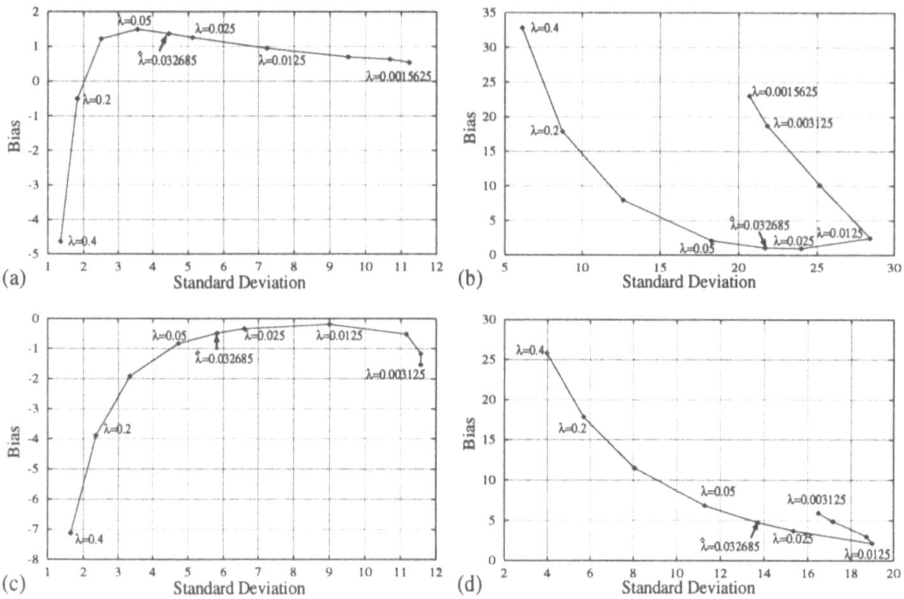

Figure 5. The effects of several different values for λ in membrane reconstructions. The same regions as in Fig. 4 were measured by using the standard definitions for regional bias and standard deviation. The arrows indicate the $\hat{\lambda}$ obtained from MPL.

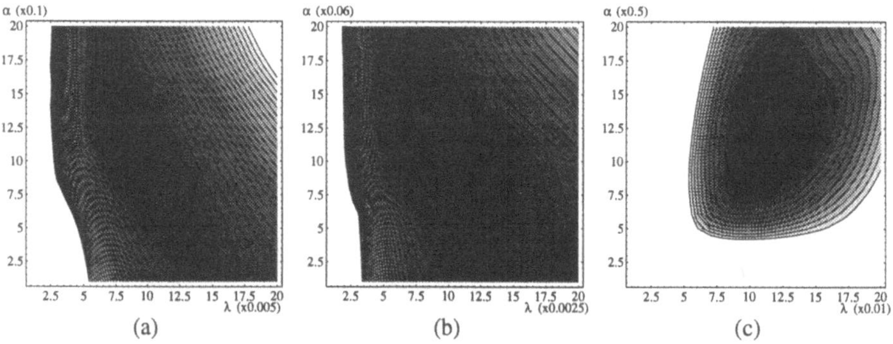

Figure 6. Contour plots for sensitivity of the hyperparameters λ (abscissa) and α (ordinate) in terms of RMSE and maximum pseudolikelihood. (a) Plot for WM. (b) Plot for WP. Each contour level in (a) and (b) indicates the equi-error within a range and the darkest areas correspond to the hyperparameters for the least RMSE's. (c) Plot for WM. Each contour level in (c) indicates the equi-pseudolikelihood within a range and the darkest area corresponds to the hyperparameters for the maximum pseudolikelihood.

MAXIMUM LIKELIHOOD HYPERPARAMETER ESTIMATION FOR GIBBS PRIORS WITH APPLICATIONS TO PET

Z. ZHOU, R. M. LEAHY AND E. U. MUMCUOGLU
Signal and Image Processing Institute,
University of Southern California,
Los Angeles, CA 90089-2564

Abstract. The choice of the parameters of the prior, *the hyperparameters*, plays a critical role in Bayesian PET image estimation and other applications. Of particular importance for the case of Gibbs priors is the global hyperparameter which multiplies the Hamiltonian. Here we consider maximum likelihood (ML) estimation of the global hyperparameter from incomplete data. For PET, the incomplete data are the measured projection data, in general, they are some noisy transformation on the unknown image. True ML estimation of hyperparameters from incomplete data is intractable for most image processing problems. Here we develop an approximate ML estimator for the global hyperparameter which is computed simultaneously with a MAP image estimate. This new algorithm relies on an approximation closely related to the mean field theory of statistical mechanics. We present the results of a Monte-Carlo study that examines the bias and variance of this estimator and shows its application to PET.

1. Introduction

Bayesian methods using Markov random field (MRF) priors have proven to be very powerful for reconstruction of high quality images. However, a major problem limiting their utility is the lack of a practical and robust method for selecting the parameters of the prior, i.e. the *hyperparameters*. Of particular importance for the case of Gibbs priors is the global hyperparameter β that multiplies the Gibbs energy function. This hyperparameter controls the degree of influence of the Gibbs prior on the solution.

Y. Bizais et al. (eds.), Information Processing in Medical Imaging, 39–51.

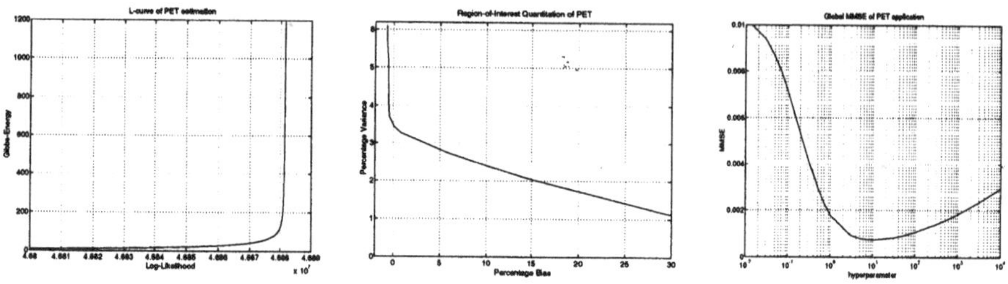

Figure 1. Illustration of the quantitative effect of the global hyperparameter (a) L-curve, (b)percentage bias versus percentage variance in ROIs for a range of hyperparameter values, (c) Global mean-squared-error of reconstructed PET image versus β.

In this work we are primarily interested in estimating an image λ from a set of data y. For example, in the PET application described later, λ is the image of mean emission rates from the source pixels and y is the coincidence sinogram. The image is estimated by maximizing over the posterior density $P(\lambda|y)$ to form a *maximum a posteriori* (MAP) estimate. The posterior density is proportional to the product of the likelihood function for the data, $P(y|\lambda)$, and a prior on the image, $P(\lambda|\beta)$. The Gibbs prior is of the form $P(\lambda|\beta) = Z_\beta^{-1} exp\{-\beta U(\lambda)\}$ where $U(\lambda)$ is the Gibbs energy, Z_β is the partition function, and β is the hyperparameter. To illustrate the effect of the hyperparameter we show three curves in Figure 1 computed for a typical application to PET image reconstruction. In Figure 1(a), we have a typical L-curve [6], which is a plot, for all valid β, of the value of the Gibbs energy $U(\lambda)$ versus the log-likelihood $lnP(y|\lambda)$, computed at the MAP solution for each value of β. Similarly, in (b), we plot the percentage bias versus variance of a MAP image estimate for the average intensity computed over a pre-selected ROI for a range of hyperparameter values. In both curves, we observe two characteristic parts, namely a flat part where the MAP solution is dominated by the prior and has high bias, and an almost vertical part, where the solution is dominated by the likelihood function and has high variance. Heuristically, the region between these two characteristic parts, i.e. the "corner", corresponds to a good balance between fidelity to the data and smoothness of the solution, or equivalently, between bias and variance in the estimate. Figure 1(c) shows a curve of the squared error in the MAP estimate for a range of hyperparameter values. It is clear from these examples that an appropriate method for selecting β is necessary to control the trade off between bias and variance.

The hyperparameter is often chosen in an *ad hoc* fashion through visual inspection of the resulting images. There are two basic approaches for choosing β in a more principled manner: (i) treating β as a regularization parameter and applying techniques such as generalized cross validation, the L-curve, and χ^2 goodness of fit tests; (ii) estimation theoretic approaches such as maximum likelihood. We restrict our attention to the latter class.

The image, λ, which is drawn from the process characterized by the parameter β, is not observed directly. Instead, we observe a second process which is conditioned on this image. Selection of the hyperparameter can therefore be viewed as a ML estimation problem in an incomplete/complete data framework and is a natural candidate for the EM algorithm. However, in most imaging applications, the high dimensionality and the complexity of the densities involved make the EM approach intractable. Geman and McClure [4] propose using a stochastic relaxation technique, such as a Gibbs sampler, to evaluate the E-step of the EM algorithm. While this approach provides a means of overcoming the intractability of the true EM algorithm, the computational cost remains extremely high. Several other estimation methods have been studied which do not share the desirable properties of true MLE but do offer computational tractability. These include generalized ML [1, 7] and the method of moment [4] (see Section 2.2).

Here we describe an approximation based on the mean field theory of statistical mechanics. The method allows simultaneous ML hyperparameter estimation and MAP image reconstruction. We present a brief summary of the problem formulation and describe the approximate ML estimation scheme below. We present the results of a Monte-Carlo study that examines the bias and variance of this estimator and show its applications to PET.

2. Hyperparameter Estimation from Incomplete Data

2.1. GIBBS PRIORS FOR IMAGE ESTIMATION

We will assume a homogeneous isotropic MRF model for the image, λ, characterized by the Gibbs distribution:

$$P(\lambda|\beta) = \frac{1}{Z(\beta)} \exp\{-\beta U(\lambda)\} \tag{1}$$

with Gibbs energy $U(\lambda)$, partition function $Z(\beta)$ and global hyperparameter β. The partition function is given by $Z(\beta) = \int_\lambda \exp\{-\beta U(\lambda)\} d\lambda$. Here we confine the Gibbs energy function to pairwise interactions:

$$U(\lambda) = \sum_j \sum_{k>j, k\in N_j} \kappa_{jk} V(\lambda_j, \lambda_k), \tag{2}$$

where N_j denotes the set of eight nearest neighbors of pixel j with κ_{jk} unity for horizontal and vertical neighbors and $1/\sqrt{2}$ for diagonal neighbors.

The defining feature of a MRF is $P(\lambda_j|\lambda_k, k \neq j) = P(\lambda_j|\lambda_k, k \in N_j)$. For the pairwise interaction model above, the conditional density has the special form:

$$P(\lambda_j|\lambda_k, k \in N_j) = \frac{\exp\{-\beta \sum_{k \in N_j} \kappa_{jk} V(\lambda_j, \lambda_k)\}}{\int_{\lambda_j} \exp\{-\beta \sum_{k \in N_j} \kappa_{jk} V(\lambda_j, \lambda_k)\} d\lambda_j}. \tag{3}$$

The specific potential functions, $V(\cdot, \cdot)$, used in this work are as follows:

$$\begin{aligned} &\text{Quadratic function:} && V_1(\lambda_j, \lambda_k) = (\lambda_j - \lambda_k)^2 \\ &\text{Huber function:} && V_2(\lambda_j, \lambda_k) = \begin{cases} \frac{1}{2\delta}(\lambda_j - \lambda_k)^2, & \text{if } |\lambda_j - \lambda_k| \leq \delta \\ |\lambda_j - \lambda_k| - \frac{\delta}{2}, & \text{otherwise} \end{cases} \\ &\text{Geman and McClure:} && V_3(\lambda_j, \lambda_k) = \frac{(\lambda_j - \lambda_k)^2}{\delta^2 + (\lambda_j - \lambda_k)^2}. \end{aligned} \tag{4}$$

All three potential functions represent stochastic models of smooth images. However, $V_1(\cdot)$ favors images with small image gradients everywhere, while $V_3(\cdot)$ favors piecewise flat images with step edges. The Huber function, $V_2(\cdot)$ represents a compromise between these two.

2.2. ML ESTIMATION OF β FROM INCOMPLETE DATA

The observed incomplete data, y, is related to the hyperparameter β only indirectly through the complete data, the image λ. A maximum likelihood estimate of β from incomplete data can be found only from the marginalized likelihood function:

$$P(y|\beta) = \int_\lambda P(y|\lambda) P(\lambda|\beta) d\lambda = \frac{\int_\lambda \exp\{\ln P(y|\lambda) - \beta U(\lambda)\} d\lambda}{\int_\lambda \exp\{-\beta U(\lambda)\} d\lambda} = \frac{Z(y, \beta)}{Z(\beta)} \tag{5}$$

where

$$Z(y, \beta) = \int_\lambda \exp\{\ln P(y|\lambda) - \beta U(\lambda)\} d\lambda \tag{6}$$

is the partition function of the posterior density. It can be shown [5] that

$$\frac{\partial \ln Z(\beta)}{\partial \beta} = -\frac{\int_\lambda U(\lambda) \exp\{-\beta U(\lambda)\} d\lambda}{\int_\lambda \exp\{-\beta U(\lambda)\} d\lambda} = -E[U(\lambda)|\beta], \tag{7}$$

$$\frac{\partial \ln Z(y, \beta)}{\partial \beta} = -\frac{\int_\lambda U(\lambda) \exp\{\ln P(y|\lambda) - \beta U(\lambda)\} d\lambda}{\int_\lambda \exp\{\ln P(y|\lambda) - \beta U(\lambda)\} d\lambda} = -E[U(\lambda)|y, \beta]. \tag{8}$$

The β that maximizes (5) is found as the β for which the gradient of the log of the marginalized likelihood (5) is zero. It follows from (5), (7) and (8) that the ML estimate of β is a root of the likelihood equation [4]:

$$E[U(\lambda)|y, \beta] = E[U(\lambda)|\beta], \tag{9}$$

where $E[.|y,\beta]$ and $E[.|\beta]$ denote expectations with respect to the posterior and prior densities respectively.

This equation can in principle be solved using an EM algorithm, but it is intractable in general because of the large dimensionality and complexity of the joint Gibbs distribution. Geman and McClure [4] note that a solution can be found using stochastic sampling from the posterior and prior densities to approximate the expectations. Due to the complexity of sampling from the posterior, the computation cost remains unacceptable. In [4], a second estimation method is also described. This method of moments (MOM) simply requires the computation of a statistic $M(y)$ of the data. The parameter β is then chosen as a root of the equation: $M(y) = E[M(y)|\beta]$, where the moment curve $E[M(y)|\beta]$ is pre-computed for a large range of β and should be monotonic to ensure the identifiability of the hyperparameter.

In [1, 7] an alternative simplified approach is taken whereby, instead of maximizing with respect to β over the marginalized density (5), β is computed with λ as the pair $\{\beta, \lambda\}$ that jointly maximize $P(y,\lambda|\beta)$:

$$\{\hat{\lambda},\hat{\beta}\} = \arg\max_{\lambda,\beta} \ln P(y,\lambda|\beta) = \arg\max_{\lambda}\{\ln P(y|\lambda) + \max_{\beta} \ln P(\lambda|\beta)\}. \quad (10)$$

Some authors have called this the generalized ML (GML) method [7]. In practice, direct maximization of $\ln P(\lambda|\beta)$ with respect to β is difficult for most priors. This step can be approximated using a maximum pseudo-likelihood approach [1]. We refer to this as the generalized maximum pseudo-likelihood (GMPL) method in the following.

2.3. MEAN FIELD APPROXIMATIONS

The intractability of true MLE of β is due to the complexity and dimensionality of the joint density $P(y,\lambda|\beta)$ - it is essentially impossible to compute the marginal density in (5) for each new data set. The problem becomes solvable if we approximate the complicated multidimensional densities with a separable joint density function equal to a product of one dimensional probability densities, one per pixel. The multi-dimensional integrals involved in computing the marginal density in (5), partition functions, or moments, can then be approximated with a product or sum of one dimensional integrals with respect to these one dimensional pixel-wise densities.

Approximating Gibbs distributions using separable joint density functions is the basis for the mean field theory in statistical mechanics [2]. The basic idea is to focus on one particular particle (in our case a pixel site) in the system and assume that the role of the neighboring particles (pixels) can be approximated by an average field which acts on the tagged particle. This approach, therefore, neglects the effects of statistical fluctuations in all pixels other than current tagged one. The corresponding joint description is simply the product of that for each individual particle or pixel.

The mean field approximation defines a separable *mean field reference system* with density $P_{MF}(\lambda)$, to approximate the original Gibbs distribution $P(\lambda)$. For simplicity of notation, we use $P(\lambda) = \frac{1}{Z}\exp\{-E(\lambda)\}$ to indicate the original Gibbs distribution. The mean field approximation can be applied to either the prior, $P(\lambda|\beta)$, or posterior, $P(\lambda|y,\beta)$, densities in Bayesian inverse problems, with $E(\lambda) = \beta U(\lambda)$ for the prior and $E(\lambda) = -\ln P(y|\lambda) + \beta U(\lambda)$ for the posterior.

Since the reference field is separable in λ, the new mean field Gibbs distribution $P_{MF}(\lambda)$ can be written as the product of a set of 1-D densities:

$$P_{MF}(\lambda) = \frac{1}{Z_{MF}}\exp\{-E_{MF}(\lambda)\} \equiv \prod_j P_j^{mf}(\lambda_j), \tag{11}$$

where the local reference densities are defined as [2, 10],

$$P_j^{mf}(\lambda_j) = P(\lambda_j|\lambda_{\partial j})|_{\lambda_{\partial j}=<\lambda_{\partial j}>^{mf}}, \text{ with } <\lambda_j>^{mf} = \int_{\lambda_j} \lambda_j P_j^{mf}(\lambda_j)d\lambda_j \tag{12}$$

and ∂j denotes the set of neighbor pixels of site j for the neighborhood system of the original field $P(\lambda)$. Note that the mean $<\lambda_j>^{mf}$ is computed with respect to the reference field. The conditional densities in (11) must be of the form

$$P_j^{mf}(\lambda_j) = \frac{1}{Z_j^{mf}}\exp\{-E_j^{mf}(\lambda_j; <\lambda_{\partial j}>^{mf})\} \tag{13}$$

where the local energy function $E_j^{mf}(\lambda_j; <\lambda_{\partial j}>^{mf})$ is a sum over all potential functions in the original Gibbs energy, $E(\lambda)$, that include λ_j, and the corresponding partition function Z_j^{mf} is given by:

$$Z_j^{mf} = \int_{\lambda_j} \exp\{-E_j^{mf}(\lambda_j; <\lambda_{\partial j}>^{mf})\}d\lambda_j. \tag{14}$$

Combining the local energy and partition functions gives the overall *mean field energy function* $E_{MF}(\lambda)$, and *mean field partition function*, Z_{MF}:

$$E_{MF}(\lambda) = \sum_j E_j^{mf}(\lambda_j; <\lambda_{\partial j}>^{mf}) \qquad Z_{MF} = \prod_j Z_j^{mf}. \tag{15}$$

Note that the mean field $<\lambda>^{mf}$ must be computed recursively as the local reference densities $P_j^{mf}(\lambda_j)$ are coupled through $<\lambda>^{mf}$. This recursive procedure can be performed using the well known iterated conditional expectation (ICE) method which has been used elsewhere for MMSE and MAP image estimation.

Clearly the specific form of the local energy must depend on the original Gibbs distribution that is being approximated. For the methods described below, we make use of mean field approximations of both the prior and posterior densities. For the Gibbs priors with pairwise interactions defined in Section (2.1), it follows from (3) and (13) that the reference local energies should be of the form:

$$E_j^{mf}(\lambda_j; <\lambda_{\partial j}>^{mf}) = \beta \sum_{k \in N_j} V(\lambda_j, \lambda_k^{PR}) \tag{16}$$

where $\lambda_j^{PR} = <\lambda_j>^{mf}$ is the mean of the mean field reference system for the prior. For the posterior density, the likelihood for the data $P(y|\lambda)$ must also be included:

$$E_j^{mf}(\lambda_j; <\lambda_{\partial j}>^{mf}) = -\ln P(y|\lambda_j, \lambda_{S\backslash j}^{PO}) + \beta \sum_{k \in N_j} V(\lambda_j, \lambda_k^{PO}) \tag{17}$$

where $\lambda^{PO} = <\lambda_j>^{mf}$ is the mean of the mean field reference system for the posterior density. Here $S\backslash j$ denotes the set of all pixels except j. Note that the introduction of the likelihood term, $P(y|\lambda)$, generally increases the size of the neighborhood system of the posterior relative to the prior. In the PET case considered below, the neighborhood graph for the posterior density becomes fully connected.

2.4. HYPERPARAMETER ESTIMATION

Above we described how a Gibbs distribution can be approximated by a mean field reference system. We now use this mean field approximation to simplify the hyperparameter estimation problem described in Section (2.2).

We can use the mean field approximated prior and posterior partition functions directly in (7) and (8). Using the definitions of the local reference field energies, (16) and (17), and the relationship between these and the reference field partition function (15) we have:

$$Z(\beta) \approx Z^{PR}(\beta) = \prod_j \int_{\lambda_j} \exp\{-\beta \sum_{k \in N_j} V(\lambda_j, \lambda_k^{PR})\} d\lambda_j \tag{18}$$

$$Z(y, \beta) \approx Z^{PO}(y, \beta) = \prod_j \int_{\lambda_j} \exp\{\ln P(y|\lambda_j, \lambda_{S\backslash j}^{PO}) - \beta \sum_{k \in N_j} V(\lambda_j, \lambda_k^{PO})\} d\lambda_j \tag{19}$$

Differentiating the logarithm of the two partition functions in β and equating them in (9) gives us the approximate likelihood equation:

$$\sum_j \frac{\int_{\lambda_j} U_j^{PO}(\lambda_j) \exp\{\ln P(y|\lambda_j, \lambda_{S\backslash j}^{PO}) - \beta U_j^{PO}(\lambda_j)\} d\lambda_j}{\int_{\lambda_j} \exp\{\ln P(y|\lambda_j, \lambda_{S\backslash j}^{PO}) - \beta U_j^{PO}(\lambda_j)\} d\lambda_j}$$
$$= \sum_j \frac{\int_{\lambda_j} U_j^{PR}(\lambda_j) \exp\{-\beta U_j^{PR}(\lambda_j)\} d\lambda_j}{\int_{\lambda_j} \exp\{-\beta U_j^{PR}(\lambda_j)\} d\lambda_j}, \quad (20)$$

where $U_j^{PR}(\lambda_j) = \sum_{k \in N_j} V(\lambda_j, \lambda_k^{PR})$ and $U_j^{PO}(\lambda_j) = \sum_{k \in N_j} V(\lambda_j, \lambda_k^{PO})$. The mean field approximated ML estimate of β (MFAML1) is a root of (20).

2.5. SELECTION OF OPTIMAL APPROXIMATION VIA THE GBF BOUND

The mean field approach described above was developed heuristically, i.e. we argued that the statistical influence of the neighbors of each site can be approximated by their means, and on this basis the separable reference field is selected. There are of course an infinite number of separable functions that could be used for the reference field. Here we provide a brief outline for selecting an optimal reference field for use in partition function approximation and hence for hyperparameter estimation. This approach is based on the *Gibbs-Bogoliubov-Feynman bound* [2] and first order perturbation theory. Unfortunately, a closed form solution to this problem is possible only for a restricted class of Gibbs energies. Due to space limitations the full method will not be presented here. A full treatment is given in [12].

Theorem 1 Gibbs-Bogoliubov-Feynman (GBF) bound
For a Gibbs distribution with partition function Z and Gibbs energy E, and any mean field reference Gibbs distribution with partition function Z_{MF} and Gibbs energy E_{MF}, we have the following inequality

$$Z \geq Z_{MF} \exp\{- < E - E_{MF} >_{MF}\}. \quad (21)$$

where $< \ldots >_{MF} \stackrel{\text{def}}{=} Z_{MF}^{-1} \int_{\mathcal{X}} [\ldots] \exp(-E_{MF}) d\mathbf{x}$.

This result leads to an optimal approximation of the partition function. Since $Z_{MF} \exp\{- < E - E_{MF} >_{MF}\}$ can never exceed the true Z, it follows that the optimal separable reference field, i.e. that which leads to the best approximation of the partition function, can be found by maximizing the right-side of (21) over all separable reference fields.

This optimization problem can be solved for a restricted class of MRFs. In cases where the Gibbs energy can be written in the form:

$$E(\lambda) = \sum_j \left[\sum_k a_{jk} \lambda_j^k + \frac{1}{2} \sum_{i \in N_j} b_{ij} \lambda_i \lambda_j \right], \tag{22}$$

an optimal reference field can be found by choosing the ΔH_j in the reference field energy:

$$E_{MF}(\lambda) = \sum_j \left[\sum_k a_{jk} \lambda_j^k + \Delta H_j \lambda_j \right] \tag{23}$$

to maximize the right hand side of the GBF bound. Examples of fields for which this approach can be applied include (i) a prior MRF with the quadratic potential function $V_1(\cdot)$, and (ii) a posterior MRF where the likelihood function $P(y|\lambda)$ is Gaussian with mean a linear function of λ and the prior is again quadratic. In the case of the Poisson likelihood function used in PET, and/or the non-quadratic priors, the optimization problem does not lend itself to a closed form solution.

In [12], we show that (i) for the case of Gaussian prior and posterior densities the optimal reference fields are identical to those described by the local energy function (16) and (17) above; (ii) the constant $< E - E_{MF} >_{MF}$ is zero for the quadratic prior but depends on the posterior mean field in the case of the posterior partition function approximation. Now consider the impact on the approximate method described above. The method described in 2.4 equates the derivatives of the prior and posterior log-partition functions. Using the GBF bound results in a modification of the derivatives, since the constants are different for the prior and posterior partition function, and hence the mean field approximated likelihood equation is modified to include this difference. This constitutes our second method (MFAML2) and is compared with the others in our Monte-Carlo experiment.

2.6. MEAN VS. MODE FIELD APPROXIMATIONS

In many imaging applications, including PET, we are more interested in computing a MAP estimate of the image than a minimum mean square error (MMSE) estimate. These correspond respectively, to the mode and the mean of the posterior densities. Therefore, rather than also computing the mean field of the posterior reference field, we replace the mean field with a *mode-field*. This mode is computed using an iterative MAP estimation procedure. Note that using the separable approximations described above, the mode of the original and reference fields are identical. In cases where the posterior density is unimodal and symmetric, mean and mode fields are equivalent. Such is the case for Gaussian data with the convex potential functions V_1 and V_2 defined in (4). For PET data, the Poisson likelihood is asymmetric and the two methods are not equivalent. However, for relatively high count data, the Poisson likelihood is well approximated by a symmetric Gaussian function (i.e. the log-likelihood function $\ln P(y|\lambda)$ is approximately symmetric). We therefore anticipate only minor differences between the mode-field and mean-field approximations in this case.

Using the approximations described above, the MAP estimate of the image and the ML estimate of the hyperparameter are jointly computed using a two step iteration:

[1] Initialize the image $\lambda^k = \lambda^0$ and hyperparameter $\beta^k = \beta^0$. Set $k = 0$.

[2] Maximize $P(\lambda|y;\beta^k)$ to find λ^{k+1}.

[3] Solve the approximated likelihood equation for β^{k+1} using λ^{k+1} as the current posterior mode field.

[4] Set $k = k + 1$, goto step [2].

In practice, neither steps 2 or 3 need be iterated to convergence before moving to the next step. We have no convergence proof for this method. However, in running the method for a wide range conditions, we have never observed a case in which the method does not converge.

3. Performances Studies

3.1. ESTIMATOR BIAS AND VARIANCE USING STOCHASTIC SAMPLING

We have compared the performance of the two mode field approximated maximum likelihood methods, (MFAML1-2), with a generalized maximum pseudo-likelihood (GMPL) technique and the method of moments (MOM) [4]. In the MOM, the statistic $M(y)$ is chosen as the same form as the Gibbs energy function of the prior, computed over the sinogram y with 8-nearest neighbor interactions.

We performed Monte Carlo studies for image restoration as follows. For each value of the hyperparameter, 50 sample images were drawn from the quadratic prior, V_1 in (4), using a stochastic Gibbs sampler. Each sampled image was then blurred by a 3×3 kernel with 0.884 at the center, 0.028 at its nearest neighbors and 0.001 at the diagonals. Pseudo-random Gaussian noise was generated to contaminate each of these images. The hyperparameters were then estimated using each of the methods listed above. Ensemble averaging over the 50 hyperparameter estimates was used to compute bias and variance of the estimators. The results shown in Table 1 are for additive Gaussian noise $N(0, 16)$ on an image with intensities in the range $(0, 100)$. Note that in all cases, both MFAML methods out-perform both of the other techniques. The differences are very clear for the cases where β is large, corresponding to the case of very smooth images. Although the results deteriorate for all estimators as β increases, the performance of

TABLE 1. Monte Carlo Test comparing performance of generalized maximum pseudo-likelihood (GMPL), the method of moments (MOM), and the mode field approximated ML methods (MFAML). Mean, percentage bias and variance were computed using 50 independent images drawn from the prior using a Gibbs sampler and then blurred and contaminated with Gaussian noise $N(0, 16)$. The prior had a quadratic Hamiltonian defined on a second order neighborhood. (* indicates algorithm fails to reach a solution. STD=Standard Deviation)

True β	0.0004	0.0010	0.0040	0.0100	0.0400	0.100
GMPL Mean	4.134e-4	1.093e-3	7.742e-3	*	*	*
GMPL Bias (%)	3.35%	9.30%	93.6%	*	*	*
GMPL STD (%)	1.74%	1.60%	6.49%	*	*	*
MOM Mean	4.039e-4	1.012e-3	4.175e-3	1.154e-2	0.0775	0.3565
MOM Bias (%)	0.97%	1.20%	4.37%	15.4%	93.7%	257 %
MOM STD (%)	2.13%	3.21%	10.9%	31.0%	340%	380 %
MFAML1 Mean	3.932e-4	9.777e-4	3.794e-3	9.889e-3	0.0306	0.0579
MFAML1 Bias (%)	-1.70%	-2.23%	-5.14%	-11.1%	-23.4%	-42.1%
MFAML1 STD (%)	1.53%	1.45%	1.63%	2.58%	7.08%	16.3%
MFAML2 Mean	4.153e-4	1.004e-3	3.977e-3	9.526e-3	0.0357	0.07842
MFAML2 Bias (%)	3.8%	0.42%	-0.572%	-4.74%	-10.7%	-21.58%
MFAML2 STD (%)	1.21%	1.37%	1.27%	2.51%	4.02%	9.02%

the MFAML methods is markedly superior to MOM and GMPL. In our experiments, we found that when the smoothing kernel becomes more ill-conditioned or the variance of the additive noise increases, both GMPL and MOM will fail to identify the hyperparameter, while the MFAML methods still produce reasonable accuracy. Of the two MFAML methods, the one using optimal partition function approximation performs better.

3.2. APPLICATION TO PET RECONSTRUCTION

The method described above is applicable to a wide range of inverse problems in image processing. Here we illustrate the application of the method to PET data. The description here is necessarily brief, see [8, 9] for a more detailed description of our approach. Realistic Poisson data was generated using a single plane of the 3-D Hoffman brain phantom. Attenuation correction factors (ACFs) were computed assuming uniform attenuation within the head boundary. A 10% uniform randoms distribution was added based on the assumption that the singles distribution was uniform across the detectors. A scatter distribution was computed by using the transmission and emission images in a Klein-Nishina based Compton scatter model (about 10%). To model variable detector sensitivity, the mean sinogram data were scaled using values for intrinsic sensitivity typical in a clinical scanner. The ACFs, randoms and scatter mean values were assumed to be known. The sinogram, detector, and image sizes were chosen to simulate the Siemens ECAT831 PET scanner. Fifty realizations of the Poisson data were generated for both high and low counts.

PET images were reconstructed for each of the three priors defined in (4). MAP images were estimated for each of the fifty sets of data for both high and low count, for each of the priors and for a range of values of β. Since we do not know the true hyperparameter for the brain phantom, we generated a curve of average total squared error in the reconstructed image versus the range of hyperparameter values (see Figure 2). We then ran simultaneous reconstruction and hyperparameter estimation for each of the data sets and for each of the priors, and marked on Figure 2 the mean value of β obtained. Here we used the MFAML1 method for hyperparameter estimation. Note that the hyperparameter value that the MFAML method produced lies very close to the smallest average squared-error for both high and low count tests. The best performance for the low count data is only slightly worse than that for the high count case. For both count rates, the standard deviation of the estimated hyperparameter was less than 10%.

References

1. J. Besag, "On the statistical analysis of dirty pictures," *Journal of the Royal Statistical Society, B,* vol. 48(3), pp. 259-302, 1986.
2. D. Chandler, *Introduction to Modern Statistical Mechanics.* Oxford Univ. Press, 1987.
3. P. Craven and G. Wahba, "Smoothing Noisy Data with Spline Function," *Numerische Mathematik,* vol. 31, pp. 377-403, 1979.
4. S. Geman and D. McClure, "Statistical Methods For Tomographic Image Reconstruction," *In Proc. of the 46th Session of the ISI, Bulletin of the ISI,* Vol.52, 1987.
5. D. Geiger and F. Girosi,"Parallel and Deterministic Algorithms for MRFs: Surface Reconstruction and Integration," *IEEE Trans.Patt. Anal. and Machine Intell.,* Vol. PAMI-12, pp 401-412, May 1991.
6. P. C. Hansen,"Analysis of Discrete Ill-posed Problems by Means of the L-curve," *SIAM Review* vol. 34, No.4,pp.561-580, 1992.
7. A. Mohammad-Djafari, "On the Estimation of Hyperparameters in Bayesian Approach of Solving Inverse Problems," *In Proc. ICASSP-93* pp. V495-498.
8. E. Mumcuoglu, R. Leahy, S. Cherry, Z. Zhou, "Fast Gradient-Based Methods for Bayesian Reconstruction of Transmission and Emission PET Images," *IEEE Trans. on Medical Imaging,* pp. 687-702, Dec., 1994.
9. E. Mumcuoglu and R. Leahy, "A Gradient Projection Conjugate Gradient Algorithm for PET Reconstruction," In *IEEE Proc. Nucl. Sci. Symp and Med. Imag. Conf. 1994.*
10. J. Zhang, "The Mean Field Theory in EM Procedures for Markov Random Fields,"*IEEE Trans. on Signal Processing.* vol. 40. No. 10. Oct. 1992.
11. J. Zhang, J. W.Modestino, and D. A.Langan,"Maximum-Likelihood Parameter Estimation for Unsupervised Stochastic Model-Based Image Segmentation,"*IEEE Trans. Image Proc.* vol. 3, pp. 404-420, July, 1994.
12. Z. Zhou,"Maximum Likelihood Hyperparameter Estimation for Gibbs Priors from Incomplete Data with Applications in Image Processing," Ph.D Dissertation, Signal and Image Processing Institute, University of Southern California, 1994.

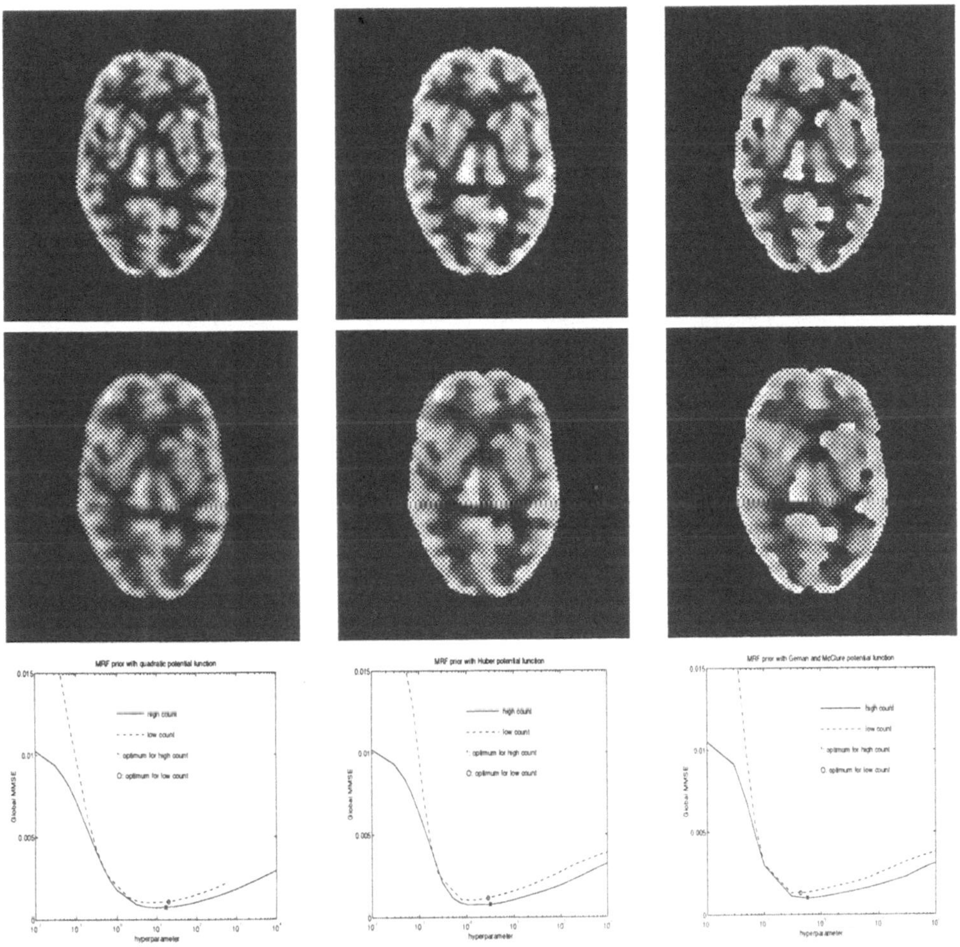

Figure 2. PET simulation experiment: The top row shows sample PET images reconstructed, with simultaneous MFAML hyperparameter estimation, from pseudo-random high-count data, using each of the three priors: left - quadratic, center - Huber, right - Geman and McClure. The middle row shows sample PET images reconstructed, with simultaneous MFAML hyperparameter estimation, from pseudo-random low-count data, using each of the three priors. The bottom row shows the total squared error in the estimated MAP images as a function of β. These errors were averaged over 50 independent realizations as described in Section 3.2. The solid line corresponds to the high count data and the dashed line to the low count data. The star indicates the mean value of β obtained using the MFAML method for high count data and the small circle for low count data. Note its close proximity to the minimum squared error.

IMPROVED QUANTITATION IN SPECT IMAGING USING FULLY 3D ITERATIVE NON-STATIONARY SCATTER COMPENSATION

FREEK J. BEEKMAN, CHRIS KAMPHUIS, PETER P. VAN RIJK

AND

MAX A. VIERGEVER

Department of Radiology and Nuclear Medicine,
University Hospital Utrecht, Heidelberglaan 100,
3584CX Utrecht, The Netherlands

Abstract. The quality and quantitative accuracy of iteratively reconstructed SPECT images strongly improves when better point spread function models of the gamma camera are used during reconstruction. A major cause of image degradation is photon crosstalk between different slices of the gamma camera resolution can be compensated for by fully 3D iterative detector and scatter response compensation. A fully 3D projector back-projector (proback) has been developed that evaluates both the distance dependent detector point spread function and the object shape and distance dependent scatter point spread function of SPECT. The proposed approach avoids the huge memory requirements of storing the full transition matrix needed for 3D reconstruction including object shape dependent scatter. In addition, the method avoids the need for lengthy Monte Carlo simulations to generate such a matrix, because the parameters of the scatter response are obtained from line source measurements. Three iterative reconstruction methods are investigated: (i) 3D reconstruction including only attenuation and 2D detector response, (ii) 2D reconstruction including attenuation, 1D detector response, attenuation and 1D scatter response, and (iii) fully 3D reconstruction (attenuation, 2D detector response and 2D scatter response). Reconstructions of a water filled cylinder containing regions of different activity levels were performed. Fully 3D reconstruction including patient shape dependent scatter response results in higher accuracy and quality than the other probacks, without increasing noise as occurs in scatter subtraction based methods.

Y. Bizais et al. (eds.), Information Processing in Medical Imaging, 53–64.

1. INTRODUCTION

Detection of of photons which have undergone scatter events in the patient is one of the main factors of distortion in SPECT imaging. Over the last two decades, many methods have been addressed this problem so as to reduce the resulting degradation of image contrast and loss of quantitation [1]. Image degradations caused by scatter can, for example, be corrected for by multiple energy windows approaches, like Compton window subtraction [2], dual photo-peak window scatter correction [3, 4] and some others e.g. [5, 6]. These relatively simple methods all amplify noise because they subtract the (noisy) estimation of the scatter distribution from the acquired data. Moreover, the estimation of the amount of scatter which has to be subtracted is not exact, for example due to inaccuracies in the scatter fraction estimation [7], as calculated from images from different energy windows. The disadvantages of multi-window scatter compensation methods can be avoided by accurately modeling the scatter response and including this model in an iterative reconstruction algorithm [8, 9, 10]; A comparison of scatter subtraction based methods with 2D MLEM incorporating scatter [11] shows a better signal-to-noise ratio in the latter approach.

In iterative SPECT reconstruction the quality and quantitative accuracy of the reconstructed image depends strongly on the accuracy of the forward modeling. This dependency calls for modeling of the complete two-dimensional (2D) point spread function of the imaging system, and is called fully 3D reconstruction. Because a significant amount of photons emitted from the slice to be reconstructed are detected in out of slice projection pixels, 2D reconstruction is insufficient for quantitative SPECT reconstruction. Fully 3D iterative reconstruction attempts to deconvolve the blurring effects caused by photon crosstalk between different transaxial slices, so as to improve image contrast, resolution, quantitation and signal-to-noise ratio compared to 2D reconstruction [12, 13, 14, 15]. In the present paper a method is developed to improve scatter compensation in terms of accuracy and signal-to-noise ratio by using a 2D scatter response (fully 3D scatter compensation) instead of the 1D scatter response (2D or single-slice scatter compensation [8, 9]).

For scatter modeling in iterative algorithms, a complicating factor is that the scatter response depends on the source depth and distance, and on the shape of the patient [16, 9, 10]. One method to include the object shape dependent scatter response in SPECT reconstruction is to compute the transition matrix using Monte Carlo (MC) simulations [8]. However, the computation time of MC simulations is excessive and has to be performed anew for each patient. Furthermore, the memory required to store the complete transition matrix is huge, especially when information from out-of-slice photons taken into account: the complete matrix contains $I \times J$ elements where J is the total number of pixels in all projections and I is the number of voxels to be reconstructed. For the -conservative- case of 64^2 projection images and 64 projection views, this amounts to a memory demand of a few tens of Gbytes (in the case of two byte integers), which is prohibitive for iterative reconstruction. In addition, generation of such a 3D matrix takes several months on a state-of-the-art work station [17].

Instead of generating and storing a complete matrix, one may settle for a projector-backprojector (proback) that models the spatially varying PSF. Previously presented probacks do not include effects of the curved shape of the patient on the scatter response [18, 19], or include object shape dependent scatter in just one dimension [8, 9, 10]. The latter approach is satisfactory only for a limited class of emitting objects (in particular objects having the same activity and attenuation distribution in every transversal slice). Moreover, the assumption of a one-dimensional response results in images with worse signal-to-noise ratio than fully 3D reconstructed images, as was shown already in a simpler case where only collimator blurring and attenuation was modeled [14].

The purpose of this paper is to develop a memory-efficient fully 3D proback based on a PSF model that takes object shape dependent scatter into account. A comparison of 2D vs. fully 3D reconstruction compensating for attenuation, detector response and scatter, will be made on the basis of measured projection data of a physical phantom.

2. MATERIALS AND METHODS

In [16] a depth, distance, and object shape dependent description of SPECT in terms of an analytical expression for the point spread function (PSF) was developed. The parameters of the PSF were obtained from line source measurements in air and one line source measurement using a triangular reservoir [20, 21]. In this way the scatter response for a rectangular reservoir (slab scatter response) was approximated and subsequently extended to include effects of the curved object shape [16, 9, 10]. The model was evaluated by line [16] and point source [21] measurements.

Figure 1 shows the coordinate system and the method used to generalize the PSF model for a source behind a slab to curved object shapes. The relative position between the projected position (x^v, y^v) of the source voxel and the projection pixel of interest at position (x^p, y^p) is denoted by $\vec{r} = (x, y)$. The distance of the source voxel to the camera is denoted by z. The depth of the source in the plane at distance z in the object is denoted by $d(x^v, y^v, z)$. The total PSF model can be written as

$$PSF = \{P\delta(r) + \pi^{-1}SsK_0(sr)\} * (u/\pi)e^{-ur^2} \quad (1)$$

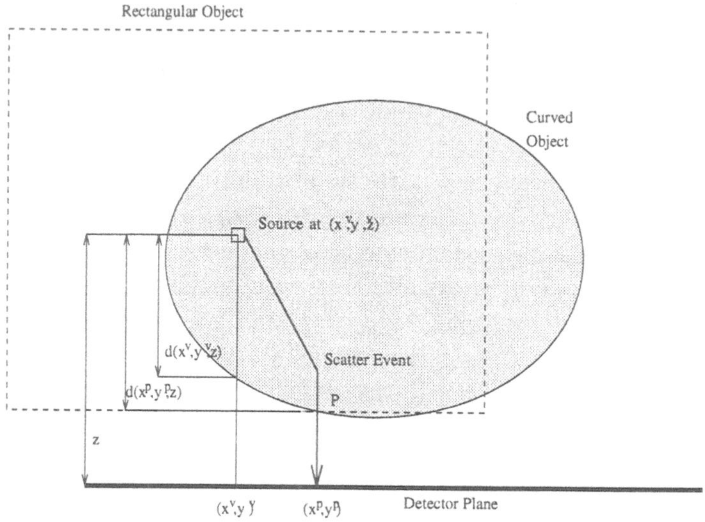

Figure 1. Coordinate system used in the formulation of the PSF model. Because of the limited acceptance angle of the collimator, photons detected at position (x^p, y^p) must have left the object near point P on the object surface, with a local depth $d(x^p, y^p, z)$. The assumption for extension of the scatter response behind a slab to curved object shapes is that photons leaving at the object at point P have the same photon history as photons leaving the front of a slab with point source depth $d(x^p, y^p, z)$. Therefore, the scatter response at point (x^p, y^p) is evaluated at $d(x^p, y^p, z)$ instead of $d(x^v, y^v, z)$ as would be correct for a point source behind a slab.

with $r = \sqrt{x^2 + y^2}$, $u = u(z)$, $P = P(d(x^p, y^p, z))$, $S = S(d(x^p, y^p, z))$, $s = s(d(x^p, y^p, z))$ and $*$ the 2D convolution operation in the x, y-plane. K_0 is the zeroth order modified Bessel function. The width of the geometric response depends on $u(z) = \frac{4ln2}{R_i^2 + (c_1 + c_2 z)^2}$, where R_i is the intrinsic resolution of the camera, and P is given by

$$P = c_3 e^{-\mu_l d} (\frac{\pi}{4ln2})^{\frac{1}{2}} \tag{2}$$

with $d = d(x^p, y^p, z)$. The linear attenuation coefficient in the narrow beam approximation (μ_l) was fixed at $0.152cm^{-1}$ [22]. The functions $s(d)$ and $S(d)$, determining the slope and amplitude of the scatter response, are given by

$$s(d) = (c_4 d + c_5 d^2)^{-1} \tag{3}$$

$$S(d) = c_6 s(d) d e^{-c_7 d} \tag{4}$$

The constants $c_i, i = 1, ..., 7$ are the model parameters which have been determined from line source measurements.

The right hand side of (1) can be interpreted as the PSF for ideal collimation (δ-like detector response) convolved with the detector PSF, normalized with respect to its 2D integral over the detector plane. Consequently, the total PSF for a point source at position (x^v, y^v, z) can be written as

$$\begin{aligned} PSF(z, \vec{d_z}, \vec{r}) &= \{P(z, d(x^p, y^p, z))\delta(\vec{r}) + \\ & \quad PSF^{sc.i}(d(x^p, y^p, z), r)\} * PSF'^{unsc}(z, r) && (5) \\ &= PSF^{ideal}(d(x^p, y^p, z), r) * PSF'^{unsc}(z, r) && (6) \end{aligned}$$

with $PSF^{sc.i}$ the scatter response for ideal collimation, PSF^{ideal} the total response (scatter + primary) for ideal collimation, PSF'^{unsc} the normalized detector PSF, and $\vec{d_z}$ the map containing all depths $d(x_p, y_p)$ of the object with respect to a plane at distance z from the detector.

2.1. A PROJECTOR/BACK-PROJECTOR INCLUDING 3D OBJECT SHAPE DEPENDENT SCATTER

In this paper we will concentrate on the Maximimum Likelihood Expectation Maximization (MLEM) algorithm, which maximizes the likelihood that the image estimate $\vec{\tilde{\lambda}}$ generates the measured projection $\vec{p}$ by means of an expectation maximization scheme, and the Method of Ordered Subsets (OS) which creates solutions close to MLEM but in a much shorter time [23, 24, 25, 26]. Let $\vec{\lambda}$ denote the image distribution, $\vec{p}$ the projection data and c_{ij} the probability (assumed by the forward model) that a photon from voxel i is detected in projection pixel j. Then the MLEM algorithm reads [27]

$$\tilde{\lambda}_i^{k+1} = \frac{\tilde{\lambda}_i^k}{\sum_j c_{ij}} \sum_j \left(\frac{c_{ij} p_j}{\sum_i c_{ij} \tilde{\lambda}_i^k} \right) = \tilde{\lambda}_i^k N_i \sum_j \left(\frac{c_{ij} p_j}{p_{sj}^k} \right) \tag{7}$$

where the subscript s in p_{sj}^k indicates *simulated* projection pixels, and $\vec{\tilde{\lambda}}^k$ the image estimate at iteration k. In the ideal case the reprojection p_{sj}^k is an accurate noiseless simulation of the projection of $\tilde{\lambda}_i^k$. The normalization factor $\vec{N}$ with $N_i = (\sum_j c_{ij})^{-1}$ is the total fraction of detected photons emitted from a voxel at position i.

By defining coordinates for every projection angle θ by x_θ^p and y_θ^p, and denoting the transition probability from voxel (x^v, y^v, z) to detector pixel (x_θ^p, y_θ^p) by the transition matrix elements $c^\theta_{(x^v,y^v,z)\to(x_\theta^p,y_\theta^p)}$ we have for $N_i = N_{(x^v,y^v,z)}$;

$$N_{(x^v,y^v,z)} = \left(\sum_{\theta,(x_\theta^p,y_\theta^p)} c^\theta_{(x^v,y^v,z)\to(x_\theta^p,y_\theta^p)} \right)^{-1}, \tag{8}$$

the reprojection of $\vec{\tilde{\lambda}}$ on projection θ is given by

$$p_s^{\theta,k}(x_\theta^p, y_\theta^p) = \sum_{(x^v,y^v,z)} c^\theta_{(x^v,y^v,z)\rightarrow(x^p_\theta,y^p_\theta)} \tilde{\lambda}^k_{(x^v,y^v,z)}. \tag{9}$$

Upon substituting (9) into (7) we obtain

$$\tilde{\lambda}^{k+1}_{(x^v,y^v,z)} = \tilde{\lambda}^k_{(x^v,y^v,z)} \quad N_{(x^v,y^v,z)} \sum_{\theta,(x^p_\theta,y^p_\theta)} \left(c^\theta_{(x^v,y^v,z)\rightarrow(x^p_\theta,y^p_\theta)} \frac{p^\theta(x^p_\theta, y^p_\theta)}{p_s^{\theta,k}(x^p_\theta, y^p_\theta)} \right) \tag{10}$$

where $p^\theta(x^p_\theta, y^p_\theta)$ is the measured projection at angle θ.

2.1.1. *Acceleration of iterative reconstruction using Ordered Subsets*
MLEM is known to converge very slowly. This is a problem, especially when accurate, proback models are used: much the more elements of the transition matrix are non-zero, the more calculations have to be performed during one projection back-projection operation. In the case of fully 3D reconstruction including evaluation of scatter this would result in prohibitively long reconstruction times. The method using "Ordered Subsets" (OS) [23] allows for acceleration by a factor of nearly 60 (in the case of 120 projections) compared to conventional MLEM, while leading to very similar solutions [23, 24, 25, 26].

The OS method uses only a few projection images (in this study two projection images) at a time to calculate a normalization map and an update of $\vec{\tilde{\lambda}}$. For every update a different subset of projections is used until all projections have been adressed. For OS, normalization maps have to be calculated for all subsets; these maps are denoted by $\vec{N}^{\mathbf{s}}$. During reconstruction one normalization map at the time is used. The other maps, together occupying more then hundred Mbytes, are reversibly compressed and stored on disc. The update of the image estimate is performed according to

$$\tilde{\lambda}^{k+1}_{(x^v,y^v,z)} = \tilde{\lambda}^k_{(x^v,y^v,z)} \quad N^{\mathbf{s}}_{(x^v,y^v,z)} \sum_{\{\theta,(x^p_\theta,y^p_\theta)\}\theta\in\mathbf{s}} \left(c^\theta_{(x^v,y^v,z)\rightarrow(x^p_\theta,y^p_\theta)} \frac{p^\theta(x^p_\theta, y^p_\theta)}{p_s^{\theta,k}(x^p_\theta, y^p_\theta)} \right) \tag{11}$$

When all subsets have been used $\vec{\tilde{\lambda}}^{\mathbf{S}}$ is obtained, where $\mathbf{S}$ is the number of subsets. This estimate is called *one iteration* OSS. When the number of subsets is for example 60 (as in the next section) this is referred to as OS60.

In all reconstructions performed in the present study, each subset contains two opposing projection images. For further details see [24, 25, 28].

2.1.2. *Inclusion of the Object Shape Dependent PSF model in a Projector Back-Projector*

To avoid interpolation during projection back-projection the image estimate and attenuation map are rotated to align with the coordinate system of the rotated detector [29]. Upon denoting the k-th image estimate by $\tilde{\lambda}^k_{(x^v_\theta, y^v_\theta, z_\theta)}$ we have for the projection operation using the object shape dependent PSF according to (9):

$$p_s^{\theta,k}(x^p_\theta, y^p_\theta) = \sum_{(x^v_\theta, y^v_\theta, z_\theta)} PSF(z_\theta, \vec{d^\theta_{z_\theta}}, x^p_\theta - x^v_\theta, y^p_\theta - y^v_\theta)\tilde{\lambda}^k_{(x^v_\theta, y^v_\theta, z_\theta)} \qquad (12)$$

where each projection has a different depth map for each different angle θ and each distance z_θ to the detector denoted by $\vec{d^\theta_{z_\theta}}$. Upon expressing the back-projection operation in terms of the object shape dependent PSF we can write the proback as

$$\sum_{\theta,(x^p_\theta, y^p_\theta)} \left(\frac{PSF(z_\theta, \vec{d^\theta_{z_\theta}}, x^p_\theta - x^v_\theta, y^p_\theta - y^v_\theta) p^\theta(x^p_\theta, y^p_\theta)}{p_s^{\theta,k}(x^p_\theta, y^p_\theta)} \right). \qquad (13)$$

With equation (6) and the commutation property of the convolution operation this becomes

$$\sum_{\theta,(x^p_\theta, y^p_\theta)} (PSF^{ideal}(d^\theta(x^p_\theta, y^p_\theta, z_\theta), x^p_\theta - x^v_\theta, y^p_\theta - y^v_\theta)$$

$$PSF'^{unsc}(x^p_\theta, y^p_\theta, z_\theta) * \left(\frac{p^\theta(x^p_\theta, y^p_\theta)}{p_s^{\theta,k}(x^p_\theta, y^p_\theta)} \right)) \qquad (14)$$

The normalization maps $N^s_{(x^v, y^v, z)}$ have to be calculated only once (by back-projecting with unit projection pixel values).

2.2. GAMMA CAMERA MEASUREMENTS

The parameterization of the detector and scatter response is described in [16, 20, 21].

In order to assess the performance of the different probacks to quantify the activity in various regions, 120 projections of 64×64 pixels were acquired of a cylindrical phantom with rectangular inserts. An average of 2.1×10^6 counts per projection were acquired using a high resolution parallel hole collimator: the pixel size was 5.34mm.

The cylinder contained three rectangular inserts made of stacked sheets of PMMA (perspex) to emulate regions of different activity concentrations (see figure 2). The sheets (60 × 60mm) with a thickness of 3.0 mm were stacked and separated by spacers with the same thickness as the PMMA sheets (see figure 2) in such a way that activity concentrations of 1/3, 1/2 and 2/3 times the concentration in the homogeneous parts of the water filled cylinder were obtained. The transaxial thickness of the lower count regions was approximately 40 mm.

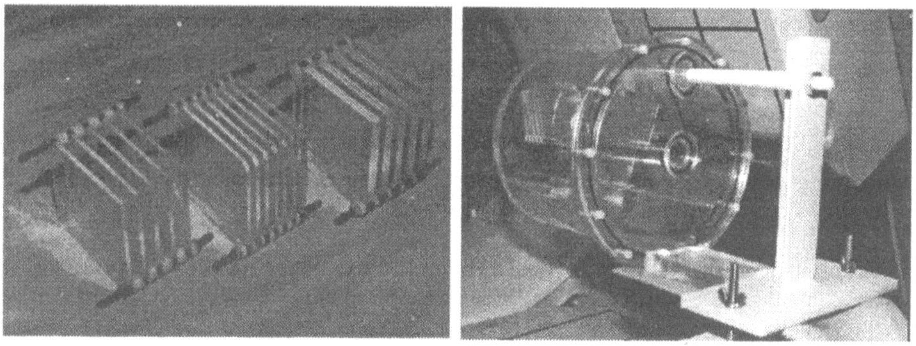

Figure 2. Left: Inserts made of stacked sheets of perspex, to emulate regions with activity levels of 1/3, 1/2 and 2/3 of the activity in the uniform water regions. Right: Cylinder with stacked sheet inserts, mounted on the head support of the SPECT bed.

3. RESULTS

In figures 3 and 4 reconstructed slices are shown of the cylinder with rectangular regions of different activity levels. A region of 10 × 10 pixels wide in the center of the cylinder (at a transaxial slice where the activity is homogeneous because only water is present) was used to normalize each reconstructed object, which enables us to determine whether the reconstruction algorithms produced the correct activity ratios. Profiles through the low activity regions in a transaxial slice show that both 2D and fully 3D reconstruction compensating for scatter (figure 3, center and right) improve quantitation compared to fully 3D reconstruction which only models attenuation and detector response (figure 3, left). Fully 3D scatter correction seems to provide slightly better results than 2D reconstruction including scatter. In "sagittal" slices (figure 4) the differences become more obvious; fully 3D reconstruction including scatter now produces results which are much closer to the desired activity distribution than 2D reconstruction. In particular the edges between regions of different activity levels are more clearly defined and the estimated activity levels are closer to the desired

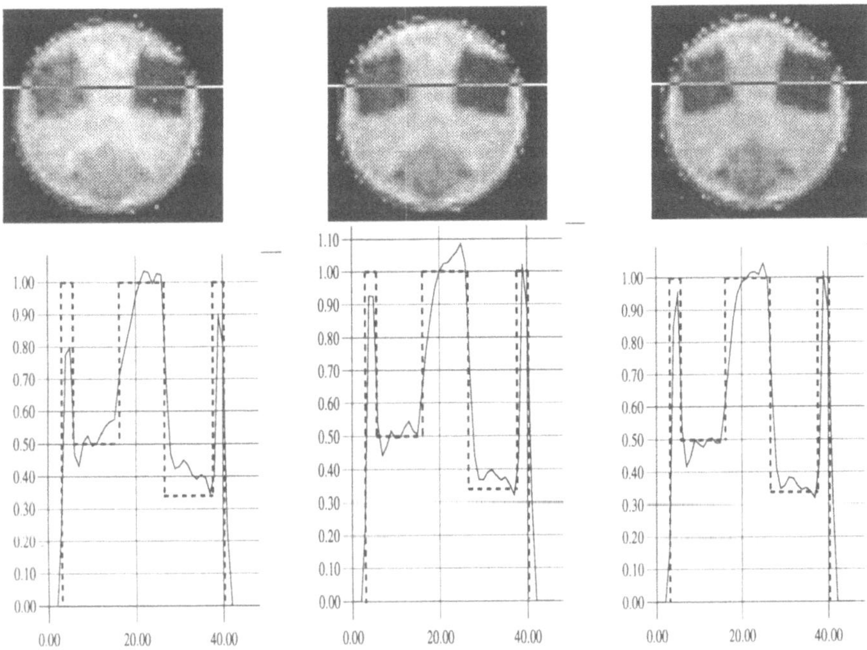

Figure 3. Transaxial slices and profiles through the cylinder with regions with different activity concentrations, obtained after 3 iterations OS60. Left: Fully 3D including attenuation and 2D detector response compensation but no scatter compensation. Center: 2D reconstruction including attenuation, 1D detector and 1D scatter compensation. Right: Fully 3D reconstruction compensating for attenuation, 2D detector and 2D scatter. Dashed lines were drawn to show the desired activity levels.

activity levels. Moreover, the images produced by the proback performing fully 3D scatter compensation appear to be less noisy.

4. CONCLUSIONS AND DISCUSSION

We have presented an efficient projector/backprojector (proback) for true 3D iterative SPECT reconstruction, based on an analytical PSF model that takes into account object shape dependent scatter. With this approach only a few Mbytes of memory is needed compared to the tens of Gbytes needed to store a Monte Carlo generated transition matrix containing a different PSF for every point in the object. Moreover, the presented strategy significantly alleviates the computational burden of calculating the transition matrix: the tables representing the object shape dependent forward model are generated in a few seconds.

Using a cylinder containing rectangular regions with different activity levels, we demonstrated that fully 3D iterative reconstruction including the full scatter response compensation results in excellent quantitation. A large

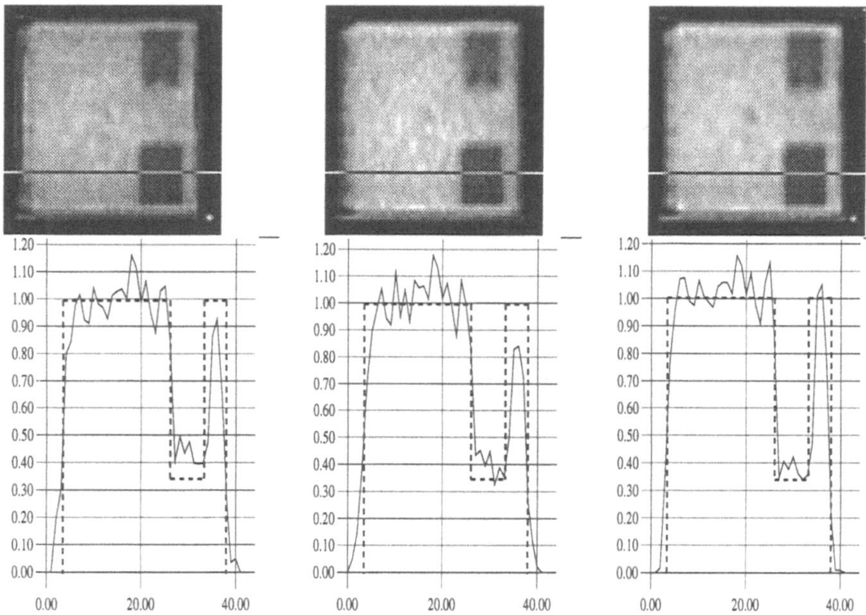

Figure 4. As figure 3, for sagittal slices and profiles.

advantage compared to subtraction based methods is that no increase of noise occurs: instead the images appear to be less noisy than reconstructions where no scatter correction is applied.

The use of a PSF model including scatter implies long reconstruction times. For example one iteration OS60 (the equivalent of approximately 60 iteration MLEM) to reconstruct an entire 3D brain phantom from simulated data ([21]) (voxel size 5.34^3) takes 54 minutes when fully 3D scatter compensation is applied. Methods to attack the problem of long reconstruction times may include (i) parallel computer architectures [30, 31] These are expected to speed up the reconstruction with a factor which is nearly equal to the number of processors; (ii) better initial image estimates than a uniform distribution used in the present study; (iii) simplifying the projection back-projection equation.

Extension of the PSF model to non-homogeneous media can be realized by using water equivalent depths for the calculation of depth maps [32, 9, 21]. The accuracy of a projector modeling line source responses in a non-uniform attenuator based on water-equivalent depths were presented in [17]. It was shown that projection data for a complex thorax phantom generated using water-equivalent depths was in relatively good agreement ($\approx 5\%$ error in the scatter-to-primary ratios) with MCS projection data. Different manufacturers provide a transmission device on their SPECT systems, which enables simultaneous transmission and emission scanning (e.g. [33]). This is expected to increase the accuracy of iterative compensation for both attenuation and scatter.

References

1. I. Buvat, H. Benali, A. Todd-Pokropek, and R. DiPaolo, "Scatter correction in scintigraphy: State of the art," *Eur.J.Nucl.Med.*, vol. 21, pp. 675–694, 1994.
2. R. Jaszczak, K. Greer, and C. Floyd, "Improved SPECT quantification using compensation for scattered photons," *J. Nucl. Med.*, vol. 25, pp. 893–900, 1984.
3. M. King, G. Hademenos, and S. Glick, "A dual photopeak window method for scatter correction," *J. Nucl. Med.*, vol. 33, pp. 605–612, 1992.
4. G. Hademenos, M. Ljungberg, M. King, and S. Glick, "A Monte Carlo investigation of the dual photopeak window scatter correction method," *IEEE Trans.Nucl.Sci.*, vol. 40, pp. 179–185, 1993.
5. K. Ogawa, Y. Harata, and T. Ichihara, "A practical method for position-dependent Compton-scatter correction in Single Photon Emission CT," *IEEE Trans.Med.Im.*, vol. 10, pp. 408–412, 1991.
6. P. Pretorius, A. van Rensburg, A. van Aswegen, M. Lotter, D. Serfontein, and C. Herbst, "The channel ratio method of scatter correction for radionuclide image quantitation," *J.Nucl.Med.*, vol. 34, pp. 330–335, 1993.
7. M. Ljungberg, M. King, G. Hademenos, and S. Strand, "Comparison of four scatter correction methods using monte carlo simulated source distributions," *J.Nucl.Med.*, vol. 35, pp. 143–151, 1994.
8. C. Floyd, R. Jaszczak, K. Greer, and R. Coleman, "Inverse Monte Carlo as an unified reconstruction algorithm for ECT," *J.Nucl.Med.*, vol. 27, pp. 1577–1585, 1986.
9. E. Frey and B. Tsui, "A practical method for incorporating scatter in a projector-backprojector for accurate scatter compensation in SPECT," *IEEE Trans. Nucl.Sci.*, vol. 40, pp. 1107–1116, 1993.
10. E. Frey, Z. Lu, and B. Tsui, "A fast projector backprojector pair modeling the asymmetric, spatially varying scatter response in SPECT imaging," *IEEE Trans.Nucl.Sci.*, vol. 40, pp. 1192–1197, 1993.
11. E. Frey, B. Tsui, and M. Ljungberg, "A comparison of scatter compensation methods in SPECT: Subtraction-based techniques versus iterative reconstruction with accurate modeling of the scatter response," *Conference Record of 1992 IEEE Nuclear Science Symposium and Medical Imaging Conference, F. Kirsten ed., IEEE Press, Piscataway NY*, pp. 1035–1037, 1993.
12. G. Zeng, G. Gulberg, B. Tsui, and J. Terry, "Three-dimensional iterative reconstruction algorithms with attenuation and geometric point response correction," *IEEE Trans.Nucl.Sci.*, vol. 38, pp. 693–703, 1991.
13. T. Hebert and S. Gopal, "The GEM MAP algorithm with 3-D SPECT system response," *IEEE Trans.Med.Im.*, vol. 11, pp. 81–90, 1992.
14. B. Tsui, X. Zhao, E. Frey, and W. McCartney, "Quantitative SPECT: Basics and clinical considerations," *Sem.Nucl.Med.*, vol. 24, pp. 38–65, 1994.
15. B. Tsui, E. Frey, X. Zhao, D. Lalush, R. Johnston, and W. McCartney, "The importance and implementation of accurate 3D compensation methods for quantitative SPECT," *Phys.Med.Biol.*, vol. 39, pp. 509–530, 1994.
16. F. Beekman, E. Eijkman, M. Viergever, G. Borm, and E. Slijpen, "Object shape dependent PSF model for SPECT imaging," *IEEE Trans.Nucl.Sci.*, vol. 40, pp. 31–39, 1993.
17. E. Frey and B. Tsui, "Modeling the scatter response function in inhomogeneous scattering media for SPECT," *IEEE Trans.Nucl.Sci.*, vol. 41, pp. 1585–1593, 1994.

18. B. Penney, M. King, and K. Knesaurek, "A projector-backprojector pair which accounts for the depth and distance dependent blurring in SPECT," *IEEE Trans.Nucl.Sci.*, vol. 37, pp. 681–686, 1990.
19. N. Rajeevan, B. Penney, and C. Byrne, "An accelerated 3D maximum likelihood SPECT reconstruction," *Conference Record of 1991 IEEE Nuclear Science Symposium and Medical Imaging Conference, F. Kirsten ed., IEEE Press, Piscataway NY*, vol. 2, pp. 1963–1968, 1992.
20. F. Beekman, E. Frey, C. Kamphuis, B. Tsui, and M. Viergever, "A new phantom for fast determination of the scatter response of a gamma camera," *IEEE Trans.Nucl.Sci.*, vol. 41, pp. 1481–1488, 1994.
21. F. Beekman and M. Viergever, "Fast SPECT simulation including object shape dependent scatter," *IEEE Trans.Med.Im., in press.*
22. S. Manglos, C. Floyd, R. Jaszczak, K. Greer, C. Harris, and R. Coleman, "Experimentally measured scatter fractions and energy spectra as a test of Monte Carlo simulations," *Phys.Med.Biol.*, vol. 32, pp. 335–343, 1987.
23. H. Hudson, B. Hutton, and R. Larkin, "Accelerated EM reconstruction using ordered subsets," *J.Nucl.Med.*, vol. 33, p. 960, 1991.
24. H. M. Hudson and R. S. Larkin, "Accelerated image reconstruction using ordered subsets of projection data," *IEEE Trans.Med.Im.*, pp. 601–609.
25. J. Li, R. Jaszczak, K. Greer, and R. Coleman, "Implementation of an accelerated iterative algorithm for cone beam SPECT," *Phys. Med. Biol.*, vol. 39, pp. 643–653, 1994.
26. C. Kamphuis, *Accelerated SPECT reconstruction including object shape dependent scatter.* MSc thesis: University of Nijmegen, 1994.
27. K. Lange and R. Carson, "E.M. reconstruction algorithms for emission and transmission tomography," *J.Comput.Assist.Tomog.*, vol. 8, pp. 306–316, 1985.
28. F. Beekman, *Fully 3D SPECT Reconstruction with Object Shape Dependent Scatter Compensation.* PhD thesis, Utrecht University, 1995.
29. G. Zeng and G. Gullberg, "Frequency domain implementation of three-dimensional geometric point response in SPECT imaging," *Conference Record of 1991 IEEE Nuclear Science Symposium and Medical Imaging Conference, IEEE Press, Piscataway NY*, pp. 1943–1947, 1992.
30. A. McCarthy and M. Miller, "Maximum likelihood SPECT in clinical computation times using mesh-connected parallel computers," *IEEE Trans.Med.Im.*, vol. 10, pp. 426–436, 1991.
31. M. Miller and C. Butler, "3D maximum a priory estimation for single photon emission computed tomography on massively-parallel computers," *IEEE Trans.Med.Im.*, vol. 12, pp. 560–565, 1993.
32. D. Gilland, B. Tsui, H. Hu, and J. Perry, *Computer simulated cardiac SPECT data for use in evaluating reconstruction algorithms.* In: Medical Images: Formation, Handling and Evaluation. A.E. Todd-Pokropek and M.A. Viergever eds., NATO ASI Series F98, Springer-Verlag, Berlin, pp. 523-536, 1992.
33. C. Tung, G. Gullberg, G. Zeng, P. Christian, F. Datz, and H. Morgan, "Nonuniform attenuation correction using simultaneous transmission and emission converging tomography," *IEEE Trans.Nucl.Sci.*, vol. 38, pp. 1134–1143, 1992.

LINEAR ITERATIVE RECONSTRUCTION ALGORITHMS: STUDY OF OBSERVER PERFORMANCE.

Craig K. ABBEY[1] and Harrison H. BARRETT[2]
Program in Applied Mathematics and Dept. of Radiology, Univ. of Arizona[1]
Optical Sciences Center and Dept. of Radiology, Univ. of Arizona, Tucson AZ, USA[2]

1. INTRODUCTION

Iterative algorithms for tomographic reconstruction are attractive, in spite of their computational expense, because they enforce agreement between the data and the reconstruction and allow easy incorporation of accurate models of the imaging process and prior knowledge of the object being imaged [1-12]. There is, however, still no general agreement in the literature as to which algorithm is best in a particular situation or even how to optimize a single algorithm. Every algorithm has one or more free parameters, such as a regularizing or smoothing parameter or the iteration number at which the algorithm is stopped. These parameters control the tradeoff between sharpness and noise in the image, but there is no general method for choosing them.

To compare or optimize algorithms, a method of assessing image quality is needed. Available methods include subjective comparison, measures of image fidelity [13], statistical criteria based on degree of agreement with the data [14], and task-based methods in which image quality is defined in terms of how well some observer can perform some specific task of clinical interest [15-21]. We adopt the latter approach in this paper.

The tasks of interest can be divided generically into classification and estimation tasks [21]. Classification tasks range from detection of a nonrandom signal on a nonrandom background to complicated multiclass pattern recognition. Estimation tasks [18, 20, 21] are concerned with extraction of numerical information from the images, such as activity within a region of interest. For both kinds of task, the goal is to design the imaging system or reconstruction algorithm in such a way as to optimize the performance of the chosen observer on the task of interest. This paper will concentrate entirely on classification tasks.

Y. Bizais et al. (eds.), Information Processing in Medical Imaging, 65–76.

For classification tasks, the observer can be a human or a mathematical model. The performance of a human observer is measured experimentally by psychophysical studies [15], a well-established but time-consuming tool in image-quality research. Mathematical-model observers include the true ideal or Bayesian observer [22-26], the ideal linear or Hotelling observer [27, 28], various models that incorporate characteristics of the human visual system [29-31], and ad hoc models chosen mainly for computational ease. The success of a model is judged by how well it predicts the outcome of psychophysical experiments.

The objective of this paper is to present a general approach to optimizing algorithms and to illustrate the approach with some preliminary results on a specific algorithm, the linear Landweber (LLW) algorithm [32]. Components of the approach include development of methods of calculating the performance of model observers, selection of a model observer based on psychophysical studies, and use of the model to select algorithmic parameters. For illustration, the parameter investigated here is the number of iterations.

2. THEORY

In image reconstruction an object is typically discretized into pixels to yield a set of linear equations described by the equation $\mathbf{g} = \mathbf{Hf} + \mathbf{n}$. The object $\mathbf{f}$ is a vector of pixel intensities and the system matrix $\mathbf{H}$ is an M × N matrix whose elements represent the contribution of an image pixel to a given detector. The purpose of reconstruction and restoration algorithms is to calculate an estimate of $\mathbf{f}$, denoted $\hat{\mathbf{f}}$ here, from the data vector $\mathbf{g}$.

Most linear iterative algorithms are special cases of a general linear iterative algorithm (GLIA) of the form [33]

$$\hat{\mathbf{f}}^{(k+1)} = \hat{\mathbf{f}}^{(k)} + \mathbf{B}\left(\mathbf{g} - \mathbf{H}\hat{\mathbf{f}}^{(k)}\right) \qquad \text{(Eq. 1)}$$

Equation (1) can be thought of as a family of reconstruction algorithms where each element of the family is identified by the choice of the reconstruction operator $\mathbf{B}$. The linear Landweber algorithm, used as a test case for the methods proposed here, is defined by $\mathbf{B} = \alpha\mathbf{H}^t$, where α is a user input acceleration parameter and the superscript t denotes matrix transpose. The GLIA can be made *more* general by allowing the backprojector to vary with iteration so that $\mathbf{B} = \mathbf{B}_k$. These variable $\mathbf{B}_k$ could be used to describe multigrid approaches [34] or iteration-dependent acceleration methods [35, 36]. The results given in this work extend to this case but the notation becomes increasingly cumbersome.

A straightforward application of induction to (Eq. 1) reveals that

$$\hat{\mathbf{f}}^{(k+1)} = \sum_{k'=0}^{k} (\mathbf{I}-\mathbf{BH})^{k'}\mathbf{Bg} \;+\; (\mathbf{I}-\mathbf{BH})^{k+1}\mathbf{f}_0 \;= \mathbf{O}_k\mathbf{g} + \mathbf{Q}_k\mathbf{f}_0 . \qquad \text{(Eq. 2)}$$

In this form, the reconstruction is seen to be a function of only the data and the initial estimate. The reconstruction operator $\mathbf{O}_k$ will be used freely throughout this work. It is important to realize that this matrix is prohibitively large in most imaging systems. The example used to demonstrate our results uses a relatively small two-dimensional tomographic system in which $\mathbf{O}_k$ is a 4096 by 4096 array. It would require 64 megabytes of memory to store this matrix in uncompressed 4-byte real format. Typical three-dimensional systems involve matrices that are on the order of 100,000 by 100,000 elements. In practice we never actually compute $\mathbf{O}_k$; rather we compute its matrix products by running the algorithm on the vector of interest. For example, if we wish to compute $\mathbf{y} = \mathbf{O}_k\mathbf{x}$, we can set $\mathbf{y}^{(0)} = \mathbf{0}$, and iterate

$$\mathbf{y}^{(k+1)} = \mathbf{y}^{(k)} + \mathbf{B}(\mathbf{x} - \mathbf{H}\mathbf{y}^{(k)}) \qquad \text{(Eq. 3)}$$

until the product we desire is attained. Similarly, to compute $\mathbf{x} = \mathbf{O}_k^t\mathbf{y}$, we can iterate

$$\mathbf{x}^{(k+1)} = \mathbf{x}^{(k)} + \mathbf{B}'(\mathbf{y} - \mathbf{H}'\mathbf{x}^{(k)}) . \qquad \text{(Eq. 4)}$$

In most medical-imaging applications, the statistics governing data collection are Poisson. The resulting measurements are subject to signal-dependent noise. Object variability can also be incorporated into the data statistics by the use of a Bayesian prior distribution on $\mathbf{f}$ [21]. The expectation and covariance of the data are then given by

$$\bar{\mathbf{g}} = \mathbf{H}\bar{\mathbf{f}}, \qquad \mathbf{K}_g = \mathrm{Diag}(\mathbf{H}\bar{\mathbf{f}}) + \mathbf{H}\mathbf{K}_f\mathbf{H}',$$

where $\bar{\mathbf{f}}$ and $\mathbf{K}_f$ are the expectation and covariance of the prior distribution on $\mathbf{f}$. Using these formulas and (Eq. 3), we find the expectation and covariance of GLIA reconstructions to be

$$\mathrm{E}\left(\hat{\mathbf{f}}^{(k+1)}\right) = \mathbf{O}_k\mathbf{H}\bar{\mathbf{f}} + \mathbf{Q}_k\mathbf{f}_0 , \qquad \text{(Eq. 5)}$$

and

$$\mathbf{K}_{\hat{f}^{(k+1)}} = \mathbf{O}_k\left(\mathrm{Diag}(\mathbf{H}\bar{\mathbf{f}}) + \mathbf{H}\mathbf{K}_f\mathbf{H}'\right)\mathbf{O}_k' . \qquad \text{(Eq. 6)}$$

The expectation and covariance will be used to define a linear observer signal-to-noise ratio.

The theory behind linear observers has been treated by Barrett [21]. In this work, a linear observer is identified by a vector or template $\mathbf{w}_k$. We add the subscript k here to indicate the potential dependence of the observer on the ensemble statistics of the reconstructions and hence on the stopping point. The object $\mathbf{f}$ is presumed to consist of either a background $\mathbf{b}$ and known signal $\mathbf{s}$ or just the background. The task in this case is detection of the signal. From (Eq. 5) and (Eq. 6), the observer signal-to-noise ratio (SNR) or detectability d is given by

$$d_{k+1}^2 = \frac{\left(\mathbf{w}_k^t \mathbf{O}_k \mathbf{H}\mathbf{s}\right)^2}{\mathbf{w}_k^t \mathbf{O}_k \left(\mathrm{Diag}\left(\mathbf{H}\left(\mathbf{b} + \frac{1}{2}\mathbf{s}\right)\right) + \mathbf{H}\mathbf{K}_b\mathbf{H}^t\right)\mathbf{O}_k^t \mathbf{w}_k}. \qquad \text{(Eq. 7)}$$

Note that, with the exception of $\mathbf{K}_b$, all products used in this expression can be evaluated by the matrix-product methods described earlier. If products involving $\mathbf{K}_b$ can be implemented without storing this matrix, then d_k^2 can be evaluated for very large systems without storing any of the matrices in (Eq. 7). We also note that (Eq. 7) is independent of the initial estimate $\mathbf{f}_0$ provided it is the same for both signal-absent and signal-present images.

Since the model observer templates $\mathbf{w}_k$ are often functions of same statistical quantities that define the SNR, the fast-product approach used above is useful in examining a number of popular model observers. We present here computational methods for evaluating the Hotelling (HOT) observer, the Nonprewhitening Matched Filter (NPW) observer, and the Channelized Hotelling observer with an arbitrary bank of channels. The Hotelling observer [27, 28] is defined as the expected signal after an initial inverse filtering to remove noise correlation. The observer is defined by

$$\mathbf{w}_k^{HOT} = \left(\mathbf{O}_k \left(\mathrm{Diag}\left(\mathbf{H}\left(\mathbf{b} + \frac{1}{2}\mathbf{s}\right)\right) + \mathbf{H}\mathbf{K}_b\mathbf{H}^t\right)\mathbf{O}_k^t\right)^{-1} \mathbf{O}_k \mathbf{H}\mathbf{s}. \qquad \text{(Eq. 8)}$$

Direct calculation of the matrix inverse in (Eq. 8) may not be feasible. However, because products involving the matrix to be inverted can be computed reasonably efficiently using (Eq. 5) and (Eq. 6), the Hotelling template can be found using a variety of inversion algorithms using only these forward products [37, 38]. The NPW observer [22] is defined by

$$\mathbf{w}_k^{NPW} = \mathbf{O}_k \mathbf{H}\mathbf{s} \qquad \text{(Eq. 9)}$$

and can be computed directly from (Eq. 5).

The Channelized Hotelling observer [30] attempts to incorporate elements of the human visual system by processing the image through a number of frequency-selective filters and then acts as a Hotelling observer on this greatly reduced set of data. The formula for the observer template is given by

$$\mathbf{w}_k^{CH} = \mathbf{T}\left(\mathbf{T}'\mathbf{O}_k\left(\mathrm{Diag}\left(\mathbf{H}\left(\mathbf{b}+\frac{1}{2}\mathbf{s}\right)\right)+\mathbf{H}\mathbf{K}_b\mathbf{H}'\right)\mathbf{O}_k'\mathbf{T}\right)^{-1}\mathbf{T}'\mathbf{O}_k\mathbf{H}\mathbf{s}. \quad \text{(Eq. 10)}$$

The matrix $\mathbf{T}$ is the bank of filters used to process the data. It is usually highly rectangular leading to a great reduction in dimensionality of the inverse. This reduction in dimensionality makes direct inversion feasible. Two kinds of visual channels are considered in the subsequent model observer studies corresponding to overlapping (DG) and nonoverlapping channel models.

3. METHODS

To test the effect of stopping point on observer SNR, a number of experiments were performed using, as a test case, a simulated two-dimensional parallel-beam emission tomography imaging system with either 250,000 or 500,000 expected counts per data set. The LLW algorithm was used to reconstruct 64×64 pixel images from 64 projection angles. In all cases the task was detection of a known 10% contrast dark disk signal of radius 3 pixels in the center of a uniform bright disk background of radius 19 pixels. The signal was nonrandom and hence all object variation came in the form of a random background $\mathbf{b}$, so $\mathbf{K}_f = \mathbf{K}_b$. In the first study, the only source of variation in the reconstruction came from the Poisson statistics of the data. We refer to this scheme as the "fixed-background" case. The object covariance in this case is $\mathbf{K}_b = \mathbf{0}$. The second study used a statistically defined or "lumpy" background where the background is modeled as a stationary random process with a Gaussian power spectral density[39]. The lumpy background is confined to the background disk region and hence is only locally stationary near the center of the image. The covariance of this windowed lumpy background is given by $\mathbf{K}_b = \mathbf{W}\mathbf{F}^{-1}\mathbf{S}\mathbf{F}\mathbf{W}^t$. Here $\mathbf{S}$ is the diagonal background power spectral density, $\mathbf{F}$ is the two-dimensional discrete-Fourier transform matrix, and $\mathbf{W}$ is a diagonal matrix specifying the support of the background disk. Parameters of the power spectral density were set so that the pixel standard deviation was again 10% of the expected background intensity and the correlation length of the noise was set to 5.0 pixels in length. Products involving this covariance can be implemented quickly using FFT routines.

Signal-to-noise ratios for a number of model observers were calculated for each kind of background using the methods described earlier. Stopping points from 1 to 128 were evaluated.

Human-observer performance metrics for the two backgrounds were measured by psychophysical studies using receiver operating characteristic (ROC) analysis [15]. A subset of seven stopping points - 1, 4, 8, 16, 32, 64, and 128 iterations - were chosen. This exponential sampling scheme captures the rapid early adjustments and slower convergence behavior of the LLW algorithm. The readers rated 150 images (75 signal present and 75 signal absent) for each stopping point and background in an environment controlled for ambient light level and screen brightness and contrast. The resulting areas under the ROC curves were transformed to human observer SNRs via the error function [40].

Pilot studies indicated a severe reduction in performance at the early stopping points due to very low signal contrast (approximately 1%). To create images with detectable contrast for the human observer studies, $\overline{\mathbf{b}}$, the expected signal background was subtracted and the result scaled to obtain a higher expected signal contrast. For the fixed-background study, all stopping points could be scaled for a 10% expected signal contrast across stopping point. In the variable-background studies, 5% and 8% expected signal contrast were used at 1 and 4 iterations respectively, to keep the intensity fluctuations from the background variation within a reasonable range of gray-levels. A 10% signal contrast was used for the higher iteration stopping points. The pilot studies also indicated generally poor human performance at expected count rates of 250,000 events. To boost performance in this case, the expected count rate was increased to 500,000 events. The fixed background images were read by 5 observers and the variable background cases were read by 8 observers.

4. RESULTS AND DISCUSSION

We have found it instructive to analyze the fixed and lumpy background cases in terms of local the local noise power spectrum (LNPS) [40, 41]. As seen in figure 1, the signal amplitude and LNPS vary enormously with iteration, but their ratio, a frequency-dependent SNR is almost constant. The invariance of this SNR is echoed in Figures 2a and 2b by the Hotelling observer and the

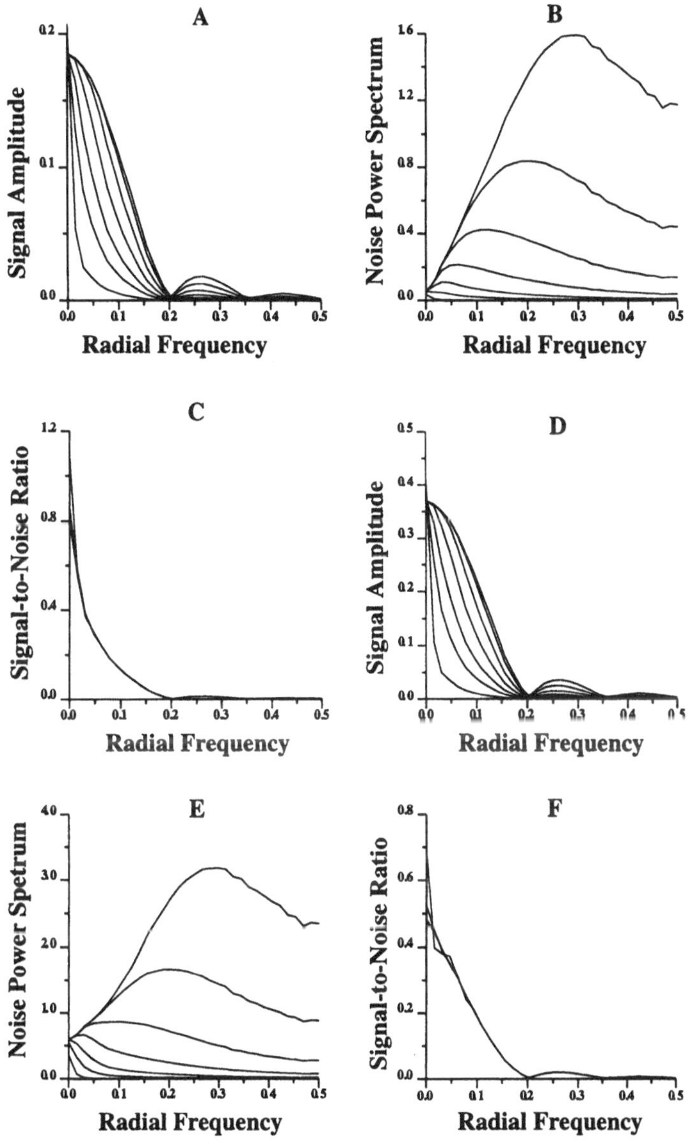

Figure 1. Local Stationarity and the effect of stopping point. Each graph contains a plot for each of the seven stoppping points from 1 to 128 iterations. Plots in the top row correspond to the known background and the bottom row correspond to the lumpy background. Plots A and D show the Fourier amplitude of the reconstructed signal while B and E track the spectral density of the noise. These combine to give radial frequency SNRs in C and F that are highly invariant over the stopping points.

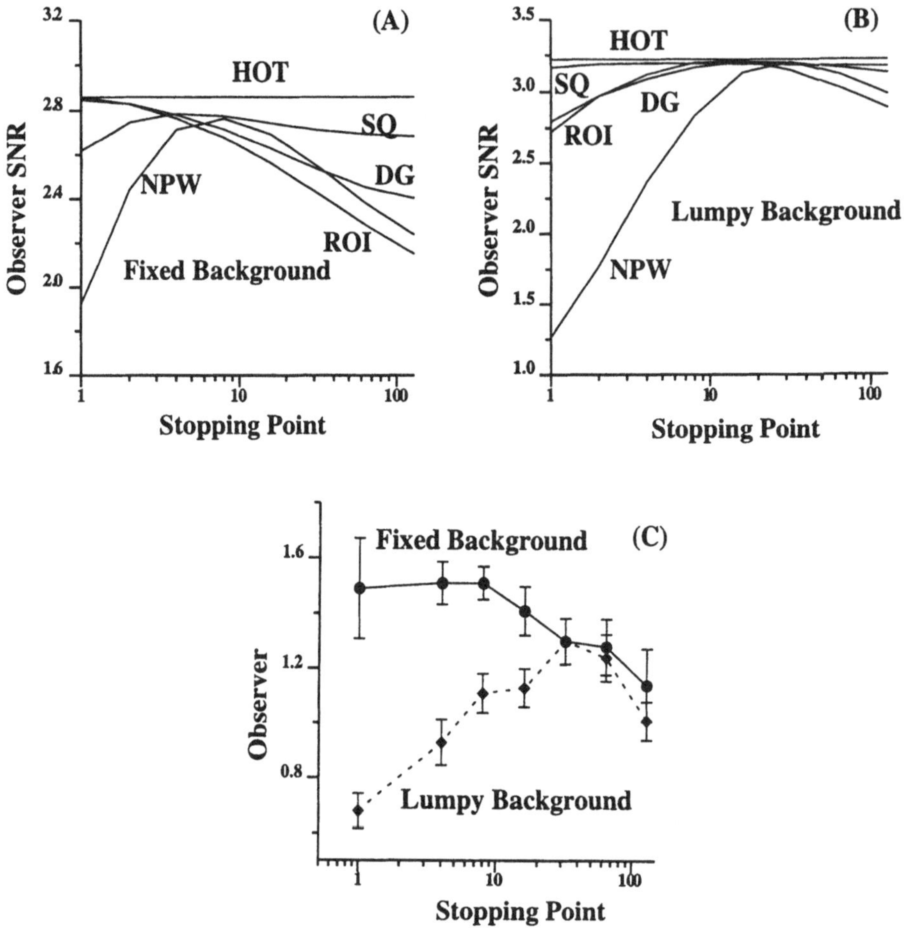

Figure 2. The effect of stopping criteria on observer performance. (A) and (B) plot model observer performance vs. stopping point. HOT is the Hotelling observer, NPW is the Nonprewhitening Matched Filter, ROI is the Region of Interest, SQ is the square Channel Hotelling, DG is the Difference of Gaussian Channel Hotelling. The general increase in observer SNR from (A) to (B) is accounted for by the increased exposure time necessary for a reasonable level of human detection efficiency. Plot (C) is the psychophysical study results.

Channelized Hotelling observer. Fairly general assumptions on the GLIA result in an invariant Hotelling observer SNR. The degradation due to processing through the frequency-selective channels present in the Channelized Hotelling observer results in little effect for these images. The remaining observers show relatively large effects with iteration number.

Human observer performance is plotted in figure 2d. Significant effects ($p < .05$) are found in all across iteration number and background variation. These data indicate that different stopping points are optimal for the different imaging tasks.

5. CONCLUSIONS

In this work we have motivated the need for an objective definition of image quality to resolve issues in the optimization of image reconstruction algorithms. Model observers play an important practical role in this definition by presenting an alternative to human-observer studies. The methods we have developed can be used to evaluate the model-observer signal-to-noise ratio of a broad class of linear reconstruction algorithms and observer models. Because of their matrix-product structure, our expressions are well suited to the analysis of large imaging systems found in practice.

Our specific studies of the linear Landweber algorithm reveal that the effect of iteration point on detection performance is dependent on both the kind of objects to be imaged and on the observer. The psychophysical studies presented here indicate that the optimal stopping point for the human observer is critically dependent on the imaging task. As the task becomes increasingly complex, realized here by an increasingly complex random background, the human observers required more iterations to achieve optimal performance.

6. ACKNOWLEDGMENTS

The authors would like to thank Elizabeth Krupinski, Ed Soares, Jack Denny, and Kyle Myers for their helpful discussions and input. This work was supported by NCI grant RO1 -CA52643-05.

7. REFERENCES

[1] M.A. King, R.B. Schwinger, P.W. Doherty, and B.C. Penney, "Two-dimensional filtering of SPECT images using the Metz and Wiener filters," *J. Nucl. Med.* 25, pp 1234-1240, 1984

[2] T.F. Budinger, S.E. Derenzo, W.L. Greenberg, G.T. Gullberg, and R.H. Huesman, "Quantitative potentials of dynamic emission computed tomography," *J. Nucl. Med* 19, pp 309-315, 1978

[3] A.R. Formiconi, A. Pupi, and A. Passeri, "Compensation of spatial system response in SPECT with conjugate gradient reconstruction technique," *Phys. Med. Biol* 34(1), pp 69-84, 1989

[4] B.M.W. Tsui, G.T. Gullberg, E.R. Edgerton, J. Glen Ballard, J.R. Perry, W.H. McCartney, and J. Berg, "Correction of nonuniform attenuation in cardiac SPECT imaging," *J. Nucl. Med* 30, pp 497-507, 1989

[5] T.E. Walters, W.Simon, D.A. Chesler, and J.A. Correia, "Attenuation Correction in Gamma Emission Computed Tomography," *J. Computer Assisted Tomography* 5(1), pp 89-94, 1981

[6] C.E. Floyd, R.J. Jaszczak, S.H. Manglos, K.L. Greer, and R.E. Coleman, "Improved SPECT reconstruction using inverse Monte Carlo," International Workshop on Physics and Engineering of Computerized Multidimensional Imaging and Processing, Newport Beach, CA, 1986

[7] G.T. Gullberg, R.H. Huesman, J.A. Malko, N.J. Pelc, and T.F. Budinger, "An attenuated projector-backprojector for iterative SPECT reconstruction," *Phys. Med. Biol* 30(8), pp 799-816, 1985

[8] E. Tanaka and H. Toyama, "A generalized weighted backprojection algorithm for single photon emission computed tomography," VIIIth Int. Conf. Information Processing in Medical Imaging, Brussels, 1993

[9] C.-T. Chen and C.E. Metz, "Application of the EM algorithm in image reconstruction and image processing," *Proc. SPIE* 767, pp 326-331, 1987

[10] R.M. Lewitt, "Alternatives to voxels for image representation in iterative reconstruction algorithms," *Phys. Med. Biol.* 37(3), pp 705-716, 1992

[11] K.F. Koral and W.L. Rogers, "Application of ART to time-coded emission tomography," *Phys. Med. Biol.* 24(5), pp 879-894, 1979

[12] D.G. Politte and D.L. Snyder, "The use of constraints to eliminate artifacts in maximum-like¬lihood image estimation for emission tomography," *IEEE Trans. on Nuc. Med.* 35(1), pp 608-610, 1988

[13] P.B. Fellgett and E.H. Linfoot, "On the assessment of optical images," *Philos. Trans. Roy. Soc. London*, A247, pp 369-407, 1955

[14] J. Llacer, E. Veklerov, L.R. Baxter, S.T. Grafton, L.K. Griffeth, R.A. Hawkins, C.K. Hoh, J.C. Mazziotta, E.J. Hoffman, and C.E. Metz, "Results of a clinical receiver operating characteristic study comparing filtered backprojection and maximum likelihood estimator images in FDG PET studies," *J. Nucl. Med.* 34(7), pp 1198-1203, 1993

[15] J.A. Swets, "ROC analysis applied to the evaluation of medical imaging techniques," *Invest. Radiol.*, 4, pp 109-112, 1979

[16] K. Hanson, "Method of evaluating image recovery algorithms based on task performance," *J. Opt. Soc. Amer. A.*, 7, pp 1294-1304, 1990

[17] R.F. Wagner, K.J. Myers, D.G. Brown, M.J. Tapiovaara and A.E. Burgess, "High-order tasks: human vs. machine performance," *Proc. SPIE*, 1090, pp 183-194, 1989

[18] S. Matej, G.T. Herman, T.K. Narayan, S.S. Furuie, R.M. Lewitt and P. Kinahan, "Evaluation of task-oriented performance of several fully 3-D PET reconstruction algorithms," *Phys. Med. Biol.*, 39, pp 355-367, 1994

[19] D.R. Gilland, B.M.W. Tsui, W.H. McCartney, J.R. Perry, and J. Berg, "Determination of the optimum filter function for SPECT imaging," *J. Nucl. Med.* 29(5), pp 643-650, 1988

[20] S.P. Mueller, M.F. Kijewski, S.C. Moore and B.L. Holman, "Maximum-likelihood estimation -- a model for optimal quantitation in nuclear medicine," J. Nucl. Med. 31, pp 1693-1701, 1990

[21] H.H. Barrett, "Objective assessment of image quality: Effects of quantum noise and object variability," *J. Opt. Soc. Am. A*, 7, pp 1266-1278, 1990

[22] R.F. Wagner and D.G. Brown, "Unified SNR analysis of medical imaging systems," *Phys. Med. Biol.* 30, pp 489-518, 1985

[23] A.E. Burgess, R.F. Wagner, R.J. Jennings, and H.B. Barlow, "Efficiency of human visual discrimination," *Science* 214, pp 93-94, 1981

[24] B.M.W. Tsui, C.E. Metz, F.B. Atkins, S.J. Starr, and R.N. Beck, "A comparison of optimum spatial resolution in nuclear imaging based on statistical theory and on observer performance," *Phys. Med. Biol.* 23, pp 654-676, 1978

[25] R. Shaw, "The equivalent quantum efficiency of the photographic process," *J. Photog. Sci.* 11, pp 199-204, 1963

[26] P.F. Judy and R.G. Swensson, "Detection of small focal lesions in CT images: effects of reconstruction filters and visual display windows," *British Journal of Radiology*, 58, pp 137-145, 1985

[27] W.E. Smith and H.H. Barrett, "Hotelling trace criterion as a figure of merit for the optimization of imaging systems," *J. Opt. Soc. Amer. A*, 3, pp 717-723, 1986

[28] R.D. Fiete, H.H. Barrett, W.E. Smith and K.J. Myers, "Hotelling trace criterion and its correlation with human-observer performance," *J. Opt. Soc. Am. A.*, 4, pp 945-953, 1987

[29] K.J. Myers and H.H. Barrett "Addition of a channel mechanism to the ideal-observer model," *J. Opt. Soc. Am. A.*, 4, pp 2447-2457, 1987

[30] J. Yao and H.H. Barrett, "Predicting Human Performance by a Channelized Hotelling Observer Model," *SPIE Math. Methods Med. Imaging*, 1768, pp 161-168, 1992

[31] A. Burgess, "Statistically defined background: performance of a modified nonprewhitening observer model," *J. Opt. Soc. Am. A*, 11(4), pp 1237-1242, 1994

[32] L. Landweber, "An iteration formula for Fredholm integral equations of the first kind," *Am. J. Math.*, 73, pp 615-624, 1951

[33] X-L. Xu, J-S. Liow and S.C. Strother, "Iterative algebraic reconstruction algorithms for emission computed tomography: A unified framework and its application to positron emission tomography," *Med. Phys.*, 20(6), pp 1675-1684, 1993

[34] T-S. Pan and A. E. Yagle, "Acceleration and filtering in the generalized landweber iteration using a variable shaping matrix," *IEEE Trans Med. Imaging*, MI-12(2), pp 278-286, 1993

[35] G.T. Herman and L.B. Meyer, "Algebraic reconstruction methods can be made computationally efficient," *IEEE Trans Med.Imaging* MI-12(3) pp 600-609

[36] K.M. Hanson, "Optimization of the constrained algebraic reconstruction technique for the performance of a variety of visual tasks," *in Information Processing in Medical Imaging*, Ortendahl and Llacer, eds, pp 45-57, Wiley-Liss, New York, 1991

[37] G.H. Golub and C.F. Van Loan, *Matrix Computations*, Second Edition, Johns Hopkins University Press, Baltimore, 1989

[38] W.H. Press, S.A. Teukolsky, W.T. Vetterling and B.P. Flannery, *Numerical Recipes: The Art of Scientific Computing*, Cambridge University Press, Cambridge, 1992

[39] J.P. Rolland and H.H. Barrett, "Effect of random background inhomogeneity on observer detection performance," *J. Opt. Soc. Am. A.*, 9(5), pp 649-658, 1992

[40] C.E. Metz, "Some practical issues of experimental design and data analysis in radiological ROC studies" *Invest. Radiol.* 24, pp 234-245, 1989

[41] E.J. Soares, "Attenuation, noise and image quality in single-photon emission computed tomography," University of Arizona Ph. D. Dissertation, 1994

[42] D.W. Wilson, "Noise and resolution properties of FB and ML-EM Reconstructed Images," University of North Carolina Ph. D. Dissertation, 1993

RECONSTRUCTION IN OPTICAL TOMOGRAPHY USING MRI BASED PRIOR KNOWLEDGE

S.R. Arridge#, M.Schweiger*

Dept. of Computer Science, University College London, UK#

*Dept. of Medical Physics, University College London, UK.**

Abstract : This paper addresses the sensitivity of Optical Tomographic reconstructions to *a priori* knowledge. We begin by formulating the problems as non-linear and ill-posed. We compare the Fréchet derivative of the forward problem with the Radon Transform, and introduce analytical expressions for simple cases, as well as Finite Element Method calculations for general cases. In particular we demonstrate the extreme non-linearity of the derivative using a software phantom derived from MRI scans. By reseting the phantom to progressively less accurate starting conditions, we investigate the feasibility of Optical Tomography in the absence of such knowledge.

1 Introduction

The development of Optical Tomography has gathered increasing interest in the last few years, but has been limited by both instrumentation difficulties in acquiring sufficient data in reasonable times, and a theoretical treatment of the reconstruction problem[1]. The fundamental difficulty is that biological tissue is a highly light scattering medium, so that the transport model deviates highly from the Radon Transform. The clinical usefulness of optical transillumination has been known for many years [2], and stems from the fact that the relative attenuation of light in tissue at different wavelengths is related to the global concentrations of certain metabolites in their oxygenated and deoxygenated states [3]. Thus an optical tomographic imaging system would provide a *functional image* of local oxygenation.

Our approach as previously reported [4,5,6] uses a Finite Element Method (FEM) that can cope with arbitrary geometries. In this paper we demonstrate the flexibility of this approach by introducing an anthropomorphic software phantom of the head that is geometrically derived from an MRI scan, and uses predefined segmentation to create regions with known optical properties. We demonstrate

Y. Bizais et al. (eds.), Information Processing in Medical Imaging, 77–88.

reconstruction of complex structures starting from incresasingly distant prior assumptions.

1.1 Data Acquisition

A variety of experimental techniques are employed to obtain the data used in Optical Tomography. These may be divided into those that use continuous intensity information, those that measure the temporal intensity distribution using a "time of flight" method that employs ultra short light pulses and a fast optical detector, and those that use a radio frequency (RF) modulated light source and phase sensitive detection of the transmitted light[7]. The time-of-flight method may be related to the non-time-resolved methods by considering the integral over time of the detected light, and to frequency domain studies via its Fourier Transform[8].

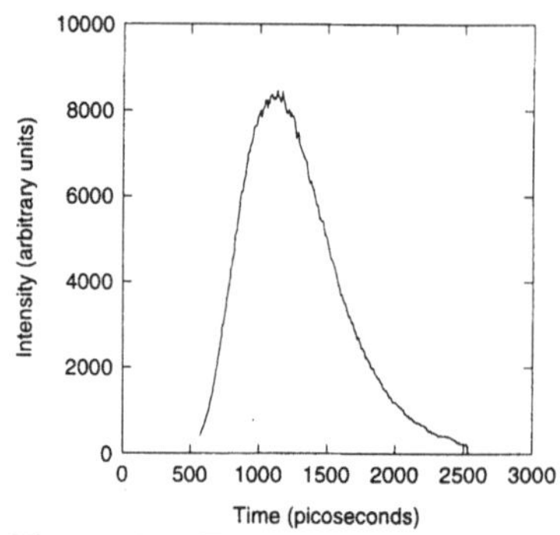

Figure 1 : Experimental time-of-flight data

Figure 1 shows an impulse response measured across 5 cm of scattering phantom emulating brain tissue using a picosecond Ti:Sapphire laser (Tsunami, Spectra Physics) and synchroscan streak camera (Hamamatsu Photonics KK Japan, C1587/M1955) with a 10 ps resolution. It illustrates the typical skewness of such impulse functions.

2 Problem Model

2.1 Photon Transport

Propagation of radiation in absorbing and scattering media may be described in general by the Radiative Transfer Equation [9]

$$\left\{\hat{s}\cdot\nabla+\mu_a(\boldsymbol{r})+\mu_s(\boldsymbol{r})+\frac{1}{c}\frac{\partial}{\partial t}\right\}\phi(\boldsymbol{r},\hat{s},t) = q(\hat{s},\hat{r},t) + \mu_s(\boldsymbol{r})\int\big(f(\hat{s}',\hat{s},\boldsymbol{r})\phi(\boldsymbol{r},\hat{s}',t)\big)\,\mathrm{d}^2\hat{s}' \qquad \textbf{(1)}$$

which describes the change of the intensity $\phi(\boldsymbol{r},\hat{s},t)$ at position $\boldsymbol{r}$ into direction $\hat{s}$ within a domain Ω, bounded by a surface $\partial\Omega$; μ_a and μ_s are the absorption and scattering coefficients respectively (dimensions of inverse length), c is the speed of light, $\boldsymbol{q}(\boldsymbol{r},\hat{s},t)$ is a source term, and $f(\hat{s}',\hat{s},\boldsymbol{r})$ is the scattering phase function characterising the intensity of a wave incident in direction $\hat{s}'$ scattered into direction $\hat{s}$.

This formulation ignores such electromagnetic wave properties as polarisation, and such particle properties as inelastic collisions, (an equivalent form for particle radiation is the Linear Transport Equation[10]), but is generally sufficient to describe the interaction of electro-magnetic radiation in tissue for many Medical Imaging modalities. For example, if μ_s and $\boldsymbol{q}$ are zero, and the system is assumed to be steady-state, then Eq.(1) becomes the differential form of the Radon Transform for CT :

$$\left\{\hat{s}\cdot\nabla+\mu_a(\boldsymbol{r})\right\}\phi(\boldsymbol{r},\hat{s}) = 0 \quad \Rightarrow \quad \phi(\boldsymbol{b},\hat{s}) = \phi(\boldsymbol{a},\hat{s})e^{-\int_0^1 \mu_a(\boldsymbol{a}+\lambda(\boldsymbol{b}-\boldsymbol{a}))d\lambda} \tag{2}$$

Similarly if μ_a and μ_s are zero and $\boldsymbol{q}$ is non zero and isotropic, we obtain the Radon Transform for SPECT :

$$\hat{s}\cdot\nabla\phi(\boldsymbol{r},\hat{s}) = q(\boldsymbol{r}) \quad \rightarrow \quad \phi(\boldsymbol{b},\hat{s}) = \int_0^1 q(\boldsymbol{a}+\lambda(\boldsymbol{b}-\boldsymbol{a}))d\lambda \tag{3}$$

If both μ_a and $\boldsymbol{q}$ are non zero, then the problem becomes the Attenuated Radon Transform, with the Exponential Radon Transform representing the case where μ_a is non zero and constant.

In Optical Tomography, scattering is the dominant process, and analytical inversions of Eq.(1) are not available except in highly restrictive geometries. Inversion schemes in Optical Tomography have depended on one the following general approaches : either restriction of the domain of measurement to those observables that give rise to straight line integrals, expansion of ϕ and f in Eq.(1) in spherical harmonics to give Partial Differential Equations (PDEs) of diffusion type, or restriction of the set of directions $\hat{s}$ to allow the problem to be modelled as a Markov Random Process with finite state space[11,12]; this last approach, although analytic, pertains to such a limited domain, and to noiseless data, that we do not consider it further here as a practical method. In the first case, the source is assumed to be a δ-function in time, and it is argued that the first photons that arrive at the measurement site have undergone almost no scattering, so that rejecting photons that arrive later than cd where d is the source-detector distance leads to a Radon Transform approximation to Eq.(1) -

i.e. only the beginning of the rising edge of the data in Figure 1 would be considered. Because the Signal-to-Noise ratio (SNR) of this approach is unusable except for unrealistically simple situations (either very low scattering, or extremely small objects) our approach has been to seek the solution of a PDE; in particular we use the Diffusion Approximation (DA):

$$\left\{\nabla\cdot\,\kappa(r)\nabla-\gamma(r)-\frac{\partial}{\partial t}\right\}\Phi(r,t) \;=\; -q_0(r,t) \tag{4}$$

where $\Phi(\mathbf{r},t)$ now represents directionless photon intensity, $\gamma(\mathbf{r}) = \mu_a(\mathbf{r})c$ and $\kappa(\mathbf{r})$ is given by:

$$\kappa(r) \;=\; \frac{c}{3(\mu_a(r)+\mu_s'(r))} \tag{5}$$

where $\mu_s'(\mathbf{r}) = (1-\bar{f})\mu_s(\mathbf{r})$ is the reduced scattering coefficient, and $\bar{f}$ is an *anisotropy factor* ($0 \leq \bar{f} \leq 1$). The values of $\bar{f}$ for biological tissue are typically of the order of 0.9, indicating strongly forward biased scattering. Use of the Diffusion Approximation (DA) is widespread, and its derivation from Eq.(1) is well-known[10]. The severe ill-posedness of inverse problems of diffusion type has to be offset against the increased SNR of the data, and the computationally efficient methods available for solution, as described below. It is possible that hybrid approaches, using less severe restrictions on the measurement domain may hold the key to more accurate methods. Such discussions are beyond the scope of this paper however.

2.2 The forward mapping and derivatives

It is convenient to think of the *Forward Problem* as a non-linear mapping $F[\mathbf{p}]:\mathbf{X}_p \rightarrow \mathbf{Y}_M$ from a space $\mathbf{X}_p$ of parameters into a space $\mathbf{Y}_M$ of measurements and the Fréchet derivative of F[.] as a mapping $F'[\mathbf{p}]:\mathbf{X}_p \rightarrow (\mathbf{Y}_M \times \mathbf{X}_p)$. A single element $\mathbf{y} \in \mathbf{Y}_M$ represents the observable M for all given source-pair geometries. Examples of M are integrated intensity E, intensity at a particular time $\Gamma(t_j)$, moments of the time-varying intensity, $\langle t\rangle,\langle t^2\rangle\ldots\langle t^n\rangle$, and phase and modulation depth in the frequency domain. The one-dimensional subspace of $F'[\mathbf{p}]$ which is the projection of the partial derivative operator corresponding to the ith component y_i of $\mathbf{y}$ represents the dependence of the measurement of a single source-detector pair on changes in $\mathbf{p}$ everywhere. In the discrete case, where $\mathbf{p}$ is reconstructed at sample points, and there are a finite number of measurements, both $\mathbf{X}_p$ and $\mathbf{Y}_M$ are of finite dimension. Then F′[**p**] is the *Jacobian* J[**p**] of the forward problem, and the subspace corresponding to y_i is a single row of its matrix representation, *J*. We have given the term *Photon Measurement Density Functions* (PMDFs) to these subspaces

Suppose we have an experiment that measures property *M* at a set of source-measurement pairs $(\xi_1, \xi_2, ..., \xi_N) \times (\zeta_1, \zeta_2, ..., \zeta_S)$, and we divide the domain Ω into L non-overlapping elemental regions $(\tau_1(\mathbf{r}'), \tau_2(\mathbf{r}'), ...\tau_L(\mathbf{r}'))$ such that $\Omega = \cup_{i=1}^{L}\tau_i(r)$, then *J* can be represented :

$$\begin{bmatrix} \Delta y(\zeta_1,\xi_1) \\ \Delta y(\zeta_1,\xi_2) \\ \cdot \\ \cdot \\ \Delta y(\zeta_i,\xi_k) \\ \cdot \\ \cdot \\ \Delta y(\zeta_S,\xi_N) \end{bmatrix} = \begin{pmatrix} J_p^{(M)}(\xi_1,\zeta_1;\boldsymbol{r}_1') & \cdots & J_p^{(M)}(\xi_1,\zeta_1;\boldsymbol{r}_j') & \cdots & J_p^{(M)}(\xi_1,\zeta_1;\boldsymbol{r}_L') \\ J_p^{(M)}(\xi_2,\zeta_1;\boldsymbol{r}_1') & \cdots & J_p^{(M)}(\xi_2,\zeta_1;\boldsymbol{r}_j') & \cdots & J_p^{(M)}(\xi_2,\zeta_1;\boldsymbol{r}_L') \\ \cdot & \cdots & \cdot & \cdots & \cdot \\ \cdot & \cdots & \cdot & \cdots & \cdot \\ J_p^{(M)}(\xi_k,\zeta_i;\boldsymbol{r}_1') & \cdots & J_p^{(M)}(\xi_k,\zeta_i;\boldsymbol{r}_j') & \cdots & J_p^{(M)}(\xi_k,\zeta_i;\boldsymbol{r}_L') \\ \cdot & \cdots & \cdot & \cdots & \cdot \\ \cdot & \cdots & \cdot & \cdots & \cdot \\ J_p^{(M)}(\xi_N,\zeta_S;\boldsymbol{r}_1') & \cdots & J_p^{(M)}(\xi_N,\zeta_S;\boldsymbol{r}_j') & \cdots & J_p^{(M)}(\xi_N,\zeta_S;\boldsymbol{r}_L') \end{pmatrix} \begin{bmatrix} \Delta \boldsymbol{p}(\boldsymbol{r}_1') \\ \cdot \\ \cdot \\ \Delta \boldsymbol{p}(\boldsymbol{r}_j') \\ \cdot \\ \cdot \\ \Delta \boldsymbol{p}(\boldsymbol{r}_L') \end{bmatrix} \qquad (6)$$

The PMDFs are rows of this matrix. The columns represent the perturbation in data due to a point inhomogeneity. Examination of *J* provides considerable insight into the rich diversity of measurement possibilities in Optical Tomography.

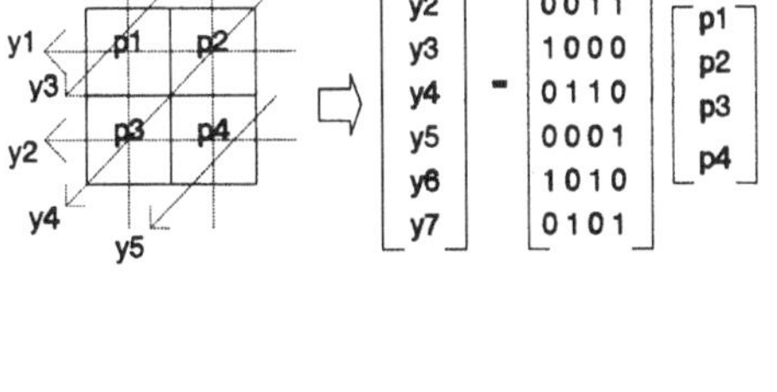

Figure 2 : Jacobian for the Radon Transform

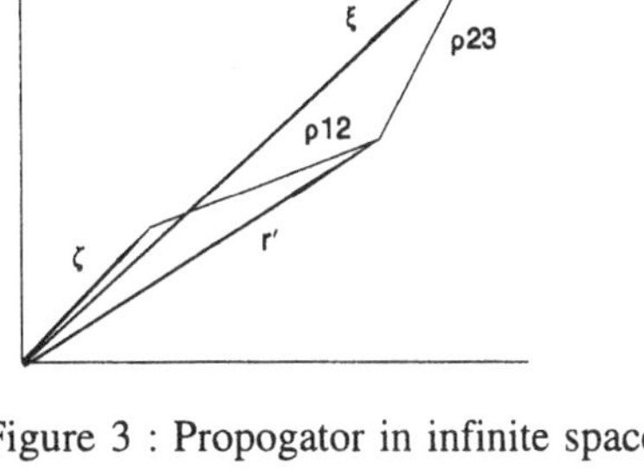

Figure 3 : Propogator in infinite space

3 Derivation of Jacobian

For the Radon Transform as given in Eq.(2), where M is the logarithm of the relative attenuation, and **p** is $\mu_a(\mathbf{r})$, the Fréchet derivative is a constant. If $\mathbf{p}(\mathbf{r})$ is sampled on a pixel basis then each row of the J is just the set of pixels intersected by the straight line from source to detector as shown in Figure 2. It is important to note that for a transport model based on Eq.(4), $\mathbf{p}(\mathbf{r})$ needs to be considered as a vector $(\gamma(\mathbf{r}),\kappa(\mathbf{r}))$. We have shown previously that non time-resolved methods are unable to distinguish between the two images simultaneously[4], which is a serious shortcoming of some approaches that only attempt reconstruction of $\gamma(\mathbf{r})$ [13].

Under the assumption of Eq.(4), the Fréchet derivative is obtained as the sum over all first order Feynman diagrams for the diffusion propagator as shown in Figure 3. For simple geometries it is analytically derivable. For the case where **p** is absorption coefficient μ_a, the time-dependent Fréchet derivative in an homogeneous infinite space is

$$J^{(\Gamma)}_{\alpha,inf}(\xi,\zeta,t;r') = [(\hat{s}(\xi)\cdot(\xi-r')]\,S^{(\Gamma)}(|\xi-r'|,|r'-\zeta|,t) \tag{7}$$

where
$$S^{(\Gamma)}(x,y,t) = \frac{1}{2(2\pi)^3}\left(\frac{1}{y^3}+\frac{(x+y)^2}{2\kappa t x y^2}\right)f(x+y,t) \tag{8}$$

and
$$f(x,t) = \left(2\kappa t\right)^{-3/2}\exp\left(-\gamma t-\frac{x^2}{4\kappa t}\right) \tag{9}$$

In the frequency domain we require just the Fourier Transform of $S^{(\Gamma)}(x,y,t)$ given by :

$$\hat{S}^{(\Gamma)}(x,y,\omega) = \frac{1}{4(2\pi)^3\kappa}\frac{1+\sigma y}{xy^2}\exp[-\sigma(x+y)] \tag{10}$$

with $\sigma = ((\gamma + i\omega)/\kappa)^{1/2}$.

Figures 4 and 5 show three-dimensional views of $J_\alpha^{(\Gamma)}$ at 50 and 200 picoseconds respectively with $\mu_a = 0.025\ \mathrm{mm}^{-1}$, and $\mu_s' = 2\ \mathrm{mm}^{-1}$. The source is at $\zeta = (0,0,0)$ and the detector at $\xi = (0,0,20)$. The function is sampled on a $64 \times$

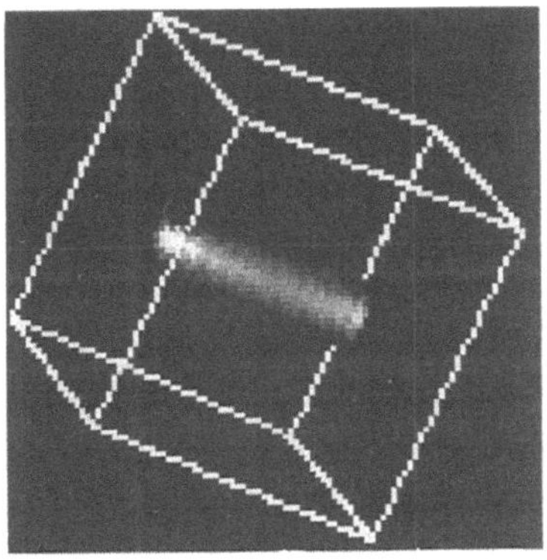

Figure 4 : $J_\alpha^{(\Gamma)}$ at 50 ps

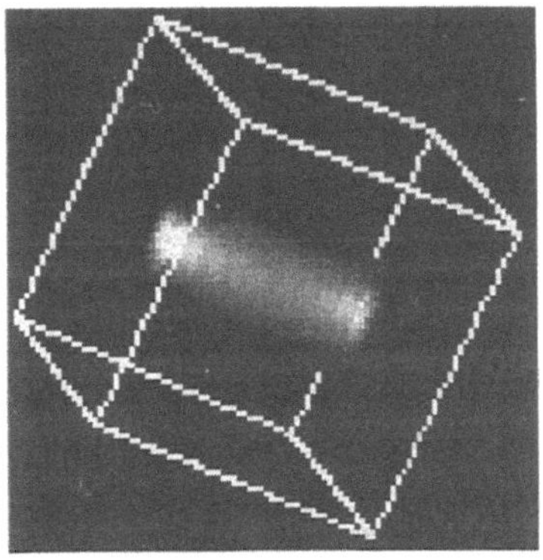

Figure 5 : $J_\alpha^{(\Gamma)}$ 200 ps

64 × 16 grid inside a 20mm × 20mm × 20mm volume centred on the source detector pair. The images are generated with Analyze™[14] using Maximum Intensity Projection mode. By contrast the Radon Transform case would be a straight line of voxels of equal weight. Two features are apparent from these figures. Firstly, the strength of the functions is concentrated near the source and detector, indicating that the predominant influence of the measurement is there. Secondly, the volume of space influencing the measurement increases with increasing time.

The above forms serve to generate all useful observables. For example, the integrated intensity is just the (ω=0) sample of the frequency domain. The moments are obtained from the Mellin Transform of Φ, by differentiation in the Fourier Domain [15] :

$$\Phi^*(s) = \mathrm{Me}[\Phi(t);s] = \int_0^\infty t^{s-1}\Gamma(t)\mathrm{d}t = (-\mathrm{i})^{(s-1)}\frac{\partial^{(s-1)}}{\partial\omega^{(s-1)}}\hat{\Phi}(\omega)\Big|_{\omega=0} \quad (11)$$

4 Finite Element Approach

In the Finite Element Method (FEM) we find not $\Phi(\mathbf{r},t)$ (the solution to Eq (4)), but $\Phi^h(\mathbf{r},t) = \sum_i^D \Phi_i(t)u_i(r) \in \mho^h$, where $\mho^h$ is a finite dimensional subspace spanned by basis functions $u_j(\mathbf{r})$, j=1..D chosen to have limited support[5]. The problem becomes one of sparse banded matrix manipulation with a large literature of computational efficiency and accuracy. The advantage of FEM over analytical approaches is its versatility and applicability to complex geometries. We demonstrate this aspect by calculating PMDFs for a head model generated from a MRI head scan. Due to limitations in computing resources all simulations are performed in 2D.

To create a 2D mesh based on an image obtained from MRI image of the head, an automated mesh generation routine has been developed, based on an adaptive mesh generation method[16], and has been extended to take into account internal boundaries of tissue types. First an edge-detection routine is applied to the image to calculate the outline of the head and all internal structures of interest as a series of closed polygons. The generation algorithm then proceeds by filling each area by triangular elements, starting from the polygonal boundaries. The absorption and scattering coefficients of the mesh elements are then set to the values appropriate for the identified tissue type. The mesh used here is shown in figure 7. The optical parameters of each element are indicated as grey scales, left μ_a and right μ_s'. The mesh consists of 29639 elements and 15025 nodes. We distinguish five different tissue types: skin, bone, cerebrospinal fluid (CSF), grey and white brain matter. The mesh may be scaled in size for simulation of different clinical situations. For simulating adult heads we use a sagittal diamter (posterior-anterior distance) of 18cms, and for neonates a diameter of 10cms. Note that the model does not claim to accurately calculate light propagation due to uncertainty of actual optical parameters, but the results still show some of the effects one may encounter when transilluminating large and inhomogeneous regions of tissue.

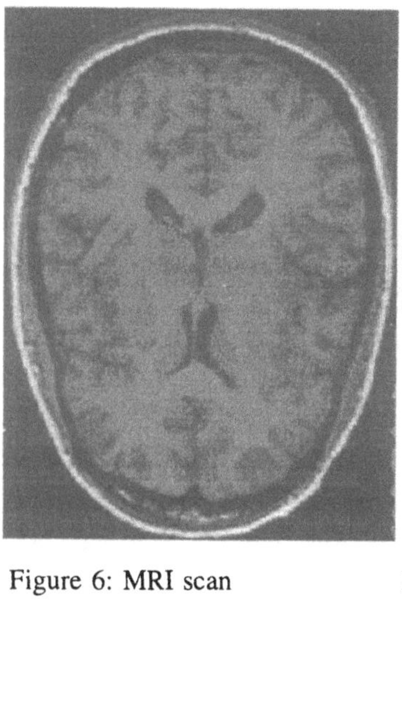

Figure 6: MRI scan

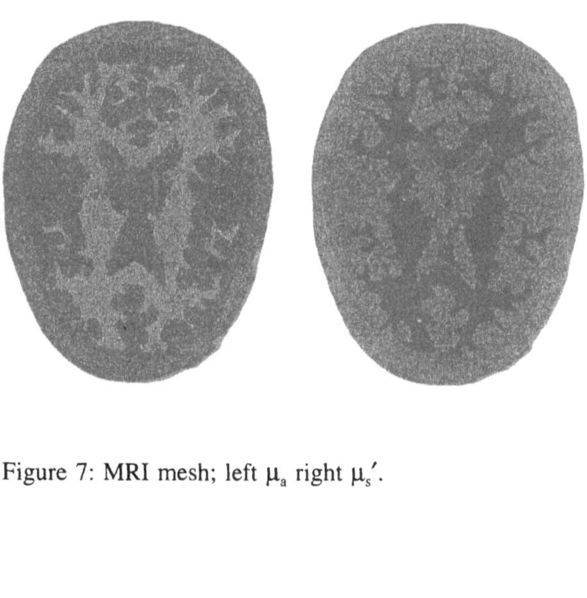

Figure 7: MRI mesh; left μ_a right μ_s'.

To illustrate the usefulness of the FEM approach we show example PMDFs, calculated on the neonatal model for apsorbtion perturbations, for integrated intensity measurements, E, in figure 8, and mean time, <t> in figure 9. The greyscale intensity displayed in the figures indicates the relative sensitivity of the measurement to a change in attenuation coefficient. The most striking feature in each case is their complexity - a simple Radon Transform approximation would yield a straight line from source to detector. The second most notable feature is that the integrated intensity measurement is dependent almost entirely on the features of the image directly under the source and detector. Almost no information is derived from the interior of the brain. In constrast the mean time measurement penetrates deeply into the brain and samples more strongly from the interior.

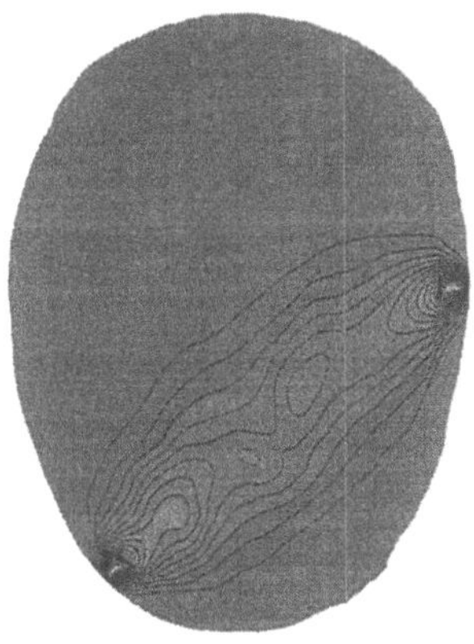

Figure 8 : Absorption PMDF for E

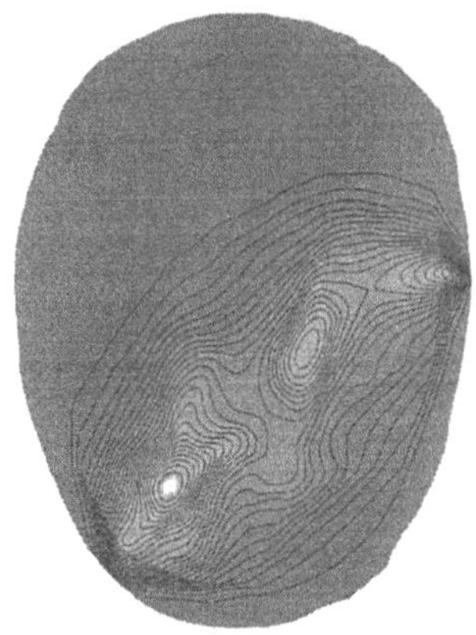

Figure 9 : Absorption PMDF for <t>

Although some caution is required due to the possible shortcomings of the model in regions of low scattering, the marked difference from the simple forms of figures 4 and 5 indicate the inadvisability of straightforward back-projection reconstruction algorithms.

5 Inverse Problem

There are many ways to approach the solution of the inverse problem[1]. By defining an appropriate error-norm, it may be expressed as a multi-dimensional optimisation problem, which we have solved using a modified Newton-Raphson approach[4]. Whereas this algorithm is feasible for limited data sets, the explicit inversion of J rapidly becomes too memory intensive for large data sets, requiring $O(N.S.L)$ storage. Since the rows of J can be computed efficiently we can investigate the use of *row-action* methods for the solution. In the following we have used the Algebraic Reconstruction Technique (ART)[17]

$$\boldsymbol{p}^{(k+1)} = \boldsymbol{p}^{(k)} + \lambda^{(k)} \frac{(y_i - \boldsymbol{J}_i \cdot \boldsymbol{p}^{(k)})}{||\boldsymbol{J}_i^T||^2} \boldsymbol{J}_i^T \qquad i = k \bmod (N.S), \qquad k = (1,2,3....) \tag{12}$$

where $\boldsymbol{J}_i$ is a $1 \times L$ row matrix that is the ith row of J and $\lambda^{(k)}$ is a relaxation parameter $0 \le \inf(\lambda^{(k)}) \le \sup(\lambda^{(k)}) < 2$. This reduces space complexity to $O(L(N+S))$, at the cost of some run-time efficiency.

5.1 Regularisation

ART is a regularising scheme where the iteration (k) takes the place of a conventional regularisation parameter (in the sense that the solution approaches the Moore-Penrose inverse as $k \to \infty$). However, we have investigated introduction of spatial constraints into ART analogous to the use of derivatives of the solution vector which prevent highly oscillatory solutions developing by controlling smoothness of the solution.

5.2 Reconstructions

We generated data with the MRI based meshes and a sagittal diameter of 45 mm. We simulated 64 equally spaced data points and 64 interleaved measurement sites, giving 4096 measuements. Included in the mesh was a small group of elements with a contrast of 500% in μ_a representing a highly localised increase in absorption. We used <t> as the only measurement and attempted reconstruction using the following starting points : prior 1 assumed the original complete distribution of parameters in the absence of the local absorber - i.e. the situation was one of *dynamic imaging* where we assume that we are only looking for differences in sequential studies; prior 2 assumed the correct distribution of $\mu_s(\mathbf{r})$, but a homogeneous distribution of $\mu_a(\mathbf{r}) = 0.022\text{mm}^{-1}$ - i.e we are attempting *absolute imaging* of the $\mu_a(\mathbf{r})$ image; prior 3 assumed a homogeneous distribution of $\mu_a(\mathbf{r}) = 0.022\text{mm}^{-1}$ and a homogeneous distribution of $\mu_s'(\mathbf{r}) = 1.0\text{mm}^{-1}$. In figure 10 we show the comparative reconstructions with 10a) being the target image in each case; all images are of $\mu_a(\mathbf{r})$. 10b) shows prior 1 and 10e) shows prior 2, (prior 3 has the same μ_a distribution). 10c) and d) show the first two iterations of ART for prior 1 - the reconstruction has clearly found the

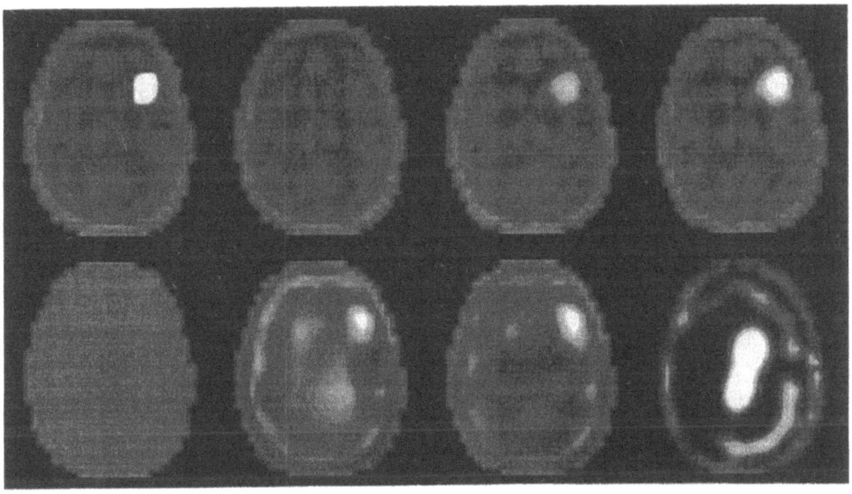

Figure 10 : Reconstruction of head images: a) target; b) prior 1; d) first two iterations of ART starting from b); e) prior 2; f),g) first two iterations of ART starting from e); h) first iteration of ART starting from no prior

absorption peak. 10f) and g) show the first two iterations of ART for prior 2 - the reconstruction has found the absorption peak but the other image features are less well resolved; the second iteration appears to correct mistakes in the first. Finally 10h) shows the image reconstructed under no prior information - it has failed completely, because it is backprojecting into regions derived from an homogeneous model, which will have a much simpler form than the behaviour demonstrated in Figures 8 and 9.

6 Conclusions

We have demonstrated that Optical Tomographic reconstruction is possible in complex geometries. At least as far as dynamic imaging goes the method seems possible, if sufficient data sampling is carried out, and reasonable knowledge is available regarding the normal distribution of parameters. The possibility of absorption imaging is also encouraging. To reconstruct absorption and scattering simultaneously from a single data type appears impossible as previously indicated[4], but incorporation of multiple data types may improve the situation.

7 Acknowledgements

This work was supported in part by the SERC, MRC, Action Research, and Hamamatsu Photonics KK.

8 References

[1] S.R.Arridge, "The forward and inverse problems in Time-Resolved Infra-red Imaging" in *Medical Optical Tomography: Functional Imaging and Monitoring*, (Eds: G. Muller et. al.,) *SPIE* (Bellingham), pp. 35-64, (1993)

[2] F.F. Jöbsis, "Noninvasive, infrared monitoring of cerebral and myocardial oxygen sufficiency and circulatory parameters," *Science*, **198**, pp. 1264-1267, (1977).

[3] M. Cope, D.T. Delpy, "System for long term measurement of cerebral blood and tissue oxygenation on newborn infants by near infrared transillumination," *Med. Biol. Eng. Comput.*, **26**, pp. 289-294, (1988).

[4] S.R.Arridge, M.Schweiger, M.Hiraoka, D.T.Delpy, "Performance of an iterative reconstruction algorithm for near infrared absorption imaging", *SPIE*,**1888**,360-371,(1993).

[5] S.R. Arridge, M. Schweiger, M. Hiraoka, D.T. Delpy, "A finite element approach for modeling photon transport in tissue," *Med. Phys.* **20(2)**, 299-309 (1993).

[6] M. Schweiger, S.R. Arridge, D.T. Delpy, "Application of the Finite-Element Method for the Forward and Inverse Models in Optical Tomography", *J. Math. Imaging and Vision*, 3, pp. 263-283 (1993)

[7] D.T. Delpy, M. Cope, P. van der Zee, S.R. Arridge, S. Wray and J. Wyatt, "Estimation of optical pathlength through tissue from direct time of flight measurement," *Phys. Med. & Biol.*, **33**, pp. 1433-1442 (1988)

[8] S.R. Arridge, M.Cope, D.T.Delpy, "Theoretical basis for the determination of optical pathlengths in tissue : temporal and frequency analysis", *Phys. Med. & Biol.*, **37**, 1531-1560, (1992)

[9] R. Chandrasekhar, *Radiation Transfer*, (Clarendon Press, Oxford, 1950)

[10] M.C. Case, P.F. Zweifel, *Linear Transport Theory*, (Addison-Wesley, New York, 1967)

[11] J.R. Singer, F.A. Grünbaum, P.D.Kohn, J.P.Zubelli, "Image reconstruction of the interior of bodies that diffuse radiation", *Science*, **248**, 990-993 (1990)

[12] S.K. Patch, "Recursive recovery of Markov transition probabilities from boundary value data", PhD Thesis, Dept. Maths and Lawrence Berkeley Laboratory, University of Berkeley, CA, (April 1994)

[13] Y. Wang, J-H Chang, R. Aronson, R.L. Barbour, H.L. Graber, J. Lubowsky, "Imaging of scattering media by diffusion tomography : an iterative perturbation approach", *Proc. SPIE*, **1641**, 58-71, (1992)

[14] R. A. Robb and C. Barillot: "Interactive display and analysis of 3-D medical images.", *IEEE Trans. Medical Imaging*, **8(3)**, 217-226 (1989).

[15] S.R. Arridge, M. Schweiger, "Direct Calculation of the moments of the distribution of photon time of flight in tissue, using a Finite Element Method", *Applied Optics*, **33**, (1994, in press).

[16] J. Peraire, M. Vahdati, K. Morgan, O.C. Zienkiewicz, "Adaptive remeshing for Compressible Flow Computations", *J. Comp. Phys.* **72**, pp. 449-466 (1987)

[17] R. Gordon, R. Bender, G.T.Herman, "Algebraic reconstruction techniques (ART) for three-dimensional electron microscopy and X-ray photography", *J. Theor. Biol*, **29**, 471-481, (1970).

EXPERIMENTAL EVALUATION OF OBSERVER MODELS FOR DETECTION OF SIGNALS IN STATISTICALLY DEFINED BACKGROUNDS

Arthur. E. BURGESS and Xing LI
CIS, Rochester Institute of Technology , Rochester, NY 14623-5604

1. ABSTRACT

Investigation of human signal detection performance for noise limited tasks with statistically defined signal or image parameters represents a step towards clinical realism. However,the ideal observer procedure is then usually non-lincar and analysis becomes mathematically intractable. Two linear, but suboptimal, observer models [the Fisher-Hotelling observer and the non-prewhitening (NPW) matched filter] have been proposed for mathematical convenience. Experiments by Rolland and Barrett [1] involving detection of signals in white noise superimposed on statistically-defined backgrounds showed that the Hotelling model gave a good fit while the simple NPW matched filter gave a poor fit. Burgess [2] showed that the NPW model could be modified to fit their data by adding a spatial frequency filter of shape similar to the human contrast sensitivity function. Results of new experiments will be presented in this paper and it will be demonstrated that neither of these two models give a satisfactory fit to the data. The results of our experiments done to determine the variation of signal detectability as a function of background spectral density can be fitted by a Fisher-Hotelling model modified to include simple circularly symmetric spatial frequency channels as proposed by Myers and Barrett [3]. However, results of our variable viewing distance experiments do not agree with predictions of the simple channelized model. The conclusion is that a more elaborate channel model must be used.

2. INTRODUCTION

Decision tasks with medical images are intrinsically noise limited. Hence, the interest in both measuring and modelling human observer performance for noise-limited, image-based tasks. Early experiments [reviewed in 4] used very simple images and tasks - detection of completely specified disks, bars, and sine waves in gaussian noise superimposed on a uniform background. These are

Y. Bizais et al. (eds.), Information Processing in Medical Imaging, 89–100.

referred to as signal known exactly (SKE) and background known exactly (BKE) tasks. The advantage of using SKE/BKE tasks is that performance for the Bayesian ideal observer is easy to calculate. The statistical efficiency of human performance for the experiments can the be determined by comparison to the ideal observer results. Some experiments were done using uncorrelated (white) noise while other were done with ramp power spectra similar to that found in computed tomography. Wagner et al [5] suggested perhaps humans might not be able to compensate for such noise correlations and proposed the simple non-prewhitening (NPW) matched filter observer model. Myers et al [6] did human observer signal detection experiments with a variety of "blue" noise power spectra of the form $f^n |H(f)|^2$ to show that the NPW model gave a much better fit to human data than the classical prewhitening matched filter. Myers and Barrett [3] subsequently demonstrated that a Fisher-Hotelling model with simple circularly symmetric spatial frequency channels whose sensitivity profiles did not overlap could also explain the "blue" noise results.

The tasks encountered in medical clinical practice tend to involve detection, discrimination or identification of signals (lesions) in non-uniform backgrounds due to normal patient structure. This situation can be simulated by using statistically defined backgrounds in synthetic images and will be referred to as the background known statistically (BKS) case . The Bayesian ideal observer would then use non-linear procedures and performance for these cases has not been evaluated. This problem can be bypassed by using suboptimal but linear observer models. Myers, Rolland, Barrett, and Wagner [7] did a theoretical analysis of the performance of the Fisher-Hotelling and the NPW matched filter observer models for BKS detection tasks. The study was a simulation of a nuclear medicine imaging system. The task was detection of a two-dimensional Gaussian signal superimposed upon an inhomogeneous background and imaged through a pinhole imaging system. For mathematical convenience the pinhole aperture was assumed to have a Gaussian transmission profile. Poisson noise corresponding to the appropriate exposure time and aperture size was added to the detected image. The lumpy background was created by randomly superimposing a random number of Gaussian functions upon a constant background. Two types of situations were studied. In the first, signal detectability as a function of pinhole aperture diameter was calculated. It should be carefully noted that the variation in aperture size causes changes in the number of gamma rays used to form the image as well as changes in image resolution. It was found that detectability reached a maximum when the aperture size was approximately equal to the signal size. In the second study, signal detectability was calculated as a function of exposure time. This gave different Poisson noise levels without any change in resolution. Rolland and Barrett [1] reported human observer performance results for these two BKS detection tasks and found that the Hotelling model gave a satisfactory fit to their human results while the simple non-prewhitening matched filter model did not. Burgess [2] showed that the Rolland and Barrett human results could be described reasonably well by a modified non-prewhitening model that includes a band-pass filter of the type attributed to the front end of the human visual system. The best choice of eye filter had the form $E(f) = f^{1.3}\exp(-cf^2)$. This will be referred to as the NPWE model. The Rolland and Barrett observer results were precise enough to distinquish between the Fisher-Hotelling and standard NPW models. However they are not sufficiently precise to allow one to clearly favour the Fisher-Hotelling model over the modified NPWE model (or vice versa). It was concluded that further experiments are necessary to settle the issue. In particular, it was suggested that experiments needed to be done with several controlled viewing distances, several different signal shapes, and several correlation distances for the lumpy background. The most important difference between the models is the adaptiveness of the Fisher-Hotelling model and it is important to determine whether the human observer can demonstrate a similar adaptability. More precise experimental data for such a variety of experiments will be presented here. It will be seen that neither the Fisher-Hotelling or the modified NPWE models give a particularly good fit to these results. Hence a third model will be considered in this paper. It is the channelized Fisher-Hotelling model of Myers and Barrett [3] and will be referred to as the FHM model for short.

3. THEORY

The images used in the experiments described in this paper are of the form

$$i(x,y) = s(x-x_o, y-y_o) + n(x,y) + b(x,y) + B \qquad \text{(Eq. 1)}$$

where $s(x-x_o,y-y_o)$ is the signal spatial function centered at position (x_o,y_o), $n(x,y)$ is white Gaussian noise , $b(x,y)$ is a spatially varying background and B is a constant bias level . The white noise has spectral density N_o and zero mean.

The lumpy backgrounds were produced by low pass filtering zero mean Gaussian noise with spectral density W_O. The constant B was set at 127 to center the image data in the 0-255 range of display memory.

The Fisher-Hotelling observer (FH) model was proposed by Fiete et al [8]. It is the optimum linear discriminator and is equivalent to the ideal observer for tasks where both signal and background are known exactly (SKE/BKE). For statistically defined tasks, the Fisher-Hotelling observer uses a cross-correlation template calculated from the covariance matrices of the signal class, noise, and backgrounds. The Fisher-Hotelling observer detectability index is given (when white noise is injected before any filter) by [7]

$$\left(d'\right)^2_{FH} = \iint \frac{|S(u,v)|^2}{[N_o + W(u,v)]^2} du dv \qquad \text{(Eq. 2)}$$

with spatial frequencies u and v , signal amplitude spectrum S(u,v), white noise power spectrum N_O, and lumpy background power spectrum W(u,v). The imaging system transfer function H(u,v) does not appear in the equation since it operates on all three image components (signal, noise and background)

The modified non-prewhitening matched filter (NPWE) model is assumed to have a fixed bandpass filter E(u,v) before the adaptive matched-filter detection stage. It was found [2] that a filter equation of the form $E(f) = f^{1.3} \exp(-cf^2)$ with circular symmetry , radial frequency f, and no DC response worked well. The NPWE model uses the filtered expected signal as a template for cross-correlation. The bandpass filter ensures that the template has negative side lobes for background estimation and subtraction. By contrast, the simple NPW model does badly for statistically defined backgrounds because it does not try to estimate the background level in the vicinity of the signal. Instead, it uses absolute amplitude values, which include the local background, as part of the decision variable. The detectability function for the NPWE observer is given by [2]

$$(d')^2{}_{npwe} = \frac{\{\iint |S(u,v)|^2 \; |E(u,v)|^2 \, du \, dv\}^2}{\iint G(u,v) \, |S(u,v)|^2 \, |H(u,v)|^2 \; |E(u,v)|^4 \, du \, dv} \qquad \text{(Eq. 3)}$$

where $\quad G(u,v) = N_o + W(u,v) \, |H(u,v)|^2$

The Channelized Fisher-Hotelling Model of Myers and Barrett(FHM) has simple circularly symmetric spatial frequency channels with non-overlapping sensitivity profiles. The output of a channel is a weighted sum of the input, so spatial frequency data within a channel cannot be decorrelated by any subsequent stage of the system. Yao and Barrett [9] have shown agreement between the FHM model and human performance of a variety of tasks. The circular symmetric allows reduction of the problem to one dimension. The detectability index for this model [8] is determined by the sum of outputs, V_i, of the integrals over the radial frequency ranges of the individual spatial frequency channels

$$\left(d'\right)^2_{FHM} = \sum V_i \qquad \text{(Eq. 4)}$$

where $V_i = \left\{2\pi\int S(f)fdf\right\}^2 / 2\pi\int [No + W\ (F)](f)fdf$

Following Myers and Barrett we will use constant Q filters and a low-frequency cut-off, f_c , to ensure no DC response. Myers and Barrett tried all combinations of Q equal to 1.5 and 2 with f_c of approximately 0.65 and 0.87 cycles per degree and found little difference in performance for the various combinations of channel parameters. Their observers were not constrained to a particular viewing distance so the values of f_c are only an estimate of average values. We also found that results were not very sensitive to values of Q and f_c. We will restrict ourselves in this paper to Q equal to 2 and f_c equal to 0.87 c/d.

4. EXPERIMENTAL METHODS

Two types of experiments were done, one with a fixed viewing distance of one meter and variable background spectral density and the other with viewing distances from 0.5 to 2.5 meters. All experiments were done using the two-alternative forced choice (2AFC) method with side-by-side presentation of the two alternative images. Each image consisted of a sample background from the ensemble of lumpy backgrounds, uncorrelated Gaussian noise, and a signal at the center of onc of the images. The two lumpy backgrounds used in a particular decision trial were uncorrelated. The signals had a known spatial profile and one of three randomly assigned amplitudes. Two signals were used, a sharp-edged disk with a radius of 4 pixels and a two dimensional Gaussian with a spatial standard deviation of 3 pixels. All experiments were done using white noise with a spectral density selected to give a noise amplitude standard deviation of 20 grey levels. The backgrounds were produced by low pass filtering white noise to give a gaussian power spectrum. The DC values of the background spectral densities (W_o) were selected to give background amplitude standard deviations of 1,2,4,8,6,and 32 grey levels. Three correlation distances (spatial standard deviations) were used (4, 8 and 16 pixels).

The images were displayed using a DEC MicroVax computer, a Peritek VCU-Q display board, and a Tektronix 634 monitor. The monitor had a mean luminance of 210 Cd/m^2. At the viewing distance of 1 meter each pixel in the display subtended an angle of 0.43 milliradians. The role of the lumpy background on observer performance is determined by the ratio of the background and noise spectral densities (W_o/N_o) so the logarithm of this parameter was used as the independent variable for analysis of results. Human performance for a given experimental condition was evaluated by determining the percentage of correct responses for each signal amplitude, calculating the corresponding value of the detectability index d' , normalizing by the signal to noise ratio (SNR) for that amplitude and averaging over the three amplitudes. Observations were done in blocks of 256 decision trials with random variation of experimental conditions between blocks. All observers did 3 training blocks for each experimental condition to gain experience before experimental data collection began. A total of 768 decisions trials were used for each condition in the experiments, giving a coefficient of variation of 5% on estimates of the d'/SNR ratio.

5. EXPERIMENTAL RESULTS

5.1. Background Spectral Density Dependence

The first set of experiments measured the variation in signal detectability with background spectral density. The results for the gaussian signal are shown in figure 1. For presentation purposes the data for the uniform background case is given the artificial W_o/N_o ratio of 0.015 (-1.8 on the logarithmic scale). The results for the disc signal are not presented here, but they show the same trend. Observer performance for the first non-zero W_o/N_o ratio is the same as the uniform background and performance then drops steadily for larger ratios. The results for the 3 models are also shown in figure 1.

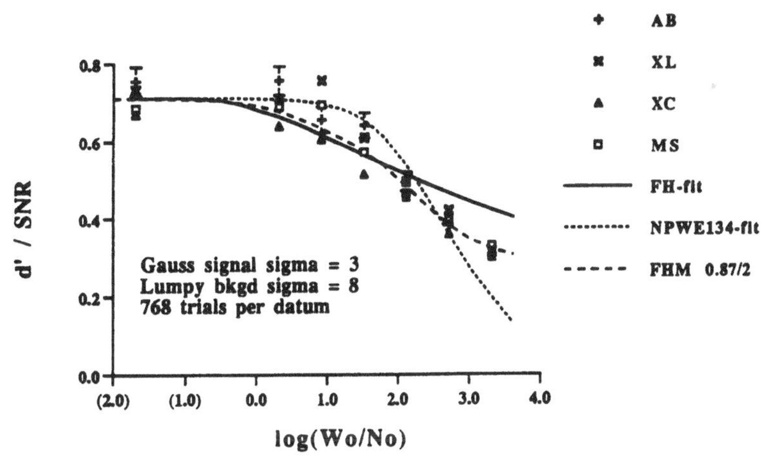

Figure 1: Observer results for detecting the Gaussian signal (s=3) with a lumpy background correlation distance of 8 pixels. The uniform background results (W_O=0) are plotted at log(W_O/N_O) equal to -1.8 for convenience. There are results for 4 human observers (AB, XL, XC, MS) and three models (FH, NPWE, FHM) with data normalized to fit the average human results for a uniform background. The error bars of d'/SNR due to sampling statistics are 5% of the data values.

The results of all models are normalized to fit the average observer results for the uniform background. The Fisher-Hotelling model agrees poorly with the human data since model performance decreases more slowly than the observer results. The NPWE model gives a slightly better fit but in most cases predicts performance lower than observers for large W_O/N_O ratios. The FHM model gives a reasonably good fit to the data.

One might wonder whether the NPWE model might fit better if a different eye filter model was used. One can address this question by an appropriate transformation of the dependent variable. The NPWE detectability index equation for a given experimental condition can be transformed so that the observer data should fall on a straight line with unit slope if any NPWE model is valid. With detectability index d'_O for the case of $W_O/N_O = 0$, the equation has the form

$$(d'_O / d')^2 - 1 = b\, W_O/N_O \qquad \text{(Eq. 5)}$$

The value of b is determined by solving the integrals in the denominator of equation 3 above. Human observer results for the appropriately transformed data are plotted in figure 2. The human results do fall on straight lines, but the slopes are wrong. They are in the range from 0.54 to 0.73. This demonstrates that no NPWE model can fit the human data since all NPWE models will give a slope of one. Changing the eye filter model will change the value of b and the theoretical line will be translated left or right, but the slope of the line will remain unity. The observer results can be fitted using the ad-hoc equation with the exponent g (with g approximately 2/3).

$$(d'_o / d')^2 - 1 = b\,[\,W_o/N_o\,]^g \qquad \text{(Eq. 6)}$$

It must be emphasized that this equation has no theoretical basis whatsoever but may be useful as a simple way of characterizing the data.

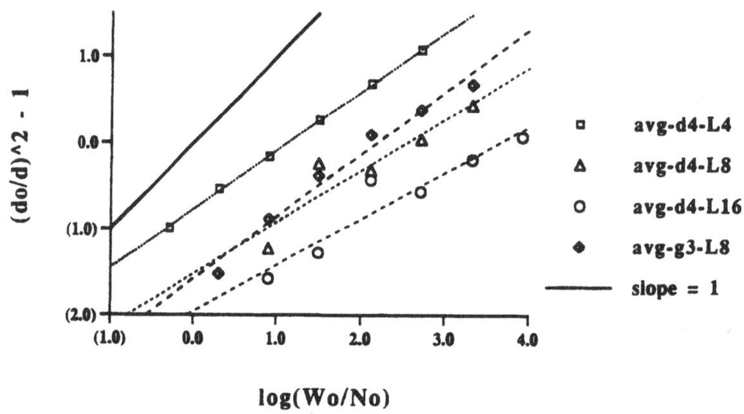

Figure 2: Transformed average results for 4 human observers, 2 signals (disc R=4 and Gaussian s=3) with 3 lumpy background correlation distances (4, 8, 16). Any modified NPWE model would give slope of one on this graph.

5.2. Viewing distance experiment

Human performance of visual tasks clearly depends on the viewing distance. For most tasks there is an optimum viewing distance that most expert observers agree upon. The basic Fisher-Hotelling model has no eye filter or channels and hence no viewing distance dependence. The other models will show a viewing dependence because the model parameters are defined in an angular spatial frequency coordinate system. As will be seen, the FHM model predicts oscillations in signal detectability as viewing distance changes. The signal detection performance of two observers was measured for 7 viewing distances

(0.5, 0.75, 1.0, 1.25, 1.5, 2 and 2.53 meters). The distance was controlled by having the observer sit in a chair with his head lightly pressed back against a head rest. Four experimental conditions were used with two signals, a disc with radius 4 pixels and a 2D Gaussian with a standard deviation of 3 pixels; and two W_O/N_O ratios, zero and 515. The observers did 3 blocks of practice trials (768 decisions) at each distance before the experimental data collection began (also 768 trials per datum).

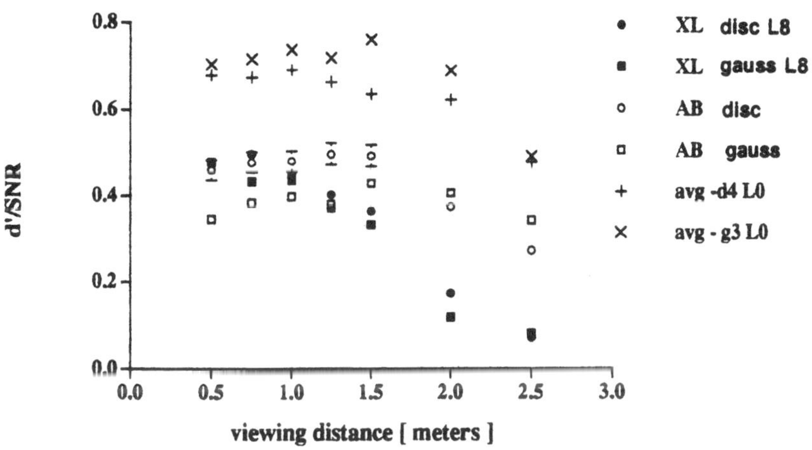

Figure 3: Observer detection results as a function of viewing distance. Two signals were used: a disc (R=4 pixels) and a Gaussian (s=3 pixels). There were two human observers (AB, XL) with 768 trials per datum. Results of the two observers and two signals are averaged for the uniform background case (L0). The individual results for the lumpy background case [L8 with s = 8 pixels and $W_O/N_O = 515$] are shown.

The experimental results are shown in figure 3. The human results for the uniform background case indicate surprisingly little distance dependence. Only the averaged results for the two observers are shown to simplify the graph but there is no evidence for oscillation in the individual observer data. For the lumpy background case, one observer (AB) shows modest variation while XL has a steady decrease in performance for distances greater than 1.5 meters. The calculated values (in relative units) of individual channel outputs and their sum for the FHM model with a lumpy background are shown in figure 4. The cause of the oscillation of performance with viewing distance is evident. The output of a given channel has a maximum at some particular viewing distance. The channel centre frequencies are separated by one octave and so the maxima of adjacent channels are at distance ratios of 2. The sum therefore oscillates. This effect could be reduced by using smaller centre frequency ratios but the overall performance would be less sensitive to the W_o/N_o ratio. In the limit of very small ratios, one would be back to the standard FH model.

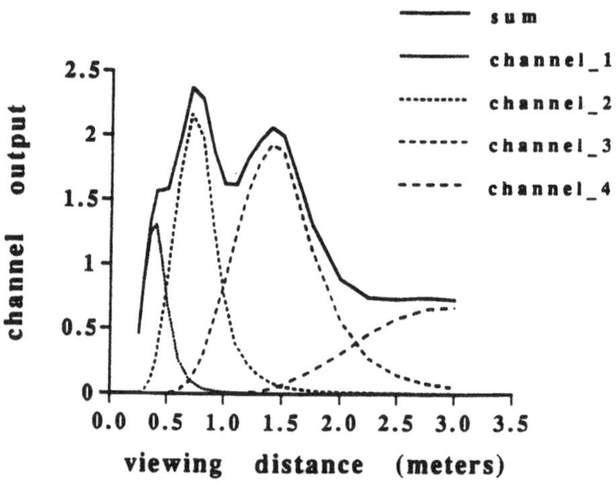

Figure 4: Viewing distance dependence for the Myers and Barrett channelized FH model with a cut-off frequency of 0.87 cycles/degree and Q equal 2. Total output $[(d')^2]$ and individual channel outputs (in relative units) for the Gaussian signal [s = 3] and a lumpy background [s = 8 , W_o/N_o = 515] as a function of distance are shown.

6. DISCUSSION

The variable background results demonstrate that the FH and NPWE models do not give good predictions of human observer performance while that the FHM model does. However the FHM model predicts oscillatory performance variation with viewing distance change. This does not agree with our human observer results or those of any previous experiments that we know of. The early human contrast sensitivity results of Lowry and DePalma [10] did show two peaks and led Campbell and Robson [11,12] to suggest visual channels. However, it has been shown consistently since then that the contrast sensitivity curve is smooth and unimodal . Many experimenters in the 1960's and 1970's obtained variable spatial frequency by varying viewing distance and none reported oscillatory detection performance behaviour. Hence one must conclude that the FHM model is unsatisfactory.

The amplitude of the oscillations in FHM model performance could be reduced by several strategms. One, as mentioned, would be to reduce the ratio of channel centre frequencies from 2 to some number closer to unity. However this approach would reduce agreement with the variable spectral density results and would not agree with many psychophysical experiments that suggest channel bandwidths of one octave. Another approach would be to assume that human observer are always working near the optimum viewing distance when not constrained by experimenters (and we believe this to be the case) and then average the model results over a reasonable range of viewing distance. These two approaches have the merit of ensuring that channel outputs are not intrinsicly correlated and hence would simplify the mathematics. We also investigated a 2D FH model with channels that were separable in cartesian coordinates and had non-overlapping sensitivity profiles. This model also gave oscillating output with varying viewing distance. The final approach would be to abandon the simplicity of non-overlapping sensitivity profiles altogether. Several channels models [13,14] with over-lapping sensitivity profiles have been used for analysis of human visual performance. These models, however, have not included an optimum linear discrimination stage after the channels. We propose to evaluate observer models which use these channels models followed by the Fisher-Hotelling optimum linear decision device.

7. ACKNOWLEDGEMENTS

The authors thank Kyle Myers and Robert Wagner for helpful discussions at all stages of this work.

8. REFERENCES

1. J.P. Rolland and H.H.Barrett, "Effect of random background inhomogeneity on observer detection performance", *J. Opt. Soc. Am.*, A9, pp 649-658, 1992
2. A.E. Burgess, " Statistically defined backgrounds: Performance of a modified nonprewhitening matched filter model ", *J. Opt. Soc. Am.*., A11, pp 1237-42, 1994
3. K.J. Myers and H.H. Barrett, "Addition of a channel mechanism to the ideal-observer model", *J. Opt. Soc. Am.*, A4, pp 2447-2457, 1987
4. A.E. Burgess, "High level visual decision efficiencies", Chapter in *Vision: Coding and Efficiency*, C. BlakemoreEdit., pp 431- 440, Cambridge Univ. Press , 1990

5. R.F. Wagner, D.G. Brown and M.S. Pastel, "Application of information theory to the assessment of computed tomography", *Medical Physics* , 6, pp 83-94, 1979
6. K.J. Myers, H.H. Barrett, M.C. Borgstrom, D.D. Patton, and G.W. Seeley,"Effect of noise correlation on detectability of disk signals in medical imaging", *J. Opt. Soc. Am.* A2, pp 1752-1759, 1985
7. K.J. Myers, J.P. Rolland, H.H. Barrett and R.F. Wagner, "Aperture optimization for emission imaging: effect of a spatially varying background", *J. Opt. Soc. Am.* , A7, pp 1279 - 1293, 1990
8. R.D. Fiete, H.H. Barrett, W.E. Smith and K.J. Myers, " The Hotelling trace criterior and its correlation with the human observer performance", *J. Opt. SocAm.,* A4, pp 945-953, 1987
9. J. Yao and H.H. Barrett, "Predicting human performance by a channelized Hotelling observer model", in *Mathematical Methods in Medical Imaging*, D.C. Wilson and J.N. Wilson eds., Proc. SPIE 1768, pp 161-168 , SPIE, 1992
10. E.M. Lowry and J.J. DePalma, "Sine Wave Response of the Visual System", *J.Opt.Soc.Am.*, 51, pp 760-766, 1961
11. F. W. Campbell and J. G. Robson, "Application of Fourier analysis to the visibility of gratings", *J. Physiol.,* 197, pp 551-566, 1968
12 M. Sachs, J. Nachmias, and J. Robson, "Spatial-frequency channels in human vision", *J.Opt. Soc. Am.,* 61, pp 1176-1186, 1971
13. A.B. Watson, "Detection and Recognition of Simple Spatial Forms", in *Physical and Biological Processing of Images*, Braddick and Sleigh eds, pp 100-114, Springer-Verlag, 1983
14. S. Daly, "The Visible Differences Predictor: An Algorithm for the Assessment of Image Fidelity", in *Digital Images and Human Vision*, A.B.Watson ed, pp 179-206, MIT Press, 1994

TOPOLOGICAL PROPERTIES OF SMOOTH ANATOMIC MAPS

Gary E. CHRISTENSEN[1], Richard D. RABBITT[2], Michael I. MILLER[3], Sarang C. JOSHI[3], Ulf GRENANDER[4], Thomas A. COOGAN[5], and David C. VAN ESSEN[5]

Dept. of Surgery, Washington Univ. School of Med., St. Louis, MO 63110, USA[1]
Dept. of Bioengineering, 2480 MEB, Univ. of Utah, Salt Lake City, Utah 84132, USA[2]
Dept. of Electrical Engineering, Washington Univ., St. Louis, MO 63130 USA[3]
Div. of Applied Mathematics, Brown Univ., Providence, RI 02912, USA[4]
Dept. of Anatomy and Neurobiology, Washington Univ. School of Med., St. Louis, MO 63110, USA[5]

ABSTRACT

A large body of work has emerged over the last decade in the area of registration of medical images and image volumes. Most of this work has focused on affine transformations which account for global translation, rotation, scale, and skew, while others are based on minimizing some quadratic energy penalty (associated with the membrane or Laplacian operator, the thin-plate spline or biharmonic operator, the linear-elasticity operator, and any polynomial of these operators). The common shortcoming of these approaches is that they do not account for the large-distance, nonlinear variation of the anatomies being registered. This paper addresses the topological issues involved with large-distance, nonlinear, volume transformations. An Eulerian reference frame is used to track the textbook as it deforms continuously into the shape of the target anatomy data set. Topological properties of transformations applied to the textbook are discussed. Specifically, it is shown that the transformation must be homeomorphic. A high-dimensional, viscous fluid model is presented for deforming the textbook that accommodates large-distance, nonlinear deformations while ensuring that the transformation is homeomorphic. 2D and 3D results are presented which map images of one macaque monkey visual cortex into another using the viscous fluid model.

1. INTRODUCTION

Unlike traditional anatomical atlases [8, 18] which describe a single anatomy, a deformable textbook [4, 7, 13, 5] is a digitized anatomic atlas that is made subject-specific by transforming its coordinate system to match that of another individual (see Fig. 1). The subject-specific textbook is called an individualized textbook and provides subject-specific information normally only known for the textbook anatomy (see Fig. 2). The terms *template* and *study* are used to denote the data sets associated with the textbook and the individual (target) anatomies, respectively.

The deformable textbook is used also as a standard anatomy for comparing function between individuals. This is accomplished by estimating the transformation that deforms the study anatomy to the textbook anatomy and then using this transformation to map the study functional images onto the textbook anatomy. Anatomical shape variation is compensated using this process because the functional data is mapped from its original

Y. Bizais et al. (eds.), Information Processing in Medical Imaging, 101–112.

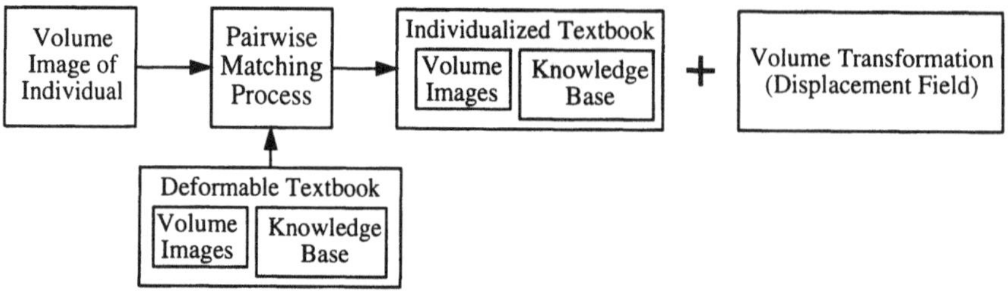

Figure 1: An individualized textbook is generated by transforming the deformable textbook coordinate system until the information in the textbook corresponds to the subject's anatomy. This is accomplished automatically by pairwise matching image volumes from the study and the textbook. Shape differences between the textbook and subject's anatomy are contained in the volume transformation.

location in the individual's coordinate system to the corresponding location in the textbook coordinate system.

Since the transformation of the textbook is used to map information from the textbook to the study or vice versa, it is crucial that the transformations be homeomorphic, i.e., continuous, 1-to-1, and onto. This ensures that the continuity and neighborhood structure of the textbook are maintained under the transformation.

The transformations applied to the textbook coordinate system need to be of high spatial dimension to accommodate the complex morphology of the brain—ventricles, cortical folds, etc. This is achieved by constraining the set of transformations to those consistent with physical deformations of real materials which accommodate large-distance, nonlinear deformations. An example constraint is the viscous fluid model described in this paper and [4, 5, 6, 7].

2. EULERIAN VS. LAGRANGIAN REFERENCE FRAME

Individualizing the anatomical textbook requires tracking the course of every material particle of the textbook from its original to its final position. Each particle can be identified on the basis of either its initial or final coordinates. If a particle is tracked with respect to its original coordinates, the reference frame is called *Lagrangian*; otherwise, if the particle is tracked with respect to its final coordinates, the reference frame is called *Eulerian*. We use an Eulerian reference frame in this work.

The Eulerian reference frame is more computationally efficient than the Lagrangian for tracking large-distance, nonlinear, volume deformations. Let N be the number of voxels in the study. In the Lagrangian reference frame, the trajectories of N particles–each originating at a different voxel location in the volume–must be tracked at each iteration. In addition, the contributions from the N particles must be interpolated onto the voxel lattice at each iteration to compute the discrete gradient of the cost function given in Eq. 5. On the other hand, when the Eulerian reference is used, the deforming continuum is tracked by computing the continuum at the N lattice sites each iteration. Because the deformed continuum is evaluated at each lattice site, no additional interpolation is required to compute the discrete gradient of Eq. 5. Evaluation of the deformed continuum on a voxel lattice is also required for visualization.

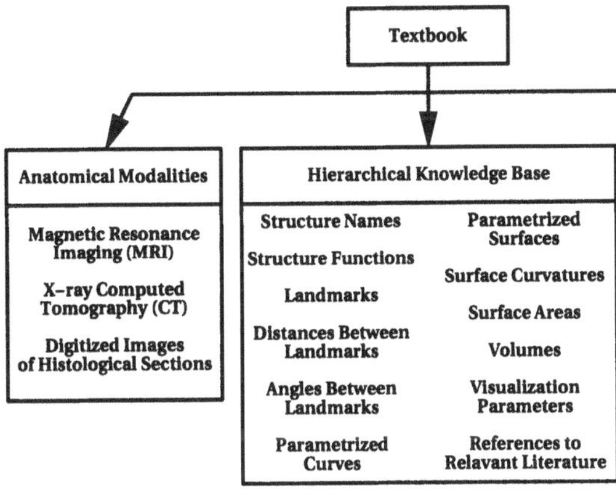

Figure 2: Overview of some of the information contained in a deformable anatomical textbook. The anatomical modalities are 3D volumetric images of the textbook anatomy; the hierarchical knowledge base is a collection of symbolic information that describes characteristics and measurements of the textbook anatomy; and the functional modalities are 3D volumetric images of function associated with the textbook anatomy. The transformation that individualizes the textbook is used to map this information onto the data set associated with the target anatomy.

3. TOPOLOGICAL PROPERTIES OF THE TEXTBOOK TRANSFORMATIONS

There are two kinds of anatomical variations which must be accommodated: normal variation between anatomies, and diseased or abnormal variation. Disease and abnormal variation other than those based on shape changes will not be addressed in this paper. To this end, we make the assumption that the template and study are topologically equivalent.

Let $\mathcal{H}$ be the set of homeomorphic transformations of the textbook coordinate system such that $\vec{h} : \Omega \rightarrow \Omega$ for $\vec{h} \in \mathcal{H}$. A homeomorphic transformation is defined to be a continuous, 1-to-1, onto mapping. These homeomorphisms are generated from translation groups applied to points $\vec{x} \in \Omega$:

$$\vec{h} : \vec{x} = (x_1, x_2, x_3) \rightarrow (x_1 - u_1(\vec{x}), x_2 - u_2(\vec{x}), x_3 - u_3(\vec{x})) . \qquad \text{(Eq. 1)}$$

The vector field $\vec{u}(\vec{x}) = [u_1(\vec{x}), u_2(\vec{x}), u_3(\vec{x})]^T$ is called the Eulerian displacement field. A time dependent displacement field is denoted by $\vec{u}(\vec{x}, t)$. It is defined as a map from points in the template to fixed observation points in the deforming continuum such that a mass particle instantaneously located at $\vec{x}$ at time t originated at point $\vec{x} - \vec{u}(\vec{x}, t)$. The time t can correspond to real or simulation time. The maps constructed from these transformations allow for the dilation, contraction, and warping of the underlying ideal coordinates of the template at a very local level.

Continuity of the transformation ensures that the local relationships between structures is maintained. That is, structures that are adjacent to each other before transformation are still adjacent after transformation even though they may have changed shape. This is due to the fact that continuous transformations map compact sets to compact sets and connected sets to connected sets.

The global 1-to-1 property of the transformation is important because it ensures that each point in the textbook anatomy is mapped to one and only one point in the deformed anatomy. When the transformation is also onto, we are assured that the entire domain of the textbook anatomy will be mapped onto the entire domain of the study anatomy. These two properties ensure that the entire knowledge base of the textbook, which is associated with coordinate locations in the textbook domain, is mapped in its entirety to the study anatomy.

A necessary condition for a transformation to be globally 1-to-1 is that it is locally 1-to-1. A transformation with domain Ω is locally 1-to-1 if about any point $p \in \Omega$ there is a neighborhood in which $\vec{h}$ is 1-to-1. Alternatively, a continuously differentiable transformation $\vec{h}$ that takes $\mathbf{R}^n$ into $\mathbf{R}^n$ is locally 1-to-1 if its Jacobian is positive, i.e.,

$$J(p) = \det\left(\vec{\nabla}\vec{h}|_p\right) > 0 . \qquad \text{(Eq. 2)}$$

If the Jacobian of the transformation is positive at each point in a given set Ω, the transformation is said to be locally 1-to-1 over Ω. If the Jacobian is nonnegative over a set Ω then open subsets of Ω are mapped to open subsets.

The next theorem from [3] give a sufficient condition to insure that a locally 1-to-1 transformation is globally 1-to-1. In practice, this theorem is used to determine whether or not a particular transformation is globally 1-to-1.

Theorem 1 *If $\vec{h}$ maps a convex set Ω onto another convex set Ω^* in such a way that the boundary of Ω is mapped exactly onto the boundary of Ω^*, and if $\vec{h}$ is locally 1-to-1 in Ω, then $\vec{h}$ must be globally 1-to-1.*

An important property of a homeomorphic transformation is that a unique inverse transformation exists and is continuous. In addition, if the homeomorphic transformation is also of class C' (continuously differentiable) then the inverse transformation is also of class C'. The existence of the inverse mapping ensures that the transformation can be used to map information not only from the textbook to the individual but also from the individual to the textbook.

4. EXAMPLE OF A TRANSFORMATION WITH SINGULAR JACOBIAN

As said previously, we assume that the template and study are topologically equivalent. Therefore the textbook transformation $\vec{h}$ must homeomorphic to ensure that the deformed textbook is topologically equivalent to the original textbook. A necessary condition that $\vec{h}$ is a homeomorphism is that the Jacobian of $\vec{h}$ for all $\vec{x} \in \Omega$ is positive. The next example demonstrates what happens when this condition is violated in both the Lagrangian and Eulerian reference frame.

Assume that $\vec{h}(\vec{x}) = (x_1 + \frac{1}{4}\sin 2\pi x_1 \sin \pi x_2, x_2)$ for $\vec{x} \in \Omega = [0,1]^2$. Although this transformation is continuous and onto in Ω, it is singular on Ω because its Jacobian $J(\vec{x}) = 1 + \frac{\pi}{2}\cos 2\pi x_1 \sin \pi x_2$ is both positive and negative on Ω. The region where $J(\vec{x})$ is less than or equal to zero is shown as the black region in Fig. 3.

If we interpret $\vec{h}$ with respect to a Lagrangian reference frame, $\vec{h}$ maps points from the undeformed coordinate system to points in the deformed system as a function of points in the undeformed coordinate system. Thus, $\vec{h}(\vec{x})$ gives the location in the deformed coordinate system to which $\vec{x}$ in the undeformed coordinate system maps.

Fig. 4 demonstrates the effect of mapping the template shown in the left panel with $\vec{h}$ with respect to a Lagrangian reference. The fact that the Jacobian of $\vec{h}$ is singular on Ω

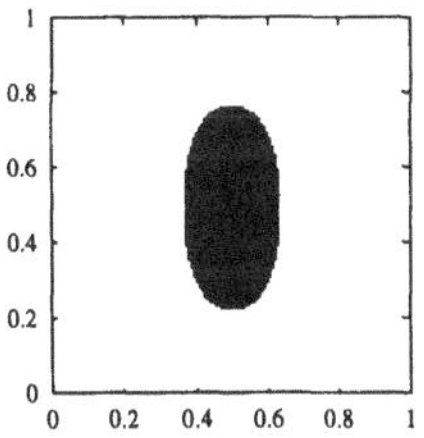

Figure 3: The black region is where the Jacobian of $\vec{h}(\vec{x}) = (x_1 + \frac{1}{4}sin2\pi x_1 sin\pi x_2, x_2)$ for $\vec{x} \in \Omega = [0,1]^2$ is less than or equal to zero.

implies that the $\vec{h}$ does not preserve the topology of the template. This is evident because the two vertical lines in the template are mapped to lines that cross over one another in the deformed coordinate system. Transformations with singular Jacobians are often referred to as a folding transformations because they "fold" the coordinate system causing the lines in this case to cross over one another.

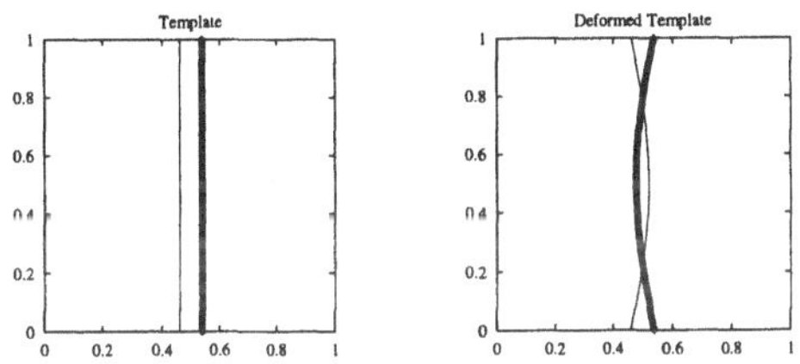

Figure 4: Two vertical lines before (left) and after (right) transformation with $\vec{h}(\vec{x}) = (x_1 + \frac{1}{4}sin2\pi x_1 sin\pi x_2, x_2)$ when $\vec{x}$ is with respect to a Lagrangian reference frame.

This example also illustrates that when $\vec{h}$ has a singular Jacobian in a Lagrangian reference frame, $\vec{h}$ is a many-to-one mapping. The two points where the deformed lines intersect are examples of the many-to-one property of $\vec{h}$.

Next we interpret $\vec{h}$ in terms of a Eulerian reference frame. In this case, $\vec{h}$ maps points from the undeformed coordinate system to points in the deformed system as a function of points in the deformed coordinate system. Thus, $\vec{h}(\vec{x})$ gives the location of the point in the undeformed coordinate system that maps to $\vec{x}$ in the deformed coordinate system.

Fig. 5 demonstrates the effect of mapping a template consisting of a vertical line using $\vec{h}$ in the Eulerian reference frame.

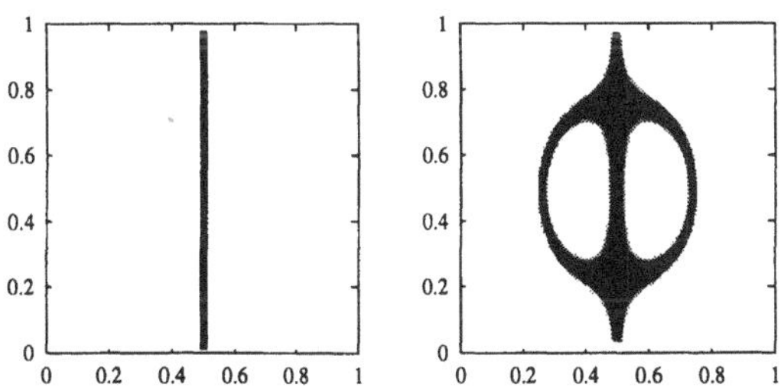

Figure 5: A template before (left) and after (right) transformation with $\vec{h}(\vec{x}) = (x_1 + \frac{1}{4}sin2\pi x_1 sin\pi x_2, x_2)$ when $\vec{x}$ is with respect to a Eulerian reference frame.

Similar to the Lagrangian case, the fact that the Jacobian of $\vec{h}$ in the Eulerian reference frame is singular implies that $\vec{h}$ does not preserve the topology of the mapped images. This is evidenced by the fact that vertical lines in the undeformed coordinate system are mapped to non-lines.

In contrast to the Lagrangian case, this example shows that a transformation $\vec{h}$ with a singular Jacobian (with respect to a Eulerian reference frame) is a 1-to-many mapping. This is seen by the fact that the single line (left) is mapped to the triple line (right) in Fig. 5.

In contrast to these examples, a homeomorphic mapping has a positive Jacobian. Therefore, this crossing of lines and mapping lines into nonlines cannot happen because topology is preserved.

5. FLUID MODEL

The viscous fluid model [7, 4, 5] was developed to overcome the shortcomings of quadratic penalty methods. Specifically, quadratic regularization penalties such as the membrane or Laplacian model [1, 17], the thin-plate or biharmonic model [17, 19, 2], and the linear elasticity model [13, 7] develop restoring forces which are proportional to the deformed distance. Except for the smallest deformations, such penalties prevent the textbook from being fully deformed into the shape of the target anatomy. Even more fundamental is the problem that quadratic base penalties are only valid for small linear deformations. This shortcoming is due to the linearization of the strain-displacement relationship in their derivations or equivalently small deformation angle assumptions.

The viscous fluid spatial transformation (VFST) is summarized from [5] as follows. At each time t, the continuum satisfies the partial differential equation (PDE)

$$\alpha\nabla^2\vec{v}(\vec{x},t) + \beta\vec{\nabla}(\vec{\nabla}^T\vec{v}(\vec{x},t)) = \vec{b}(\vec{x} - \vec{u}(\vec{x},t)) \qquad \text{(Eq. 3)}$$

where $\vec{v} = [v_1(\vec{x},t),\ \ v_1(\vec{x},t),\ \ v_3(\vec{x},t)]^T$, $x \in \Omega \subset \mathrm{R}^3$, $t \in [0,1]$ is the instantaneous velocity of the deformation field $\vec{u}$. The gradient operator is defined by $\vec{\nabla} = [\frac{\partial}{\partial x_1},\ \ \frac{\partial}{\partial x_2},\ \ \frac{\partial}{\partial x_3}]^T$ and $\nabla^2 = \vec{\nabla}^T\vec{\nabla}$ is the Laplacian. The body force per unit volume $\vec{b}$ is the driving force that deforms the textbook into the shape of study anatomy. The $\nabla^2\vec{v}$ term in (Eq. 3) is called the viscous term of the PDE. This term constrains neighboring particles of the displacement field to deform with roughly the same velocity by spatially smoothing the velocity field. The $\vec{\nabla}(\vec{\nabla}^T\vec{v})$ is called the mass source term. This term can be used to determine how much the textbook had to grow and shrink in mass to match the study. The coefficients α and β are the viscosity coefficients.

As the textbook continuously deforms over time and space, points (particles) in the textbook flow through fixed observation points–one located at the center coordinate of each voxel in the study. At any instant of time, the transformation $h(\vec{x}) = \vec{x} - \vec{u}(\vec{x})$ gives the original position of a particle as it passes through the observation point located at x. In the Eulerian reference frame, the velocity, $\vec{v}$, of a particle is related to its displacement, $\vec{u}$, by

$$\vec{v}(\vec{x},t) = \frac{\partial\vec{u}(\vec{x},t)}{\partial t} + \vec{v}(\vec{x},t)^T\vec{\nabla}\vec{u}(\vec{x},t). \qquad \text{(Eq. 4)}$$

The term $\vec{v}(\vec{x},t)^T\vec{\nabla}\vec{u}(\vec{x},t)$ accounts for the kinematic nonlinearities of particles as they pass through the observation points.

5.1. The body force

The body force appearing in the PDE associated with the momentum Eq. 3 follows directly from the Bayesian view of the PDE as the variational minimizer of the posterior distribution. The body force is given directly from the potential term in the posterior distribution associated with the sensor, linking the data to the template. The body force becomes the gradient of the total Lagrangian, call it C, associated with the log-likelihood of the data source. This is consistent with our previous work as described in [1, 9, 13]. For a Gaussian sensor model, the cost function is

$$C(T(\vec{x}), S(\vec{x}), \vec{u}) = \frac{1}{2}\int_{\Omega} |T(\vec{x} - \vec{u}(\vec{x}, t)) - S(\vec{x})|^2 d\vec{x} \ . \qquad \text{(Eq. 5)}$$

where T and S correspond to the textbook and study images being matched. For MRI and high signal-to-noise photographic images this appears to be appropriate [12, 15, 14]. Although for PET image reconstruction, Poisson models [16] and their associated variation would be more appropriate for the forcing function.

5.2. Solution on a finite spatial grid

The viscous-fluid PDE is solved on a discrete lattice associated with Ω. For large curved deformations, the transformation evaluated at these spatial grid points becomes singular over time even though the transformation evaluated on the continuum would not. To circumvent this problem, a method of regridding the template is used by generating a new, or propagated template, whenever the discretized transformation approaches local singularity. This is accomplished as follows. The discretized displacement field is propagated through time using using a discretized version of Eq. 4 (see [5]) until the Jacobian of the transformation indicates that the transformation is approaching singularity. When the magnitude of the Jacobian drops below a threshold (0.5 for the results shown later), the computation is stopped and a new propagated template is generated equal to the deformed template at the previous instant. The algorithm is restarted using the new template. When mapped back to the original template, this regridding approach is equivalent to defining a new nonhomogeneous computation grid on the undeformed continuum. Initial conditions for the propagated template are set to match the final state of the previous template, i.e., the instantaneous velocity remains the same and the displacement field corresponding to the new template is set to zero. The total transformation is continuously tracked via the concatenation of the displacement fields associated with all of the propagated templates.

6. RESULTS

This section shows some examples of the deformable textbook. Other examples this approach can be found in [5, 7, 4, 10]. The data used in these experiments are digitized images of sections of macaque monkey visual cortex. The data were collected by slicing a block from the occipital lobe into 100 μm sections using a microtome, and taking pictures of the exposed blockface with a CCD camera. Fig. 6 shows 3D volume renderings of the textbook and study monkey visual cortex used in this experiment.

The dimension of the template and study were $320 \times 240 \times 64$ and and $320 \times 240 \times 54$, respectively. These volumes were symmetrically zero-padded by background to generate equal $320 \times 240 \times 74$ volumes. The study volume was preprocessed by histogram-matching its histogram to that of the template volume. The domain of the textbook and study were assumed to be the unit cube. The intensities of the template and the study were scaled from zero to one.

A multi-resolution approach was used to deform the textbook because it improves the convergence times and it avoids many local minima, i.e., erroneous matches.

The first step in the registration was to generate a global registration between the template and the study. This was accomplished by matching nine major sulci in the textbook and the study using the technique described in [11]. Nine corresponding sulci were labeled by hand

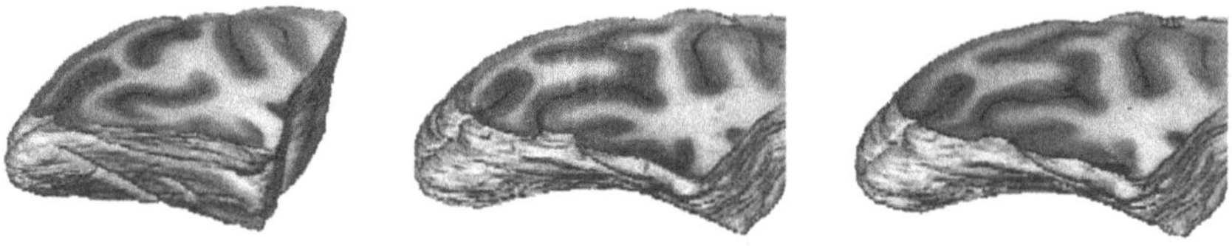

Figure 6: 3D rendering of the textbook, study, and deformed textbook monkey visual cortex cut at slice 53.

in the textbook and the study. Each sulcal line was approximated by an equal number of equally spaced landmarks. The transformation that matched the corresponding landmarks was then computed as in [11].

The final step is to refine the transformation generated by matching the sulci using the 3D viscous fluid model. This step matches the anatomy between the sulcal lines. The fluid transformation is estimated solely from the template and study data. The sulcal line information was not used in this step. The advantage of not using the sulcal lines in the final estimation step is that the data drives the deformation. Thus, sulcal lines can be then drawn quickly and do not have to be precise. On the other hand, if the sulcal lines are drawn precisely, it is reasonable to use them in conjunction with the fluid model.

Figure 7 shows slice 18 from the textbook, study, 3D sulcal deformation, and the 3D fluid deformation. Notice that the 3D sulcal line deformation has only partially transformed the template to match the shape of the study, and that the correspondence with the study is much improved after the 3D fluid deformation.

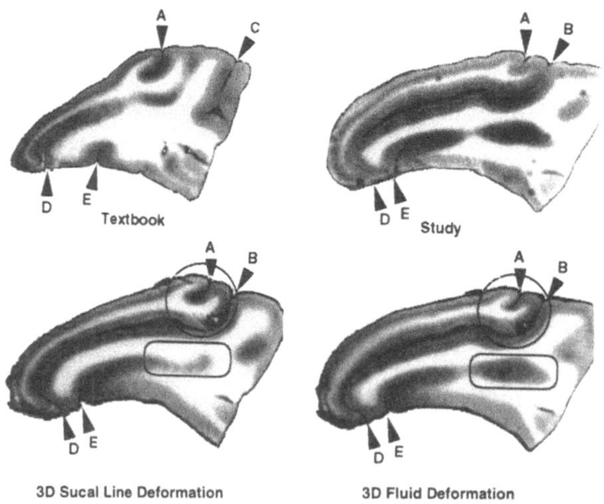

Figure 7: Slice 18 from the textbook (upper right), study (upper left), globally deformed textbook generated by the 3D sulcal transformation (lower left), and locally deformed textbook generated by the 3D fluid transformation (lower right). The 3D sulcal deformation was used as an initial condition for the 3D fluid deformation. Notice the excellent shape correspondence between all of the structures in the study and the 3D fluid deformation. Arrows show the cross section of some of the nine sulcal lines that where used. Note that not all sulcal lines are present in any one 2D slice. The ovals and circles show two obvious areas in this slice where the 3D fluid transformation improved the global transformation.

Fig. 8 shows the result of a 2D experiment. The dimensions of the textbook and study were 213 × 160. The study image was preprocessed by histogram-matching its histogram to the textbook image. The textbook was automatically deformed into the study using a 2D elastic basis transformation [13] followed by a 2D fluid transformation. The state of the template after each step is shown in the bottom row of Fig. 8.

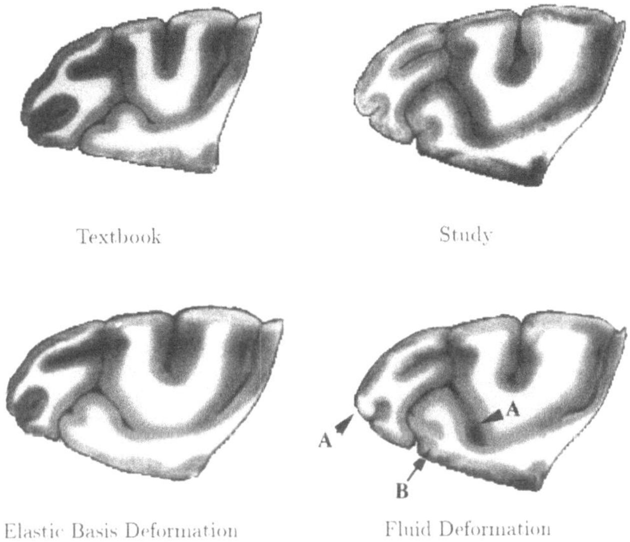

Figure 8: The top row shows a 2D image of the visual cortex of the textbook (left) and the study (right). The bottom row shows the elastic basis transformation (left) and the fluid transformation (right) of the textbook. The elastic basis transformation was used as an initial condition for the fluid transformation. Notice how similar the shape of the fluidly deformed textbook is to the study. Places marked by arrows denote local areas were the fluidly transformed textbook did not match the textbook.

Notice the good local correspondence of nearly all of the structures in the fluid transformed textbook and the study, i.e. the contours of the grey matter and white matter match nearly every where in the fluidly transformed textbook and the study. However, this experiment points out some of the problems of using a driving function (see Eq. 5) that only depends on pixel intensities. The arrows marked with the letter A (lower right panel) show two places where the intensity mismatch in the textbook and the study caused a locally incorrect registration. The first mismatch, caused by unequal lighting and inherent differences in intensity of the grey matter, resulted in the white matter in the textbook matching locally to the grey matter in the study. The second mismatch in intensities resulted in the expansion of the gap between sulci.

The problems caused by intensity mismatch can be corrected using a different driving function that depends on other characteristics besides grey levels and by better data collection in which the lighting is more uniform. However, since lighting will never be perfect and there will be critical data sets with bad lighting, it is natural to improve the intensity-only driving function.

Note that the 3D transformation Fig. 7 was more robust than the 2D transformation because the unequal lighting and highlighting effects are slice dependent. Information from the intensities in the planes above and below helped to reduce this problem.

The arrow marked by the letter B in Fig. 8 (lower right) points out the third place where the fluid transformed textbook did not locally match the study. The sulcus in this box flowed in an anatomically incorrect way to match the corresponding sulcus in study. Including landmark information [11, 2] as in the 3D experiment avoids this type of problem.

Fig. 9 shows the grids associated with the undeformed textbook (left), the elastic basis deformed textbook (middle), and the fluidly deformed textbook (right) for the 2D experiment. The grid in the right panel shows some of the local translations that were required to transform the textbook. These grids show that the topology was maintained, because the grid lines are solid continuous lines, despite being strongly deformed in places.

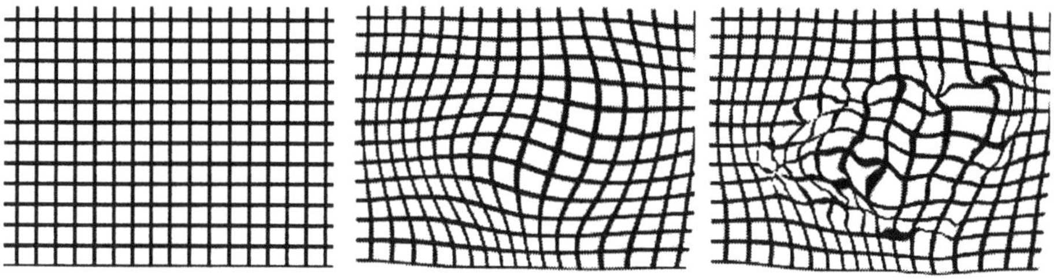

Figure 9: The grid before transformation (left), after the elastic basis transformation (middle), and after the fluid transformation (right). These grids show how the elastic basis transformation accommodated the global shape variation and how the fluid transformation accommodated the local shape variation.

7. PARALLEL COMPUTER IMPLEMENTATION

The 2D and 3D fluid transformation algorithms were implemented on the massively-parallel DECmpp 12000Sx/Model 200 (MasPar). This computer is a 128×64 mesh-connected single-instruction-multiple-data (SIMD) architecture and is well suited for solving partial differential equations such as Eq. 3.

The 2D and 3D fluid transformations were run for 100 real-time time steps and took roughly 2 minutes and seven hours to compute, respectively. This time difference is due to the additional computations that are required to solve Eq. 3 in 3D than in 2D and that the continuum was tracked at 167 times more observation points in 3D experiment than in the 2D experiment. This computation was performed at the rate of approximately 1.6Gflops, which is half of the 3.2Gflop rating of the DECmpp 12000Sx/Model 200. This estimate presumes that half of the time the processors were computing while the other half of the time the data was being transferred to and from the processors. The final transformation required the composition of seven propagated template transformations in both the 2D and 3D experiments. Eq. 3 was solved numerically using 100 iterations of Successive Overrelaxation in 3D and 250 iterations in 2D with checkerboard update [4] at each real-time time step.

8. SUMMARY

This paper presented an approach for accommodating both global and local shape variation present in anatomy while maintaining topology. For the deformed textbook to have anatomical relevance, it is crucial that the topology be maintained. Information contained in the textbook can then be projected onto the coordinates of the study, creating an individualized textbook. Functional data from individuals can be mapped back to the common coordinate system of the textbook for comparison. Accommodating local shape variation is important because it allows for a much more precise comparison between two anatomies than accommodating global shape variation alone. The transformations of the cortex from one monkey to another demonstrate that the viscous fluid model is able to accommodate the complex local shape variation of anatomical morphology in the primate brain. The human brain, with its even more highly convoluted cerebral cortex, is also well suited to this approach [4, 7, 13, 5].

9. ACKNOWLEDGMENTS

G.E.C. was supported by the Craniofacial Imaging Laboratory, St. Louis Children's Hospital, Washington University Medical Center. R.D.R, M.I.M, and S.C.J. were supported by the NIH-NCRR-RR01380, and a grant from Digital Equipment Corporation; U.G. was supported by ARO-DAAL03-92-G-0015 and ARPA-MDA972-93-1-0012; T.A.C., D.V.E., R.D.R., and M.I.M. received joint funding from NIMH, NASA, and NSF under the Human Brain Project.

References

[1] Y. Amit, U. Grenander, and M. Piccioni. Structural image restoration through deformable templates. *J. American Statistical Association*, 86(414):376–387, June 1991.

[2] F.L. Bookstein. *Morphometric Tools for Landmark Data*. Cambridge University Press, New York, 1991.

[3] R. C. Buck. *Advanced Calculus*. McGraw-Hill Book Company, St. Louis, 3 edition, 1978.

[4] G. Christensen. *Deformable Shape Models for Anatomy*. Ph.D. Thesis, Department of Electrical Engineering, Sever Institute of Technology, Washington University, St. Louis, Mo. 63130, Aug 1994.

[5] G. E. Christensen, R. D. Rabbitt, and M. I. Miller. Deformable templates using large deformation kinematics. *Submitted to the IEEE Transactions on Image Processing*, October 1994.

[6] G. E. Christensen, R. D. Rabbitt, and M.I. Miller. A deformable neuroanatomy textbook based on viscous fluid mechanics. In Jerry Prince and Thordur Runolfsson, editors, *Proceedings of the Twenty-Seventh Annual Conference on Information Sciences and Systems*, pages 211–216, Baltimore, Maryland, March 24-26 1993. Department of Electrical Engineering, The Johns Hopkins University.

[7] G. E. Christensen, R. D. Rabbitt, and M.I. Miller. 3D brain mapping using a deformable neuroanatomy. *Physics in Medicine and Biology*, 39, 1994.

[8] S.J. DeArmond, M.M. Fusco, and M.M. Dewey. *Structure of the Human Brain: A Photographic Atlas*. Oxford University Press, New York, third edition, 1989.

[9] U. Grenander, M.I. Miller, and G. Christensen. Deformable anatomical data bases using global shape models: A position paper for the 1992 electronic imaging of the human body workshop. In *Proceedings of the Cooperative Working Group on Whole Body 3-D Electronic Imaging of the Human Body*, Administered by the Crew System Ergonomics Information Analysis Center -Wright Patterson Air Force Base, March 9-11, 1992.

[10] John W. Haller, Gary E. Christensen, Sarang C. Joshi, Mohktar Gado, Michael I. Miller, and Michael W. Vannier. Precision and accuracy of a high dimensional transformation and segmentation of mr images of the hippocampus. In *Proceedings of the XIVth International Conference on Information Processing in Medical Imaging (IPMI95)*, Ile de Berder, France, June 26-30 1995.

[11] Sarang C. Joshi, Michael I. Miller, Gary E. Christensen, Thomas A. Coogan, and Ulf Grenander. The generalized dirichlet problem for mapping brain manifolds. *Monograph of the Biomedical Computer Laboratory (BCL-580)*, 1995.

[12] E.R. McVeigh, R.M. Henkelman, and M.J. Bronskill. Noise and filtration in magnetic resonance imaging. *Medical Physics*, 12 (5):586–591, 1985.

[13] M.I. Miller, G.E. Christensen, Y. Amit, and U. Grenander. Mathematical textbook of deformable neuroanatomies. *Proceedings of the National Academy of Science*, 90(24), December 1993.

[14] M.I. Miller, T.J. Schaewe, C.S. Bosch, J.J.H. Ackerman, R.G. Jost, A.S. Greene, and S.N. Nadel. A mathematical signal model for phase and frequency encoded magnetic resonance imaging data. *Monograph of the Electronics Signals and Systems Research Laboratory*, (ESSRL-91-47), 1991.

[15] T.J. Schaewe and M.I. Miller. A maximum likelihood reconstruction algorithm for magnetic resonance imaging. In R.H. Schneider, editor, *Medical Imaging IV : Image Formation*, pages 188–194. SPIE 1231, 1990.

[16] D. L. Snyder, Jr. L. J. Thomas, and M. M. Ter-Pogossian. A mathematical model for positron emission tomography systems having time-of-flight measurements. *IEEE Transactions on Nuclear Science*, NS-28:3575–3583, 1981.

[17] R. Szeliski. *Bayesian Modeling of Uncertainy in Low-Level Vision*. Kluwer Academic Publisher, Boston, 1989.

[18] J. Talairach and P. Tournoux. *Co-Planar Stereotatctic Atlas of the Human Brain*. Beorg Thieme Verlag, Stuttgart, 1988.

[19] Grace Wahba. *Spline Models for Observational Data*. Society for Industrial and Applied Mathematics, Philadelphia, 1990.

PROBABILISTIC MATCHING OF BRAIN IMAGES

J. C. GEE[1], L. LE BRIQUER AND C. BARILLOT
Laboratoire SIM
Faculté de Médecine, Université de Rennes I
35043 Rennes Cedex, France

AND

D. R. HAYNOR
Department of Radiology
University of Washington
Seattle, WA 98195, USA

Abstract. Image matching has emerged as an important area of investigation in medical image analysis. In particular, much attention has been focused on the atlas problem, in which a template representing the structural anatomy of the human brain is deformed to match anatomic brain images from a given individual. The problem is made difficult because there are important differences in both the gross and local morphology of the brain among normal individuals. We have formulated the image matching problem under a Bayesian framework. The Bayesian methodology facilitates a principled approach to the development of a matching model. Of special interest is its capacity to deal with uncertainty in the estimates, a potentially important but generally ignored aspect of the solution. In the construction of a reference system for the human brain, the Bayesian approach is well suited to the task of modeling variation in morphology. Statistical information about morphological variability, accumulated over past samples, can be formally introduced into the problem formulation to guide the matching or normalization of future data sets.

[1]Current address: Department of Computer and Information Science, University of Pennsylvania, Philadelphia, PA 19104, USA.

Y. Bizais et al. (eds.), Information Processing in Medical Imaging, 113–125.

1. Introduction

Modern imaging modalities routinely provide a variety of information—ranging from highly detailed, three-dimensional (3-D) pictures of cerebral anatomy to maps of functional activity within the brain—that have become indispensable in the investigation of the neurological integrity of an individual. These images, however, are often difficult to interpret because of ambiguity in the precise location and extent of the structure(s) of interest. This uncertainty can arise in several ways: intrinsic limitations in the resolution of the scanner; disparate angles at which image sections were obtained by the different scanners or at different times; or natural variability in both the gross and local morphology of the brain among normal individuals. The goal of our work is to ameliorate the localization of anatomical structures by developing semi-automated methods based on the use of a labeled *atlas*. An early survey of atlas-based methods for cerebral localization can be found in [1], in which advantages unique to the approach are discussed. The fundamental assumption of these methods is that the topology of cerebral structures is invariant among normal individuals. To the degree that this assumption holds, the localization problem can be solved by determining the mapping from the brain image of a given individual to some labeled representation[1] or atlas of normal neuroanatomy.

The first comprehensive effort aimed at recovering the local differences between a pair of brain images was made by Broit [5]. Broit proposed modeling the atlas as an elastic object so that it could be "physically" deformed to match the brain image of an individual. A multiresolution version of this "elastic matching" technique was implemented by Bajcsy and Kovačič [6] to speed its convergence and to handle large-scale variations in anatomy. Physically-based modeling is now widely applied in image analysis to characterize shape and its deformation. A number of global object models have been proposed for recognition tasks but these are unsuited to the atlas problem because they cannot describe local changes in shape. Even the hybrid schemes [7, 8, 9], which combine lumped and distributed parameter models, and many of the distributed parameter models [10, 11] are generally not appropriate because only the surface or contour of an object is modeled, whereas fully volumetric representations are the most useful for the atlas problem.

In the present paper, we present a Bayesian decision-theoretic formulation of the volumetric image matching problem. The Bayesian approach (to regularization) represents a generalization of many physically-based methods, such as elastic matching, and is appealing for additional important rea-

[1]This representation can take many forms but is usually derived from either a brain image of an individual or a textbook (see, for example, the atlases in [2, 3, 4]).

sons. First, because we are dealing with *uncertain* quantities, such as noisy data, from which information or decisions are derived (that must themselves then be uncertain), it is natural to adopt a probabilistic approach. Second, Bayesian analysis formally embodies the use of *prior* information that we may have about the problem. In matching, the prior serves to constrain our mappings by favoring certain configurations, in effect, regularizing the problem. Among the many operational advantages of Bayesian analysis, the most relevant to our work is that the result is a *posterior distribution* for the unknown mapping, which expresses the probability of any mapping given the observed images. The existence of this distribution makes possible a range of analyses, including the estimation of the *variance* or reliability of the estimated mapping. This information about the uncertainty in the estimate better informs the investigator in his or her interpretation of the results and indicates to the user where the addition of information may improve the results (in terms of reducing its uncertainty). In the construction of a reference system for the human brain, the Bayesian approach can provide information about the variability inherent in the normalization procedure.

2. Elements of Decision Theory

Statistical decision theory is concerned with the problem of decision making in light of relevant statistical information other than that provided by the sample [12]. One important kind of such nonsample information is prior knowledge that we may have about the unknowns. We usually also have some understanding of the possible consequences of making a particular decision, and indeed its quantification will be necessary to perform estimation. This is achieved by determining a *loss function* which measures the cost incurred for making a particular decision given some value of the unknown. In our matching problem, the unknown state of nature or *parameter* θ is the "true" displacement mapping relating the pair of images; the images constitute the sample; and the decision or *action* concerns the value to which the mapping is assigned. To evaluate the effect of taking action a_1 when the true state of nature is θ_1, the loss function is used to calculate the cost $L(\theta_1, a_1)$ that is incurred.

As with knowledge about the consequences of making particular decisions, the prior information related to θ must similarly be quantified in order to be useful. This is done by considering the unknowns to be random variables. The prior knowledge is then expressed as a probability distribution π on the parameter space Θ.

In order to actually make a decision, first note that the actual incurred loss is itself a random quantity; that is, it will never be known with certainty.

A sensible approach to take under such circumstances is to base the choice of a decision on its "expected" loss:

$$E^{\pi^*} L(\theta, a) = \int_{\Theta} L(\theta, a)\, d\pi^*(\theta),$$

where $\pi^*(\theta)$ is the final (posterior) probability distribution of θ (see Section 5). Decisions can then be made by choosing the action which minimizes the expected loss.

3. Prior Information

The use of prior information is a fundamental element of Bayesian decision theory. In many problems, the introduction of prior knowledge substantially improves the solution by, for example, reducing noise or ruling out "impossible" answers. But for image matching and other ill-posed problems, prior information in the form of constraints are essential to their solution.

In the absence of specific prior information about a given problem, constraints on the smoothness of its solution, as imposed by standard regularization, have turned out to be useful for solving a wide range of ill-posed problems—see [13, 14] for examples in early vision. In the study of biological shape, the imposition of C^1 continuity has been advocated in performing morphometrics, based on the observation that spatial variation in the proportions of shape tend to be graded [15]. Nevertheless, it will be beneficial to admit mappings of C^0 continuity, which have proved satisfactory for the atlas problem when used in combination with a separate global affine registration stage [6, 16].

The task of designing priors which reflect our expectations about the smoothness of the unknown mapping is straightforward: the prior probability corresponding to a particular stabilizing functional can be modeled by the Gibbs distribution whose energy function is the same stabilizer. For the linear elastic strain energy functional we have used [17], the Gibbs energy function is quadratic: our prior distribution $\pi(\theta)$ is therefore Gaussian[2]. Szeliski [19] has observed that the spectrum of the membrane model—to which our linear elastic prior is related—coincides in form to that of a Brownian fractal. One implication is that our priors have no natural scale and hence are suitable for modeling data of any size.

[2]It should be emphasized that smoothness constraints compose only one kind of prior information useful for image matching and that the particular prior we have chosen is not essential to the present Bayesian formulation of the matching problem—see [18] for additional discussion on prior models suitable for atlas matching.

4. Likelihood Functions

The likelihood $f(\boldsymbol{z}|\theta)$ expresses the probability of observing the data $\boldsymbol{z}$ given any particular value of the unknowns θ. Its definition requires a model for predicting the observations from given values of the unknowns as well as knowledge about the statistical characteristics of the measurement noise. The likelihood function therefore provides us with the means to explicity model the process by which our observations are degraded.

The observations encountered in image matching may be of several kinds, each corresponding to a different feature space. This is different from the usual case in which the observations would be degraded samples of the unknowns. Instead, the likelihood of a mapping is determined by measuring the similarity between pairs of features. The uncertainty in these measurements can be modeled by imagining that the similarity values are obtained from a virtual sensor. The procedure is described in [18] but the general idea is as follows: an error surface is first estimated locally about the current displacement estimate of the given point; this surface is approximated by a quadratic that we then use to define a Gaussian representing the likelihood function for the point.

5. Posterior Analysis

We began by requiring that, in the absence of additional information, a mapping should be smooth and that the likelihood of its occurrence increased with its smoothness. When data is presented to us in the form of "observations" made of the unknown mapping $\theta = \boldsymbol{\delta}$, the likelihood of an arbitrary mapping no longer depends only on its smoothness but must take into account its probability of actually having produced the observations. Our revised beliefs about the mappings $\boldsymbol{\delta}$ in view of the observed data $\boldsymbol{z}$ is simply the conditional distribution of $\boldsymbol{\delta}$ given $\boldsymbol{z}$:

$$\pi(\boldsymbol{\delta}|\boldsymbol{z}) = \frac{f(\boldsymbol{z}|\boldsymbol{\delta})\pi(\boldsymbol{\delta})}{\int_\Theta f(\boldsymbol{z}|\boldsymbol{\delta})\, d\pi(\boldsymbol{\delta})},$$

where f and π are the likelihood and prior, respectively. We determined earlier that an optimal action is one which minimizes the expected loss of the problem. We recognize now that the expectation should be taken with respect to the posterior distribution $\pi(\boldsymbol{\delta}|\boldsymbol{z})$.

5.1. DISPLACEMENT FIELD ESTIMATION

The most common estimate used in Bayesian analysis is the value of the unknown which maximizes the posterior distribution. This *maximum a pos-*

teriori (MAP) estimate is of interest because it represents the maximum likelihood estimate derived from the posterior distribution.

Consider the situation in which the similarity of an image pair is determined by some similarity measure S and the prior is the Gibbs distribution with energy U_{prior} equal to the linear elastic strain energy. Because the likelihood function is similarly a Gibbs distribution, the posterior is Gibbsian with energy $U(\boldsymbol{\delta}, \{I_{\text{subject}}, I_{\text{atlas}}\})$ equal to:

$$-\int_{\Omega_{\text{atlas}}} S(I_{\text{atlas}}(\mathbf{x} + \boldsymbol{\delta}), I_{\text{subject}}(\mathbf{x}))\, d\mathbf{x} + U_{\text{prior}}(\boldsymbol{\delta}), \tag{1}$$

where I_{atlas} and I_{subject} represent the images of the atlas and brain of the individual, respectively, and S is computed over all points $\mathbf{x}$ in the domain of the atlas Ω_{atlas}. For Gibbs distributions, MAP estimation reduces to the problem of minimizing energy U. Note that the energy functional in (1) is the same one that appears in the variational formulation of elastic matching. Thus, the current MAP estimate and the standard-regularized solution to the elastic matching problem are equivalent.

Our optimization algorithm to estimate the MAP solution is described in [20], and its performance along with that of the *minimum mean squared error* (MMSE) estimator were recently examined in [18]. The MMSE estimate is the optimal action for the *squared error loss* function: $L(\theta, a) = (\theta - a)^2$. It is equal to the mean of the posterior and can be estimated through Monte Carlo integration using the Gibbs sampler described next.

5.2. VARIANCE ESTIMATION

Although Bayesian estimates represent "optimal" interpretations of the information contained in the prior and the sample data, a potential fault with their use—and with all other methods in the brain image matching literature—is that there is no indication of the uncertainty in their values. The uncertainty encodes information about the influence of the prior on the estimate. In addition, the uncertainty in the sample data is reflected in the final reliability of the solution.

Consider a Gibbs distribution p with quadratic energy:

$$-\log p(\boldsymbol{Z} = \boldsymbol{z}) = \frac{1}{2}(\boldsymbol{z} - \boldsymbol{z}^*)^{\mathrm{T}}\mathbf{K}(\boldsymbol{z} - \boldsymbol{z}^*) + k,$$

where $\boldsymbol{z}^*$ is the minimum energy value and k a constant. This is Gaussian and, therefore, the mean and covariance completely characterize the distribution. To obtain the covariance matrix, it is only necessary to invert the matrix $\mathbf{K}$ associated with the Gibbs energy function. For energies that are not convex, the definition of the corresponding Gibbs distribution requires higher order statistics. We can nonetheless approximate the variance near an estimate by fitting a quadratic to the posterior energy function at the estimate. The matrix inversion needed to obtain the covariance matrix $\mathbf{K}^{-1}$ can be performed using a number of techniques. However, for the problem domains that are normally encountered in image matching, the covariance matrix is very large. We therefore considered only the diagonal elements of the covariance matrix, which represent the variance of the estimate at each site.

A finite element approximation to $\boldsymbol{\delta}$ was used. In the finite element method [21], the problem domain is subdivided into regions or *elements*. Neighboring elements share discrete *nodal* points along their interface. The only unknowns are those at the element nodes. The displacement $\boldsymbol{\delta}$ at any other point within an element is interpolated from the nodal values: $\delta^j = \sum_{n \in \mathcal{N}(e)} \delta^j_n N_n$, where $\mathcal{N}(e)$ is the set of nodes which delimit element e, N_i are the interpolating functions, and we write δ^i_n for the value of the ith component of the displacement at node n. It is straightforward to verify that the finite element representation of an arbitrary Gibbs energy function leads to a Markov random field (MRF) model. Since our energy functions are quadratic, the local characteristics of the MRF are Gaussian. It is thus possible to build an efficient continuous state space Gibbs sampler for our problem [22]. To illustrate its construction, consider the following quadratic (finite element) energy:

$$\begin{aligned} U(\boldsymbol{\delta}) &= \frac{1}{2}[\boldsymbol{\delta}]^{\mathrm{T}}\mathbf{K}\boldsymbol{\delta} + [\boldsymbol{\delta}]^{\mathrm{T}}\mathbf{f} + k \\ &= \frac{1}{2}\sum_{nm}[\boldsymbol{\delta}_n]^{\mathrm{T}}\mathbf{K}_{nm}\boldsymbol{\delta}_m + \sum_{n}[\boldsymbol{\delta}_n]^{\mathrm{T}}\mathbf{f}_n + k, \end{aligned}$$

where $\mathbf{K}$ is a positive definite matrix, $\mathbf{f}$ a constant vector, and k a constant. The indices n and m range over the set of all element nodes. Define the lattice of sites $\mathcal{S}$ to be the set of finite element nodes. The neighborhood system $\mathcal{G}$ at each node n comprises those nodes m which potentially exert a "force" on it: $\mathcal{G}_n = \{m | m \neq n \text{ and } m \in \mathcal{N}(\mathcal{E}(n))\}$, where $\mathcal{E}(n)$ is the set of elements containing node n. Finally, define the clique potentials as follows:

$$\begin{array}{lll} V_C(\boldsymbol{\delta}) = \frac{1}{2}[\boldsymbol{\delta}_n]^{\mathrm{T}}\mathbf{K}_{nn}\boldsymbol{\delta}_n + [\boldsymbol{\delta}_n]^{\mathrm{T}}\mathbf{f}_n & \text{if } C = \{n\}, & \\ V_C(\boldsymbol{\delta}) = [\boldsymbol{\delta}_n]^{\mathrm{T}}\mathbf{K}_{nm}\boldsymbol{\delta}_m & \text{if } C = \{n, m\}, \text{ where } n \neq m, & (2) \\ V_C(\boldsymbol{\delta}) = 0 & \text{otherwise.} & \end{array}$$

It is evident that $U(\boldsymbol{\delta}) = \sum_C V_C(\boldsymbol{\delta})$. We now establish that the local characteristics of this MRF are Gaussian or, equivalently, show that the local energies are quadratic. By definition,

$$p(\boldsymbol{\delta}_n | \boldsymbol{\delta}_m, m \neq n) \propto \exp - \sum_{\{C | n \in C\}} V_C(\boldsymbol{\delta}). \tag{3}$$

Substituting the clique potentials defined in (2) into (3), the local energy $U_n(\boldsymbol{\delta}) = \sum_{\{C|n \in C\}} V_C(\boldsymbol{\delta})$ can be rewritten as:

$$U_n(\boldsymbol{\delta}) = \frac{1}{2}[\boldsymbol{\delta}_n]^{\mathrm{T}}\mathbf{K}_{nn}\boldsymbol{\delta}_n + [\boldsymbol{\delta}_n]^{\mathrm{T}}\mathbf{f}_n + \sum_{m \in \mathcal{G}_n} [\boldsymbol{\delta}_n]^{\mathrm{T}}\mathbf{K}_{nm}\boldsymbol{\delta}_m.$$

This expression can be factored into the following quadratic form:

$$U_n(\boldsymbol{\delta}) = \frac{1}{2}(\boldsymbol{\delta}_n - \mu_n)^{\mathrm{T}}\mathbf{K}_{nn}(\boldsymbol{\delta}_n - \mu_n) - \frac{1}{2}[\mu_n]^{\mathrm{T}}\mathbf{K}_{nn}\mu_n,$$

where

$$\mu_n = -\mathbf{K}_{nn}^{-1}\left[\mathbf{f}_n + \sum_{m \in \mathcal{G}_n} \mathbf{K}_{nm}\boldsymbol{\delta}_m\right].$$

The implementation of our continuous state space Gibbs sampler at iteration k, where site n is being updated, can be summarized as follows:

- Calculate the mean vector $\mu_n^{(k)} = -\mathbf{K}_{nn}^{-1}\left[\mathbf{f}_n + \sum_{m \in \mathcal{G}_n} \mathbf{K}_{nm}\boldsymbol{\delta}_m^{(k)}\right]$.
- Set the new displacement $\boldsymbol{\delta}_n^{(k+1)} = \mu_n^{(k)} + \mathbf{b}_n$, where $\mathbf{b}_n$ is Gaussian noise with zero mean and covariance equal to $\mathbf{K}_{nn}^{-1}$.

6. Experimental Results

To demonstrate some of the manifold features of the Bayesian approach, it will be enough to use the two-dimensional data set displayed in Figure 1. The left image in the figure represents an MRI section, extracted at the level of the AC-PC plane. Its slice dimensions are 128×128 with pixel size equal to 1.87mm $\times$ 1.87mm. This image was then distorted with a known thin-plate spline to produce the image shown in the center of Figure 1. The object of the experiment was simply to recover the mapping from the original brain section to its distorted version. To make the problem more challenging, the test image depicted in the right of the figure was created by adding white Gaussian noise ($\sigma = 15$) to the spline-distorted image after first blurring it with a Gaussian ($\sigma = 1$ pixel).

Figure 2 illustrates some of the results obtained when the match between a pair of points was rated according to the similarity of their underlying tissue constituents [18]. In the current case, the MRI volume from which our section was extracted had previously been segmented into its gray and white matter components [23]. To interpet the displacement mappings in the figure, focus on their central portion and ignore those parts which displace only the background pixels—the discrepancy in the background motion is partially due to the zero displacement boundary condition that we imposed

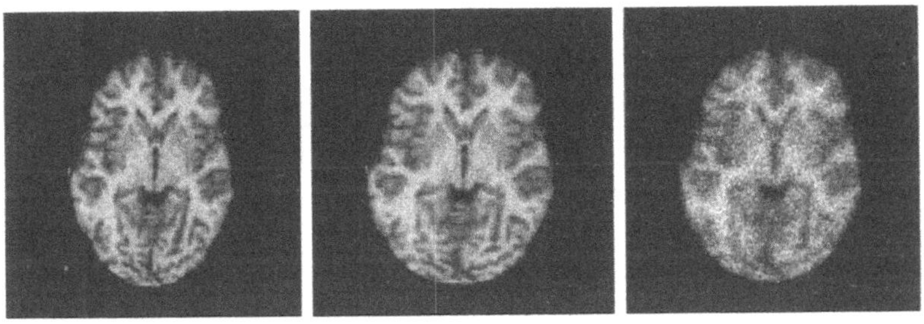

Figure 1. Original MRI section (left) and its distorted versions.

at the image periphery in lieu of first performing a global affine registration of the images. The MMSE estimate has a smoother appearance than that of the MAP solution and, as such, more closely resembles the truth in this case. In spite of the perceived disparity between their motion fields, the two estimates produce similar registrations of the gray and white matter components in terms of the amount of relative overlap between their matched and distorted versions. The specific values are listed in the second and third columns of Table 1 along with the overlap values prior to matching. Note that the MAP estimate actually results in a slightly better match of the tissue components.

Displayed to the right of the estimated mappings in Figure 2 are images of the magnitude of the variance estimates. Compare the variance of the MAP estimate (maximum = 1.1 pixels) with the original MRI section: the underlying structural anatomy is visible, reflecting the fact that the uncertainty is reduced at the gray-white interfaces but is relatively large within those portions of the image that lack informative features. In contrast, the posterior variance (maximum = 2.7 pixels) plainly depicts the complexity of the entire posterior distribution. In this case, for example, it reveals information that would otherwise be unavailable about portions of the mapping which have an exceptionally large variance—in the present experiment where the truth is known, these large uncertainties are also manifest as mismatches by the mean estimate. Unfortunately, it is not possible to visually qualify the extent of the differences between the two variance estimates in the figure because their gray-level values were not scaled in the same way.

The results for the noise-corrupted test image are presented in Figure 3. In this experiment, the normalized cross-correlation between two given points was used to measure their similarity [18]. The good visual agreement between the estimates and the truth is supported by the accurate registration that the estimates produce for the gray and white matter as indicated in Table 1. The motion fields of the two estimates are remarkably similar. Even more striking is the resemblance between their variance

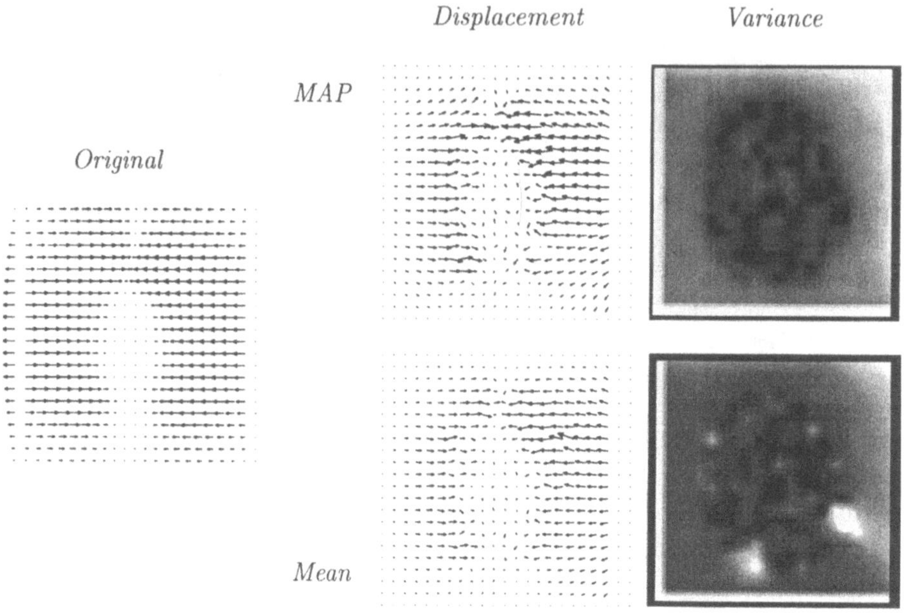

Figure 2. Original and estimated motion fields for the distorted but noise-free test image, where the match between a pair of points was rated according to the similarity of their underlying tissue constituents. Displayed to the right of each displacement estimate is the estimated (magnitude of the) variance of its values.

estimates—both have maximum values equal to 2.1 pixels. The gray-white interfaces are not only readily discernible in the variance estimate about the MAP solution but are equally apparent in the posterior variance. The implication is that the matching problem for the current experimental set-up based on cross-correlation is actually fairly "well posed." Reducing the

TABLE 1. Relative overlap between the estimated and true regions corresponding to the gray and white matter components. The overlap was defined as the ratio between the intersecting area of the two regions under comparison and the area of their union.

		Classification		Correlation	
	Original	MAP	Mean	MAP	Mean
Gray Matter	0.66	0.91	0.87	0.89	0.83
White Matter	0.68	0.91	0.87	0.90	0.85

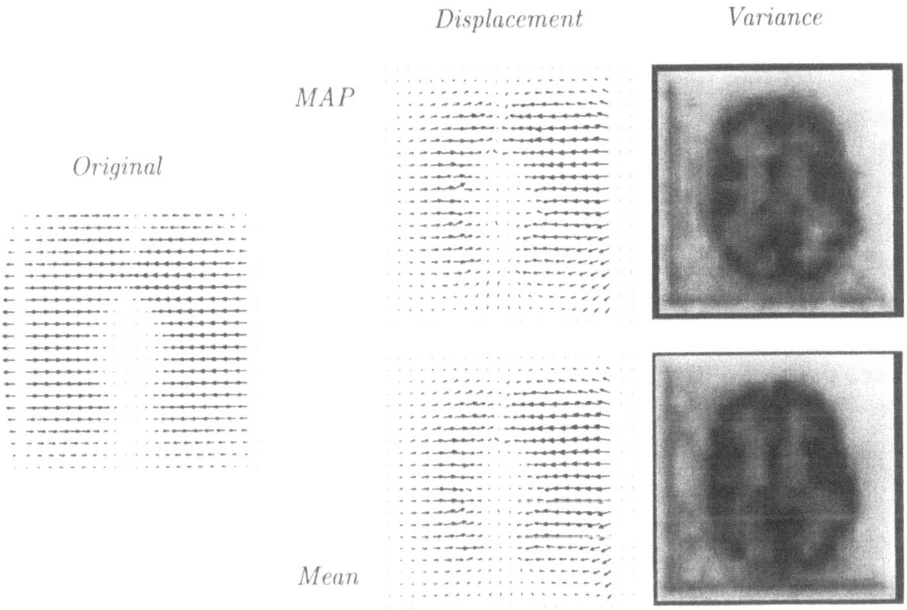

Figure 3. Original and estimated motion fields for the distorted and noise-corrupted test image, where the normalized cross correlation between two given points was used to measure their similarity. Displayed to the right of each displacement estimate is the estimated (magnitude of the) variance of its values.

"strength" of the prior by an order of magnitude results in a decrease of only approximately 4 percentage points in the final overlap values for the gray and white matter registration. However, as the influence of the prior is further reduced, the solution begins to deteriorate and eventually becomes singular.

7. Discussion

The basic elements of a Bayesian approach to the problem of matching brain images has been described. The method has important advantages that are particularly useful in the construction and application of a probabilistic reference system for the human brain. Since the method is inherently probabilistic, it is well suited to the task of modeling variation in morphology, one of the principal features of such a reference system. Prior statistical information about morphological variability, accumulated over past samples, can be formally introduced into the problem formulation to guide the matching of future data sets. Equally important is that the Bayesian methodology facilitates a principled derivation of the likelihood model, which together with the prior make up a particular matching (posterior) model. In the evaluation of any matching algorithm then, its essential assumptions can be made explicit by determining the models and the estimator effectively in use by the method. For example, the solutions obtained using variational regularization correspond to the MAP estimate. In contrast, the specification of the loss function is *part* of the problem definition in Bayesian estimation.

The task of properly building a Bayesian model is a difficult one, but the process encourages careful deliberation of the problem at hand to arrive at the assumptions used to form the components of the model. For brain image matching, a number of effective similarity measures exist, whereas there is comparatively little known about the kinds of prior models that are specifically suited to the problem (and its applications) or how their influence on the solution, relative to the likelihood, should be determined. The resolution of these issues directly impacts the viability of any matching procedure, Bayesian or otherwise, but is best addressed within a Bayesian setting. The potential benefits that follow from the Bayesian approach, including the capability to estimate and explore the uncertainty of a Bayesian solution [24], are compelling reasons for adopting the method in brain image matching.

8. Acknowledgments

The authors are grateful to the Region Council of Brittany for their financial support of this work. We thank Pr. J. M. Scarabin and the Radiology Department of the University Hospital of Rennes for their assistance in obtaining the MRI data sets.

References

1. Evans AC, Beil C, Marrett S, Thompson CJ, Hakim A. Anatomical-functional correlation using an adjustable MRI-based region of interest atlas with positron emission tomography. *J Cereb Blood Flow Metab* 1988;8:513–530.
2. Talairach J, Tournoux P. *Referentially oriented cerebral MRI anatomy: an atlas of stereotaxic anatomical correlations for gray and white matter*. Stuttgart, Germany:Thieme, 1993.
3. Tiede U, Bomans M, Höhne KH, Pommert A, Riemer M, Schiemann T, Schubert R, Lierse W. A computerized three-dimensional atlas of the human skull and brain. *Amer J Neuroradiology* 1993;14(3):551–559.
4. Nowinski WL, Fang A, Fang S, Nguyen BT, Bryan RN, Raghavan R. Three-dimensional electronic atlas of human cerebral deep structures. To appear: Bizais Y, Barillot C, Di Paola R, eds. *Information Processing in Medical Imaging*. Dordrecht:Kluwer Academic Publishers, 1995.
5. Broit C. *Optimal registration of deformed images*. Doctoral dissertation, Department of Computer and Information Science, University of Pennsylvania, PA, 1981.
6. Bajcsy R, Kovačič S. Multiresolution elastic matching. *Comput Vision, Graphics,*

Image Process 1989;46:1–21.
7. Nastar C, Ayache N. Non-rigid motion analysis in medical images: a physically based approach. In: Barrett HH, Gmitro AF, Eds. *Information Processing in Medical Imaging: Proceedings of 13th International Conference, IPMI '93*. Berlin:Springer-Verlag, 1993;LNCS 687:15–32.
8. Worring M, Smeulders AWM, Staib LH, Duncan JS. Parameterized feasible boundaries in gradient vector fields. In: Barrett HH, Gmitro AF, Eds. *Information Processing in Medical Imaging: Proceedings of 13th International Conference, IPMI '93*. Berlin:Springer-Verlag, 1993;LNCS 687:48–61.
9. Vemuri BC, Radisavljevic A, Leonard CM. In: Barrett HH, Gmitro AF, Eds. *Information Processing in Medical Imaging: Proceedings of 13th International Conference, IPMI '93*. Berlin:Springer-Verlag, 1993;LNCS 687:62–76.
10. Amini AA, Owen RL, Anandan P, Duncan JS. Non-rigid motion models for tracking the left ventricular wall. In: Colchester ACF, Hawkes DJ, eds. *Information Processing in Medical Imaging*. Berlin:Springer-Verlag, 1991;LNCS 511:343–357.
11. Cohen LD, Cohen I. Finite element methods for active contour models and ballons for 2-D and 3-D images. *IEEE PAMI* 1993;15:1131–1147.
12. Berger JO. *Statistical Decision Theory and Bayesian Analysis*. New York:Springer-Verlag, 1985.
13. Poggio T, Torre V, Koch C. Computational vision and regularization theory. *Nature* 1985;317(26):314–319.
14. Terzopoulos D. Regularization of inverse visual problems involving discontinuities. *IEEE PAMI* 1986;8(4):413–424.
15. Bookstein FL. Principal warps: Thin-plate splines and the decompostion of deformations. *IEEE PAMI* 1989;11(6):567–585.
16. Dann R, Hoford J, Kovačič S, Reivich M, Bajcsy R. Evaluation of elastic matching system for anatomic (CT, MR) and functional (PET) cerebral images. *J Comput Assist Tomogr* 1989;13:603–611.
17. Timoshenko S, Goodier JN. *Theory of elasticity*. New York:McGraw-Hill, 1951.
18. Gee JC, Le Briquer L, Barillot C, Haynor DR, Bajcsy R. Bayesian approach to the brain image matching problem. To appear: Loew MH, ed. *Medical Imaging 1995: Image Processing* Bellingham, WA:SPIE, 1995;2434.
19. Szeliski R. *Bayesian Modeling of Uncertainty in Low-level Vision*. Norwell, MA:Kluwer Academic Publishers, 1989.
20. Gee JC, Haynor DR, Reivich M, Bajcsy R. Finite element approach to warping of brain images. In: Loew MH, ed. *Medical Imaging 1994: Image Processing* Bellingham, WA:SPIE, 1994;2167.
21. Zienkiewicz OC. *The Finite Element Method in Engineering Science*. New York:McGraw-Hill, 1971.
22. Geman S, Geman D. Stochastic relaxation, Gibbs distribution, and the Bayesian restoration of images. *IEEE PAMI* 1984;6(6):721–741.
23. Lachmann F, Barillot C. Brain tissue classification from MRI data by means of texture analysis. In: Loew MH, ed. *Medical Imaging VI: Image Processing* Bellingham, WA:SPIE, 1992;1652:72–83.
24. Hanson KM, Cunningham GS. Exploring the reliability of Bayesian reconstructions. To appear: Loew MH, ed. *Medical Imaging 1995: Image Processing* Bellingham, WA:SPIE, 1995;2434.

TOWARDS AUTOMATED LABELLING OF THE CEREBRAL CORTEX USING A DEFORMABLE ATLAS

STEPHANIE R. SANDOR AND RICHARD M. LEAHY
Signal and Image Processing Institute,
Dept. Electrical Engineering-Systems
University of Southern California, Los Angeles, CA

Abstract. We describe a computerized method to automatically find and label the cortical surface in 3-D magnetic resonance (MR) brain images. The approach we take is to model a pre-labelled brain atlas as a physical object and give it elastic properties, allowing it to warp itself onto regions in a preprocessed brain image. Preprocessing consists of boundary-finding and a morphological procedure which automatically extracts the brain and sulci from an MR image and provides a smoothed representation of the brain surface to which the deformable model can rapidly converge. Our deformable models are energy-minimizing elastic surfaces that are parameterized with 3-D bicubic, B-spline surfaces.

1. Introduction

Much of the anatomical structure of the brain surface is delineated by complex patterns of cortical sulci. These sulcal patterns are routinely used as landmarks for predicting areas of functional localization within the cortex. Since there is not a simple relationship between individual cortical topology and functional localization, activation studies using PET, fMRI or magnetoencephalography must be used to more accurately localize these functional areas. An automated method for extraction of the cortical surface, and subsequent labelling of the major sulci could provide a valuable research tool for the study of the relationship between cortical anatomy and functional localization in individual subjects, and studies of inter-subject variability.

Y. Bizais et al. (eds.), Information Processing in Medical Imaging, 127–138.

This automated procedure could also be used to provide anatomical constraints for use in imaging or localization of functional activity. For example, magnetoencephalography (MEG) provides measurements of the magnetic fields produced by neural activity in the brain. These measurements can be used as the basis for imaging, or localizing, the neural activity which produces these fields. It is generally accepted that the primary current sources which give rise to these evoked fields are confined to cortex and oriented normal to the surface. Hence knowledge from MR images about the location and orientations of cortical folds can be used to improve the MEG current source estimate [1]. Labelling of the cortical surface offers the potential for utilization of probabilistic information about the relationship between cortical anatomy and functional localization.

The goal of our work is to develop a technique that is automatic and accurate in labelling the convoluted regions of the cerebral cortex. From a volume MR image of an anatomically normal brain we have developed a B-spline surface model of that brain and attached anatomic labels to all points on the surface. We match the atlas model to an image by first registering the spline model to a subject's MR image and treating the atlas as a deformable model's initial position. The model is then automatically warped to match the subject's image, and labels are transferred from the model to the image.

2. MR Image Preprocessing

The deformable models which we use in our brain labelling method converge rapidly to unambiguous image features, such as smooth image boundaries. However, matching these models to images containing complex shapes or several neighboring regions is an underconstrained problem, since image contours that are in close proximity to one another can leave the model with many possible local solutions to the boundary-finding task. Thus, in order to constrain the model to warp to the brain surface, we preprocess the MR images.

Our preprocessing consists of a series of morphological operations performed on closed regions detected in an MR image. To find these closed regions we apply a 3-D Marr-Hildreth operator to the gray-scale images. Our experience is that the spatial resolution of the Marr-Hildreth filter can be chosen so that the operator is effective in finding a closed brain surface which is a necessary requirement for the input to the morphological procedure described below. However, close examination of the extracted surface, reveals that the boundary tends to wander between the true outer brain surface (grey/csf boundary) and the inner cortical surface (grey/white boundary). This is primarily due to the partial volume effects that occur in deep cortical folds in such a way that the two sides of the sulcus appear to merge. Since the average thickness of the cortex is on the order of 2-3mm, we anticipate that the boundary mislocation will be tolerable for many applications. Such is the case when the cortical surface is used to constrain inverse solutions for magnetoencephalographic imaging due to the inherently limited resolution of the modality.

To automatically extract only the brain from MR head images our algorithm takes as input a binary edge detected image and performs a series of morphological operations on this image. A morphological erosion shrinks the brain within the skull and severs surrounding anatomical connections. The eroded brain is then extracted from the image by tracing connected regions. A morphological closing followed by set difference operations restores the brain to its original shape and size with all other head anatomy removed from the image. Also, by finding all openings in the binary brain image and then detecting which openings intersect the brain surface, the algorithm is able to accurately detect sulci locations in the image. For a more detailed discussion of our morphological algorithm see [2].

3. Cortical Anatomy and Atlas Construction

The convoluted cerebral cortex can be divided into regions or lobes whose boundaries are for the most part defined by major cortical fissures. Except for major convolutions, cerebral structures on the surfaces of the hemispheres can be difficult to recognize, due to many secondary folds and interruptions which alter the shape of cortical convolutions. However, basic recurring sulcal patterns can be identified and categorized. For example, the work of Ono *et. al.* gives an in-depth description of sulcal patterns of twenty-five specimen brains [3]. Certain trends can be found which cause particular sulci to be more readily identified. For instance, from data gathered on these brain specimens it was found that in 100% of the cases, the Sylvian fissure was never interrupted and this was true for the central sulcus in 92% of the specimens. Also, the general layout of the cortex is relatively constant among subjects, and once the main sulci are found, other important sulci can be found since they occur in a particular location and orientation with respect to the major sulci.

To develop the atlas model we used a 3-D MR volume image of a male volunteer whose brain did not show any visible anatomic abnormalities. The images were acquired on a 1.5 Tesla GE scanner in a volume acquisition mode. A spoiled grass pulse sequence was used ($T_R = 40$ msec and $T_E = 13$ msec) with flow compensation to eliminate artifacts due to moving fluids. The images were isotropically sampled with voxel dimensions of 1.09mm in all three directions. The morphological algorithm for extracting a brain from a head image was applied to this MR data.

With morphology we also can find regions of a brain image corresponding to the cortical surface and to sulci openings. By morphologically closing a binary brain image, we obtain a smoothed representation of the brain surface, and it is the outer contour of this smooth surface that we use as our model of the cortex. Also, by taking the intersection of the outer boundary of the closed brain with the set of sulci openings, we find points on the closed surface which correspond to locations of different sulci. We then select and label surface points corresponding to the sulci we want included in our atlas, and points lying between these sulci are labelled according to the lobe in which they occur. By this procedure the atlas surface is described by a list of points representing the outer surface of the brain as it appears on each slice of the closed brain image.

A B-spline surface representation provides us with a closed form expression for the atlas surface and also requires fewer points than the original atlas description. The 3-D surface is approximated by summing scaled versions of basis functions, $B_{i,j}(u,v)$, which are piecewise in two parameters. Control vertices, $V_{i,j} = (X_{i,j}, Y_{i,j}, Z_{i,j})$, are arranged in a rectangular topology called a control mesh. If $Q(u,v)$ is an approximated surface, then it can be expressed as a double summation:

$$Q(u,v) = (x(u,v), y(u,v), z(u,v)) = \sum_i \sum_j V_{i,j} B_{i,j}(u,v) \tag{1}$$

In order that $B_{i,j}(u,v)$ have rectangular parametric regions of support, we form basis functions as products of univariate B-splines:

$$B_{i,j}(u,v) = B_i(u) B_j(v). \tag{2}$$

These 3-D basis functions are called tensor product B-splines.

In order to obtain a B-spline control mesh for our atlas, we sample the contours, which correspond to the smoothed brain surface, one slice at a time (see figure 1 (a)). These surface sample points serve as the control mesh for a B-spline atlas surface. Figure 1 (b) shows a lateral view of the atlas control mesh. Once we have a control mesh, we evaluate the spline surface $Q(u,v)$ at points in the u, v parameter space, and we obtain a precise description of the surface. We attach anatomical labels to all surface points, which in this case numbered close to 5,000 points. Specifically, in the atlas presented here we label the lobe points corresponding to the frontal, parietal, and temporal lobes, and sulci points corresponding to the central, lateral and longitudinal fissures. The occipital lobe was not labelled, due to the difficulty in accurately determining its boundaries, but the lobes and sulci included in the atlas are sufficient to demonstrate our technique of atlas labelling. Figure 1 shows two views of the final labelled atlas surface.

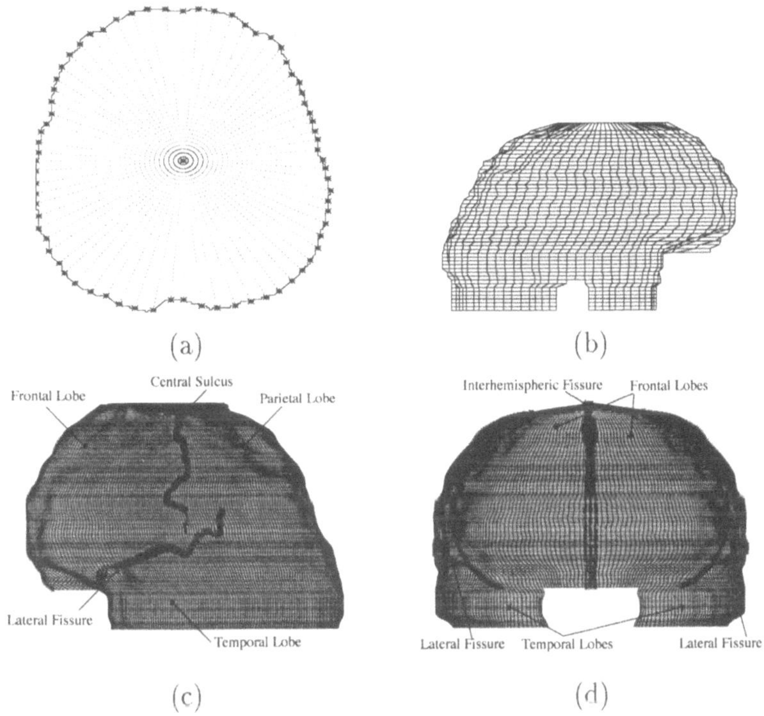

Figure 1. (a) Example of the sampling method used to obtain the B-spline control mesh. The contour represents one slice of a closed brain image, and the asterisks are our sample points which occur at the intersection of the curve with lines drawn from the center of the contour. (b) Lateral view of the atlas' B-spline control mesh obtained with the sampling method shown in (a). (c) A lateral view of the labelled atlas surface. The spline surface has been evaluated and each surface point has an anatomical label attached to it. The asterisks indicate sulci positions on the surface. (d) A frontal view of the labelled atlas surface.

4. DEFORMABLE MODELS FOR BRAIN LABELLING

The models which we incorporate into our labelling system are deformable surfaces based on "snakes", active contour models developed by Kass, Witkin, and Terzopoulos [4]. Our models are energy-minimizing elastic surfaces which search for local minima instead of a global solution. These models rely on higher-level processes to place them near a desired solution. In our atlas matching procedure we define a novel energy function that is designed to ensure that sulci points on the brain model are attracted to sulci points in the subject's image.

We have extended 2-D snakes to 3-D deformable spline models by modifying the snake energy equations and parameterizing a surface with tensor product B-splines. Let $Q(u,v) = (x(u,v), y(u,v), z(u,v))$ be a parametric description of a surface model, as in equation 1, and $\bar{Q}(u,v)$ be a parametric description of an atlas model. The total energy can be written:

$$E = \int_0^1 \int_0^1 (E_{int}(Q(u,v)) + E_{ext}(Q(u,v)))dudv \tag{3}$$

Internal energy is given by:

$$\begin{aligned} E_{int}(Q(u,v)) &= \omega_{1,0}\left|\frac{\partial Q}{\partial u}\right|^2 + \omega_{0,1}\left|\frac{\partial Q}{\partial v}\right|^2 + 2\omega_{1,1}\left|\frac{\partial^2 Q}{\partial u \partial v}\right|^2 \\ &+\omega_{2,0}\left|\frac{\partial^2 Q}{\partial u^2}\right|^2 + \omega_{0,2}\left|\frac{\partial^2 Q}{\partial v^2}\right|^2, \end{aligned} \tag{4}$$

where coefficients specify mechanical properties of a surface such as elasticity (ω_{10}, ω_{01}), rigidity (ω_{20}, ω_{02}), and twist (ω_{11}) [5]. In our application we are interested in matching a brain atlas to individual anatomy. We define the external energy term to reflect the distance between the deformed atlas and the subject's extracted brain surface:

$$E_{ext}(Q(u,v)) = C(Q(u,v)), \tag{5}$$

where $C(Q(u,v))$ is the Chamfer distance (an approximation of the Euclidean distance of each voxel to the nearest edge) between the deformable contour and image features. The discrete form of the energy equation is:

$$\begin{aligned} E &= \sum_k \sum_l [\omega_{10}(\sum_i \sum_j V_{ij} \frac{\partial B_{ij}}{\partial u}(u_k, v_l))^2 + \omega_{01}(\sum_i \sum_j V_{ij} \frac{\partial B_{ij}}{\partial v}(u_k, v_l))^2 \\ &+ 2\omega_{11}(\sum_i \sum_j V_{ij} \frac{\partial B_{ij}}{\partial u \partial v}(u_k, v_l))^2 + \omega_{20}(\sum_i \sum_j V_{ij} \frac{\partial^2 B_{ij}}{\partial u^2}(u_k, v_l))^2 \\ &+ \omega_{02}(\sum_i \sum_j V_{ij} \frac{\partial^2 B_{ij}}{\partial v^2}(u_k, v_l))^2 + C(Q(u_k, v_l)) \end{aligned} \tag{6}$$

The energy equation is a cost function which must be minimized in order for deformable models to warp onto image features. Since there is no closed-form solution to this optimization problem, we use numerical methods for energy minimization. In our work, we are concerned with demonstrating that deformable model matching is a method for introducing anatomical information into our labelling process, and we do not place any constraints on image data. Therefore, in general E_{ext} is not convex; however, we assume that when a deformable model's initial position is close to desired image features the function is locally convex and using a conjugate gradient method for minimization of E will find an acceptable local minimum.

5. MATCHING A 3-D ATLAS MODEL TO THE CORTICAL SURFACE

The principal goal of this work is to carry out 3-D atlas guided labelling of the cortical surface. In 3-D automatic brain labelling with deformable models, the major problem is to find a method such that the deformable model will converge with high probability to the desired regions. The solution we found is to generate an atlas, which is the initial position of a deformable model, from a smoothed representation of an MR image. We then warp the deformable model to a subject's MR image that has been smoothed with morphological operations equivalent to those used to create the atlas. This avoids the problem of trying to directly match highly convoluted regions in the atlas to highly convoluted regions in the individual MR images. Moreover, we are able to achieve accurate labelling with the smoothed representation of the atlas and subject's brain, using the following procedure: 1) propagating label information from the original atlas brain to its smoothed representation, 2) using a novel external energy function that causes sulci in the atlas to warp to sulci in the smoothed individual brains, and 3) propagating the labels from the warped atlas back to the high-resolution MR brain surface.

To perform 3-D image labelling we use the atlas described in section 3 (see figure 1). First, the subject's image is resliced along the same planes as the image that was used to create the atlas, then a gross registration is performed by scaling and translating the atlas. These atlas registration operations are performed on the spline control mesh and not on the surface itself, since the spline surface's shape is invariant with respect to translation, rotation, and scaling of the control mesh. The registered atlas serves as the deformable model's initial position.

In order to warp the B-spline surface, we specify the parameters which control the model's bending. Since the deformable model is initialized in a shape and location close to its desired final configuration, we weight a model's internal energy terms by parameters computed directly from its initial position. For example, let $\bar{Q}(u,v)$ be a parametric description of an anatomical atlas, which is the initial position of a surface model, then we set model parameters as:

$$\omega_{1,0}(u,v) = \frac{1}{\left|\frac{\partial \bar{Q}(u,v)}{\partial u}\right|^2 + \epsilon}, \; \omega_{0,2}(u,v) = \frac{1}{\left|\frac{\partial^2 \bar{Q}(u,v)}{\partial v^2}\right|^2 + \epsilon},$$

and similarly for $\omega_{0,1}(u,v)$, $\omega_{1,1}(u,v)$, and $\omega_{2,0}(u,v)$ where ϵ is an offset term which prevents a parameter from becoming infinite when a partial derivative goes to zero. This weighting of internal energy terms normalizes a model's bending with respect to an atlas. Thus, if a portion of an atlas is highly twisted, then the corresponding region of a deformable surface can become quite convoluted without incurring a large cost in internal energy. The opposite is true for flat areas of the anatomical atlas, since small movements are subject to a large cost in internal energy. In other words, the model's internal energy weighting parameters are automatically computed from the atlas.

Since the atlas was generated from a smoothed representation of an MR image we warp the deformable model to a subject's smoothed brain surface. This type of representation is necessary because without user interaction deformable models can not be guaranteed to converge to complex and convoluted image features. The subject's smoothed brain surface is extracted using the morphological processing described in section 2. Recall that this processing produces a smoothed representation which still retains information as to where cortical convolutions occur on the brain surface.

A Chamfer distance image serves as an external energy function. We use a Chamfer distance function which includes information about sulci locations. We calculate both the Chamfer distance of each pixel to the nearest point on the brain surface (C_{brain}) and the Chamfer distance of each pixel to the nearest sulcus point on the outer brain surface (C_{sulcus}). If we let S be a set of sulci points on the model, the external energy function can be expressed as:

$$E_{ext}(Q(u,v)) = \begin{cases} C_{sulcus}(Q(u,v)) & if\ Q(u,v) \in S \\ C_{brain}(Q(u,v)) & otherwise \end{cases} \tag{7}$$

We substitute the external energy term in equation 7 into the model's discrete energy function (equation 6), and we use the Polak-Ribiere conjugate gradient method for minimizing the model's energy, deforming the model to fit an extracted brain. In this way, model sulci points are attracted to image sulci points, and remaining model points are attracted to the brain surface.

The final step in the labelling procedure is the transfer of anatomical labels from the model to the original (unsmoothed) brain surface. As described in section 3, the B-spline surface representing the atlas is evaluated at a number of points in the u, v parameter space, and each of these points has an anatomical label associated with it. Once the model is warped, these same points in parameter space are the locations at which we evaluate the deformable B-spline surface, and thus each point on the deformed model is anatomically labelled. Finally, for each point on the extracted brain surface, we search for the closest point (in Euclidean distance) on the deformed surface and assign its label to the brain surface point.

In figures 2 and 3 we illustrate the application of this atlas matching method to 3-D MR head images. Figure 2 (a) shows a surface rendering of

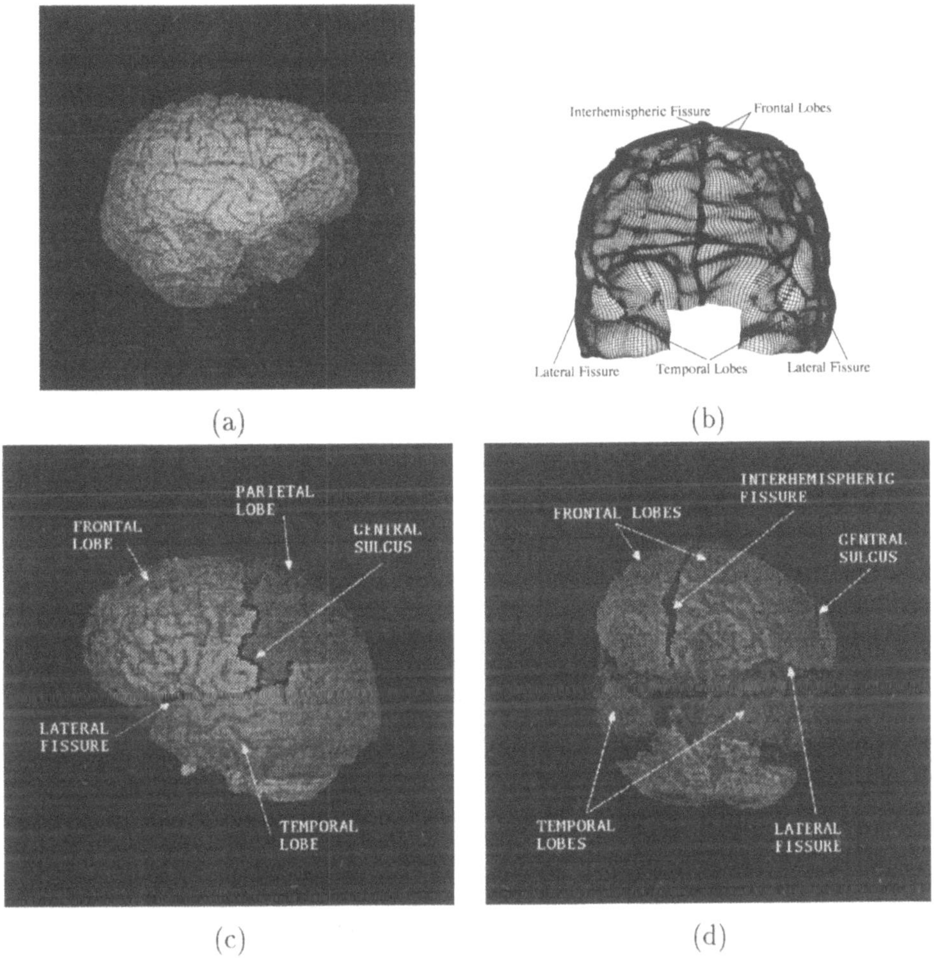

Figure 2. (a) Surface rendering of the subject's brain. The deformable model is warped to the morphologically closed version of this brain, and the anatomical labels are transferred to this high-resolution surface.(b) Frontal view of the 3-D deformable model after it has been warped to match the subject's brain surface. (c) Lateral view of the final labelled brain surface. The different gray levels correspond to different anatomical labels. (d) Frontal view of the final labelled brain surface.

the subject's morphologically extracted brain. We used the atlas described in section 3 to label regions on the brain surface corresponding to the frontal, parietal, and temporal lobes and the central, lateral, and inter-hemispheric sulci (see figure 1). The original spline surface was globally registered to match the subject brain's coordinates. The model was then allowed to deform, and figure 2 (b) depicts a view of the warped spline model after it was matched to the morphologically closed brain surface.

Labels were transferred from the deformed model to the original brain surface. Figure 2 depicts renderings of different views of the final labelled brain surface, where different gray levels correspond to different anatomical labels. Figure 3 shows a 2-D slices through the labelled brain, demonstrating that not only the visible outer surface has been labelled but also every other surface point of the morphologically extracted brain. Thus, all points which lie on the surfaces of the cortical convolutions are labelled.

6. DISCUSSSION AND CONCLUSIONS

There are two major problems with using deformable models for 3-D brain labelling. Deformable models can not be guaranteed to automatically converge to the surface of the brain in raw MR images. Also, without user interaction, deformable models can have difficulty converging to complicated object boundaries. Mathematical morphology can overcome these problems. Thus, we have developed a morphological algorithm for preprocessing these images. Once boundaries in an MR head image are detected, we apply a specific sequence of morphological operators to extract the brain as a binary object from the head image. Moreover, the morphological processing identifies points which correspond to sulci locations on the brain's surface. By extracting solely the brain we overcome one of the difficulties of matching a deformable model to these images, because we eliminate other anatomical boundaries in the image to which a deformable model may incorrectly converge.

With the images resulting from morphological processing we also overcome the problem of matching a deformable model to a convoluted surface. We generate the atlas, which is the initial position of a deformable model, from a smoothed representation of an MR image. Then we warp the deformable model to an extracted brain surface that has been processed with the equivalent smoothing operations as those used in the creation of the atlas. Therefore, the deformable model converges to the smooth outer boundary of the brain's surface. Also, since our morphological processing finds sulci locations on the smoothed brain's surface, we constrain sulci locations on the model to converge to sulci locations on the subject's brain.

Although our approach does not depend on a particular boundary-finding technique, the algorithm is highly dependent on information provided by a low-level processor, and accurate detection of region boundaries is important to all further processing. While boundary-finding results may be improved using a more sophisticated edge detector, such as the Canny-Deriche method [6], partial volume effects make it essentially impossible for any edge-based segmentation method to accuractely find the outer brain surface without smoothing over the deeper sulci. If very accurate localization of the outer brain surface is required, this may be achieved either by

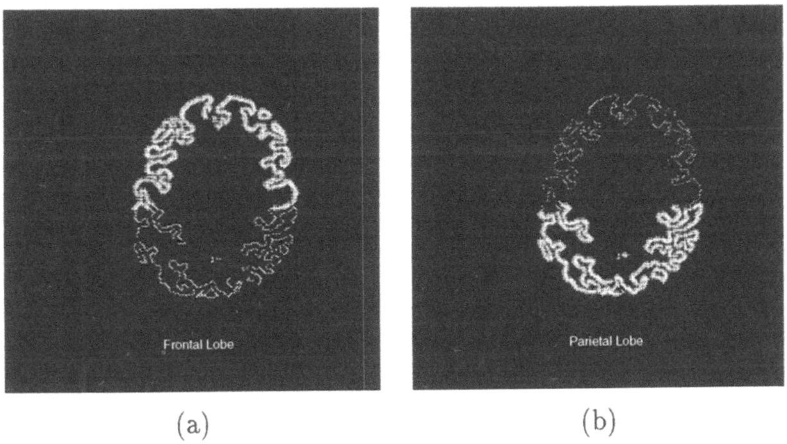

Figure 3. One 2-D slice through the brain surface showing the labelling of interior sulcal points. The anatomical labels were transferred from the warped model to the surface points (a) shows the points that were labelled as Frontal lobe. (asterisks highlight the different anatomical labels that were attached to the surface points), and (b) shows Parietal lobe points.

using an intensity based segmentation scheme [7] or by iterative relaxation of the brain surface boundary after it is extracted using our morphological method.

A modification we can make to our model is to include more anatomical features in the atlas. One way to incorporate additional sulcal patterns into our labelling technique is to extract more information from the subject's morphologically processed brain image before performing deformable atlas matching. By tracing connected sulci points which are extracted through morphological processing, it may be possible to determine whether these points are part of large, uninterrupted sulci and also to find the approximate orientation of the sulci. We could then add further terms to the deformable model's external energy function, thereby allowing model sulci points and image sulci points with similar orientations to be attracted to one another. While we have demonstrated that it is possible to automatically label the major fissures of the brain using a deformable model approach, it is an open question as to whether this approach will also work for those sulci that exhibit a greate degree of intersubject variability.

Finally, the performance of our image labelling technique can be improved by changing our global model to image registration method. Since deformable model behavior is dependent on the model's initial position with respect to image features, a more accurate initial registration technique can improve the model's performance and ensure a correct labelling.

For example, the orthogonal proportional grid system of Talairach and Tournoux allows a localization of major structures in the brain, and this method could be used as an initial registration procedure [8]. An extension of the cross-modality image registration of [9] to intersubject MR image registration can also be used for this purpose.

ACKNOWLEDGEMENT

We thank George Zubal of Yale University and Cheryl Aine of Los Alamos Research Lab for giving us access to volume MR data. We also thank Scott Grafton of USC for discussions about cortical anatomy and labelling. This work was supported by grants from the National Cancer Institute (RO1 CA579794) and the National Institute of Mental Health (R01 MH53213) and by the TRW Inc. Doctoral Fellowship program.

References

1. A. M. Dale and M. I. Sereno, "Improved localization of cortical activity by combining eeg and meg with mri cortical surface reconstruction: A linear approach," *Journal of Cognitive Neuroscience*, vol. 5, pp. 162–176, 1993.
2. S. Sandor and R. Leahy, "A 3-d morphological algorithm for automated labelling of the cortex in magnetic resonance brain images," in *Technical Report of the AAAI Symposium on Applications of Computer Vision in Medical Image Processing*, March 1994.
3. M. Ono, S. Kubik, and C. D. Abernathey, *Atlas of the Cerebral Sulci*. Stuttgart: Georg Thieme Verlag, 1990.
4. M. Kass, A. Witkin, and D. Terzopoulos, "Snakes: Active contour models," *International Journal of Computer Vision*, vol. 1, pp. 321–331, January 1988.
5. N. Ayache, I. Cohen, and L. D. Cohen, "Introducing new deformable surfaces to segment 3d images and infer differential structures," Technical Report 1403, Institut National de Recherche en Informatique et en Automatique, 1991.
6. R. Deriche, "Using canny's criteria to derive a recursively implemented optimal edge detector," *International Journal of Computer Vision*, pp. 167–187, 1987.
7. G. Gerig, J. Martin, R. Kikinis, O. Kubler, M. Shenton, and F. Jolesz, "Automating segmentation of dual-echo mr head data," in *Proceedings 12th International Conference on Information Processing in Medical Imaging*, July 1991.
8. J. Talairach and P. Tournoux, *Co-planar stereotaxic atlas of the human brain*. Stuttgart: Georg Thieme Verlag, 1988.
9. R. P. Woods, S. R. Cherry, and J. C. Mazziotta, "MRI-PET registration with automated algorithm," *Journal of Computer Assisted Tomography*, vol. 17, pp. 536–546, 1993.

AUTOMATIC 3D SEGMENTATION OF NEURO-ANATOMICAL STRUCTURES FROM MRI

D.L. COLLINS, A.C. EVANS, C. HOLMES AND T.M. PETERS
Montreal Neurological Institute, McGill University
McConnell Brain Imaging Centre
3801, University St., Montreal, Canada H3A 2B4

Abstract: This paper addresses a specific problem of model-based segmentation; namely, the automatic identification and delineation of gross anatomical structures of the human brain based on their appearance in magnetic resonance images (MRI). The approach developed in this paper depends on a general, iterative, hierarchical non-linear registration procedure and a 3-D digital model of human brain anatomy that contains both volumetric intensity-based data and geometric atlas data that co-exist in a brain based stereotaxic coordinate system.
Structure delineation is achieved by inverting the traditional segmentation strategy. Instead of matching geometric contours from an idealized atlas directly to the MRI data, segmentation is achieved by identifying the spatial transformation that, under certain constraints, best maps corresponding intensity-based features between the model and a particular MRI brain volume. When completed, atlas contours defined in the model are mapped through the recovered transformation to segment structures in the original data set and identify them by name.
Experiments for registration and segmentation are presented using simple phantoms, a realistic digital brain phantom as well as human MRI data. Validation against manual methods shows that the procedure performs well, is objective and its implementation robust.

keywords: model-based segmentation, non-linear deformation, MRI, human brain.

1. Introduction

Modern imaging modalities such as computed tomography (CT) and magnetic resonance imaging (MRI) have facilitated the investigation of complex spatial relationships between multiple structures within the brain. An essential step in the analysis of these images is the identification of anatomical structures and their sub-units. For qualitative analysis, a human investigator implicitly segments the data into its structural components based on personal knowledge of anatomy derived from textbooks and atlases. Our

Y. Bizais et al. (eds.), Information Processing in Medical Imaging, 139–152.

interests require quantitative analysis of specific regions for the mapping of functional neuroanatomy of sensorimotor and cognitive processes [13, 12, 4] as well as the analysis of neuro-anatomical variability among normal brains [11].

Therefore, explicit segmentation is needed to separate and identify structures. This is often achieved using a brain atlas as a guide for anatomical localization and functional interpretation (e.g., [18]). Unfortunately, manual outlining of structures is tedious, difficult and time consuming. Errors are due to subjectivity in atlas slice selection, structure interpretation, poor software interface design, poor eye-hand coordination, low tissue contrast, image degradation caused by artefacts and noise, and edge blurring due to partial volume effects (tissue mixing within a single voxel) and, perhaps most importantly, to the fact that tomographic slices are often not scanned at the same orientation as the atlas. These difficulties must be overcome to achieve robust, objective quantitative analysis.

1.1. PREVIOUS WORK

Present classification techniques applied to the segmentation problem can identify gross tissue types such as cerebro-spinal fluid, white or grey-matter (see [3] for a good review). However, these methods cannot identify individual neuro-anatomical structures. While computer-vision based or image-processing based algorithms delineate elementary regions from the image data based on measures of local intensity properties through scale space (e.g., [17]), these regions are not anatomically labeled nor do their borders always correspond to the outline of a cerebral structure. External domain-specific information must be used to interpret the features present in an image.

In manual techniques using an atlas and in automatic model-driven approaches, it is assumed that the target (atlas or digital model) and the data to be segmented are topologically equivalent but that internal structures are mutually deformed. This means that even though the absolute positions of structures may vary between subjects, their relative positions remain (loosely) stable and that there exists a one-to-one mapping between structures in the data and those in the atlas. Expert system techniques represent this information in the form of semantic rules to manipulate previously extracted features [1, 10]. However, some of these rule-based methods become complex and unwieldly when dealing with many structures and most are highly sensitive to the nature of a pre-segmentation step required to extract the features.

In more related work, iconic models such as idealized geometric structures have been used to represent the spatial relationships between structures of interest, and thus better constrain the segmentation problem. Zachmann has matched a given dataset to an iconic model (a voxelated volume, where the value in each voxel represents the probability of existence of a structure) for classification of the fluid spaces of the brain [20]. Kamber [14] has used a similar technique to identify grey/white-matter and ventricles in the context of multiple sclerosis lesion identification, however the model she used was based on MRI data from 12 subjects instead of the single data set used by Zachmann. These two authors used only linear transformations to match the anatomical model to the given data set.

Since there exists a significant amount of morphometric variability between subjects, nonlinear transformations must be used to account for fine structure differences persisting after linear registration. Bajcsy *et al.* [2] have used 3D elastic matching of image to model, based on a coarse-to-fine estimation of local deformations needed to register a pre-defined atlas to CT raster data. Their atlas consists of geometrical structure outlines, manually drawn on photographs of macro-tomed brain slices. A similar method has been described by Zhengping *et al.* [21] to segment MR data, where after linear registration between a voxelated model and data, the model is warped to fit the data. Unfortunately, the method is applied on 2-D slices only, and therefore it can never fully recover the complete non-linear deformation required to match the two volumes together.

1.2. OVERVIEW OF METHODOLOGY

The following example forms the basis for the segmentation strategy developed in this paper. Suppose a segmented dataset, **A**, exists, and that each of its voxels are tagged with some neuro-anatomical label. In order to segment another brain volume, **B**, in an equivalent manner, the non-linear spatial warping transformation that represents the one-to-one mapping from **A** to **B** must be recovered. Each of the labels in **A** is then mapped through the recovered transformation onto **B**, thus segmenting **B**. Hence, the segmentation problem is expressed as a registration problem. Implicit in this analysis is the assumption that the first data set is perfectly segmented and that the one-to-one mapping exists. Even though the latter is not strictly true (e.g., one brain may have sulcal patterns that another does not) the assumption does not lead to catastrophic deformations of one structure onto another in practice.

While this approach is similar to that of Bajcsy and of Zhengping, our technique differs significantly in a number of respects. The segmentation task presented in this paper turns the conventional approach upside down. Instead of matching atlas contours directly to the image data (as in [2]), a model was created that contains both volumetric raster data as well as geometric atlas data. Both components are registered within a standard brain-based coordinate system know as *stereotaxic space*[1] The segmentation strategy is then separated into two stages: a registration step followed by a delineation step. The goal of the registration process is to find the optimal 3D spatial transformation (unlike the 2D method of [21]) between the prior stereotaxic raster image model and the individual MRI data set by maximizing the match (or overlap) of features derived from the data set and those in the model. Structure delineation is then achieved by applying the inverse of the recovered transformation to contours defined in the model, thereby outlining structures in the native data.

This organization yields two important benefits over previous methods: 1) Since raster features from the data are being matched to similar features in the model, the resulting approach circumvents matching problems that confound traditional model-based segmentation techniques that attempt to fit one form of data representation, geometric contours of the atlas, directly to another, raster data of the image. The two representations are inherently different in contrast, noise and gradient properties. 2) More importantly, the segmentation strategy described here results in an atlas-independent segmentation. The geometric contours defined in the atlas are not used in any way to determine the match between data and model. Therefore, any atlas defined on the target MRI brain volume can be used for segmentation, thereby allowing for the co-existence of multiple atlases, each of which is simultaneously mappable to the native MR image volume.

2. Methods

2.1. REGISTRATION ALGORITHM

Registration of a single source brain volume to the target model is accomplished in two optimization steps: the first accounts for the linear component of the transformation function and the second, the non-linear component. The linear transformation is found automatically by estimating the best affine transformation matrix that maximizes the correlation of invariant features extracted from both source and target volumes [7].

For both linear and non-linear registration procedures, the geometrically invariant features of image intensity and image gradient magnitude are used so that the features derived from the volumetric data are independent of the original position, scale and orientation of one data set with respect to

[1] *Stereotaxic* refers to the representation of 3-D brain structures in a standardized brain-based coordinate system [18] such that all brains have the same orientation, position and size in the three orthogonal directions. Mapping volumetric data to this coordinate system permits comparison of brains from different individuals as well as to the published Talairach atlas.

the other. These features are calculated by convolution of the original data with zeroth and first order 3D isotropic Gaussian derivatives. Convolution with such an operator maintains linearity, shift-invariance and rotational-invariance in the detection of features [19, 15]. Unfortunately, this operator is not scale-invariant since the Gaussian kernel is dependent on the standard deviation, σ, a parameter which can be considered to be a measure of spatial scale.

To address the scale-invariance, the optimization used to recover the best transformation is accomplished in a hierarchical multi-scale fashion where the registration is performed at different spatial scales, starting with very blurred data and increasing detail at each step by using less blurred images, refining the registration at each stage. As well as increasing the speed of calculation with respect to performing all operations at the smallest scale (highest resolution), this strategy has the important added benefit of avoiding local minima in the objective function hyper-surface.

The non-linear registration technique is a straightforward extension of the linear registration method, except that the target volume is a small neighbourhood of the whole brain, recursively selected by stepping through the entire target volume in a 3D grid pattern [9, 6]. The global non-linear warp needed to map the data set onto the model is composed of a set of local linear deformations, with one deformation vector being defined for each grid position. Each deformation is based on the estimation of the best local linear transformation that maximizes the correlation of features from the neighbourhood surrounding the given voxel to its corresponding neighbourhood in the target data set.

Since the deformation at a single node is part of a continuous global warp, it affects not only that node, but all neighbouring nodes as well. Therefore, the correlation values for those neighbouring nodes will also change as a result. Consequently, an iterative approach is used, where a fractional value of the estimated deformation is stored for each node at each iteration. Experimentally, we have found that a fractional value less than 0.3 yielded very robust results, however too many iterations were required to be practical. While values greater than 0.75 converged quickly, an incorrect match at one node could pull neighbouring nodes from their correct location. We have selected a value of 0.6 as a compromise between robustness and speed of convergence. At this level, we have found that only 4 iterations are needed at each scale.

Initial experiments [8], using only features derived from image intensity and gradient magnitude with correlation as the residual to be minimized, showed promise when applied without explicit constraints on the deformations (i.e., plastic warping). However, occasional unrealistic mappings (edge of scalp mapped onto edge of cortex) indicated that constraints must be included in the deformation process. A simple yet effective constraint was implemented that limited the deformation magnitude at each node to two times the current scale measure σ. This eliminated all major correspondence errors.

2.2. SEGMENTATION

The segmentation procedure described in this paper depends on a two-part design for the target model volume, regardless of whether the target is a mathematical phantom or an MRI data volume. The target contains 1) the model image data with its associated invariant features and 2) a set of labels that define the structures of interest on this volume.

The features consist of blurred intensity data and gradient magnitude data, estimated at the scales required for the registration process. Note that the registration is accomplished using only information derived from the original image data. Segmentation is achieved by mapping the labels defined on the target volume through the inverse of the spatial transformation recovered by the registration procedure.

3. Experiments and results

3.1. ELLIPSOID PHANTOM

A simple volumetric brain phantom was created to demonstrate the feasibility of the segmentation strategy. The phantom contains three concentric ellipsoids representing the skin surface, cortical surface and grey-white interface, and two smaller ellipsoids to represent the ventricles. The parallelepiped represents an arbitrary structure within the phantom. The long axis of the phantom is 100mm in length, and it was generated in a volume with a 128x128x128 matrix, with 1mm^3 voxel size. Fig. 1 shows a 3-D rendering of the phantom. A thin-plate spline (TPS) [5] warp was used to deform the phantom and create a warped target. The non-linear registration algorithm was then applied to fit the original volume onto its warped counter-part. The geometric contours defined on the original phantom were mapped through the recovered non-linear transformation and overlaid on the target data set. Figure 1 shows the resulting segmentation for the linear transformation and the first, third and last scale steps (24mm, 8mm and 2mm) of the nonlinear transformation. One can see that there is a significant improvement in segmentation evident at the first non-linear result when compared to the linearly transformed contours. The successive steps in scale refine the segmentation even further, finishing with an almost perfect segmentation of the phantom.

Two quantitative error measures are used to evaluate the quality of the

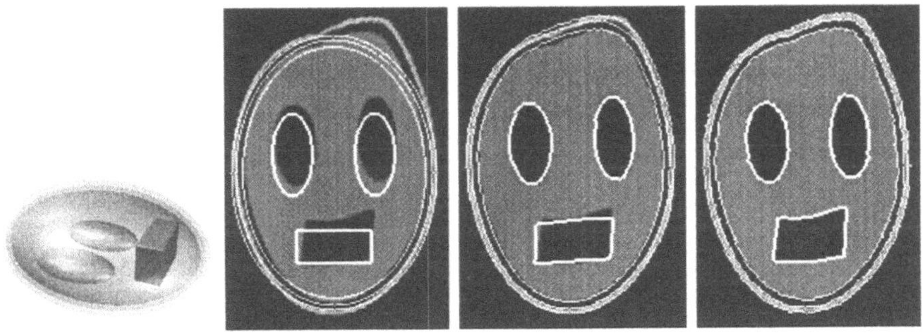

Figure 1. Segmentation of ellipsoid phantom.
From left to right: the 3D rendered phantom (a); the model contours mapped through the linear transformation (b) and the first (c) and last (d) scale steps (24mm and 2mm) of the nonlinear transformation.

segmentation. The first simply measures the percent difference in absolute volume between the true and the segmented structure. This value is defined as:

$$\delta(T,S) = \frac{V_T - V_S}{V_T} \times 100.0\%, \tag{1}$$

where δ is the percent difference in absolute volume, V_T is the volume of the true structure and V_S is the volume of the automatically segmented structure. While this value indicates differences in size between the segmented and true structures, it does not measure the quality of the segmentation, since two structures can have the same volume and not be the same shape, nor be in the same position. The second error measure is the percent volume overlap, Δ, defined as:

$$\Delta(T,S) = \frac{|T \cap S|}{|T|} \times 100.0\%, \tag{2}$$

where T is the set of voxels in the true structure, S is the set of voxels in the automatically segmented structure, $\cap$ indicates volumetric intersection and $|\cdot|$ returns the volume of the set of voxels.This value equals 100.0 when the segmented and true structures completely overlap. In fact, if the segmented structure is larger than the true structure, Δ can still equal to 100.0. Therefore, one measure forward $\Delta_f = \Delta(T,S)$ and one measure backward $\Delta_b = \Delta(S,T)$ were calculated, and the minimum of the two was used to estimate the segmentation results.

The five objects used in the definition of the ellipsoid phantom were also employed to test their segmentation. For the segmentation described above, the values of δ and Δ were calculated for each object. The results in Table 1 show that on average the estimated segmented structure volume has less than 1% error and structure overlap is better than 99%, indicating that the objects are well delineated by the contours mapped through the deformation field recovered by the non-linear registration method.

TABLE 1. Segmentation results for ellipsoid phantom.

	linear		non-linear	
structure	δ	Δ	δ	Δ
skin:	7.1%	92.8%	-0.2%	99.9%
brain:	6.1%	93.4%	0.6%	99.6%
ventricle (R):	1.7%	84.8%	0.5%	99.3%
ventricle (L):	-2.1%	77.3%	0.5%	99.1%
rectangle:	23.0%	71.1%	-1.9%	98.3%
average mag:	**8.0%**	**83.9%**	**0.7%**	**99.2%**

3.2. BRAIN PHANTOM

A more realistic evaluation of the automatic segmentation employed a digital brain phantom created by identifying different structures and tissue types from an MR data set of a single volunteer. The volume is sampled on a $1 \times 1 \times 1\text{mm}^3$ matrix, and covers the entire head. The resulting labeled volume could then be endowed with the appropriate contrast for each tissue type, and noise added, to resemble a true MR acquisition. This 3D data set provides a gold-standard with which to evaluate the linear registration algorithm. It should be noted that it does not matter that some voxels from the original MRI data may be mis-labeled. The labeled volume is now a phantom defined as truth and is no different in principle than the arbitrarily defined ellipsoid phantom. It has the single advantage that it is a complex simulated data set which closely mimics the convoluted nature of a real brain.

A set of landmarks was chosen to produce a TPS transformation that was applied to the 3D T1-weighted phantom data set to produce a topologically equivalent, but spatially warped target data set. Ten percent noise was added to the source volume.

Five anatomical structures were defined volumetrically on the original unwarped phantom data using a voxel painting feature within a software program, developed at the MNI, that allowed simultaneous tri-plane (transverse, coronal and sagittal) viewing of volumetric data. These structures included the head of caudate (left and right) as well as the lateral and third ventricles.

The manually segmented structures were mapped through the applied transformation to create the set of known structures in the warped data

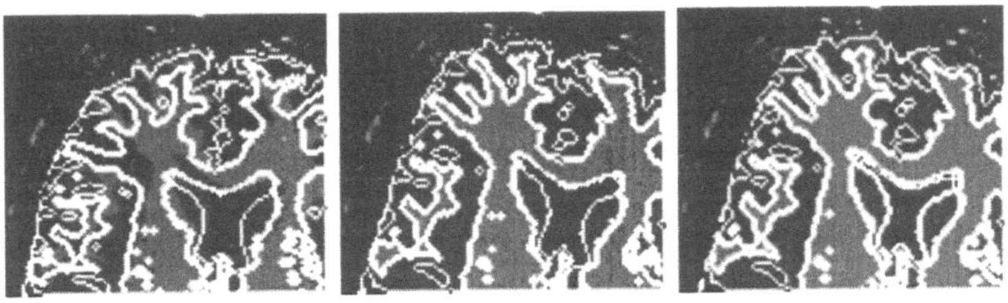

Figure 2. Segmentation of digital brain phantom.
Transverse images of the 3D digital brain phantom through the ventricles at the level of the basal ganglia with overlaid structure contours. On the left, contours are mapped through the linear transformation only. In the center are the true contours, generated by mapping the original atlas contours through the applied transformation. On the right, are the contours resulting from the segmentation algorithm. Almost no difference is visible between the two sets of contours.

volume that must be segmented by the registration/delineation procedure. These structures were used to evaluate the segmentation algorithm using the percent difference and the percent overlap measures defined above. Figure 2 shows the result of the segmentation procedure. The contours defined by the painted structures were mapped back through the inverse of the recovered transformation and overlaid on the warped data. The structures in the segmented data are outlined well by the warped contours.

Quantitative evaluation of the brain phantom segmentation is summarized in Table 2. Since a classified data set was used to create the brain phantom, the set of grey-matter and white-matter voxels were known. These too were mapped through the applied and recovered transformation for evaluation and were included in Table 2. On average, δ was reduced by more than half, from 4.2% to 1.7%, and Δ was improved 38%, from 70.4% to 97.0% when using non-linear as opposed to linear matching. Therefore, not only is there an improvement in the estimated structure volume, but the structures are properly delineated as well. The results for these structures indicate that the segmentation method functions on realistic data. For example, simulations show that $\Delta = 96.4\%$ corresponds to a deliberate misregistration of less than 1mm of a perfectly segmented caudate.

3.3. HUMAN MRI DATA

Eleven normal volunteers were scanned in a Philips Gyroscan ACS 1.5 Tesla superconducting magnet system for the experiments described below. The data were acquired with a T1-weighted 3-D spoiled gradient-echo acqui-

TABLE 2. Segmentation results for brain phantom.

	linear		non-linear	
structure	δ	Δ	δ	Δ
caudate:	1.3%	65.7%	4.9%	96.4%
ventricle:	7.0%	77.6%	-0.0%	99.1%
grey:	3.4%	72.0%	1.0%	97.0%
white:	5.0%	66.1%	-0.7%	95.6%
average:	**4.2%**	**70.4%**	**1.7%**	**97.0%**

sition with sagittal volume excitation (TR=18, TE=10, flip angle=30°, 140-180 sagittal slices). This sequence achieves a high grey/white matter contrast with good signal-to-noise characteristics (15:1).

Testing of the segmentation algorithm on real data requires a set of gold-standard structures to be defined by hand on the 11 data volumes. The voxel painting program described above was used by a trained neuro-anatomist to identify individual anatomical structures in these brain volumes. The voxels labeled by this procedure serve as the gold-standard definition for each structure on each brain, and were used to evaluate the automatic segmentation result. The left and right head of caudate were chosen for testing, because of their inherent importance in many aspects of current brain research and because they present a relatively well-defined object for manual labeling.

Without loss of generality, the segmentation algorithm allows one brain to serve as a target and structures identified on it to define the anatomical atlas that is used for segmentation of all other data sets. Therefore one brain volume, and its associated voxel labels, were selected as the model and the other ten data sets segmented automatically with it.

After application of the non-linear registration algorithm to each of the individual/model pairs, the inverse of the recovered transformation was used to resample the structure labels of the target volume onto the individual data set. Figure 3 shows three transverse slices through the ventricles and basal ganglia of one of these segmented data sets. Contours corresponding to the head of caudate are overlaid on the images for both the manual and automatic segmentation. Comparison of the two sets of images shows that the segmentation algorithm succeeds in identifying the boundaries of the structure, based on its definition in the model brain.

Table 3 shows that the overlap increases from 75.6% to 88.6% on average over the ten caudates. The percent volume difference of the right caudate is

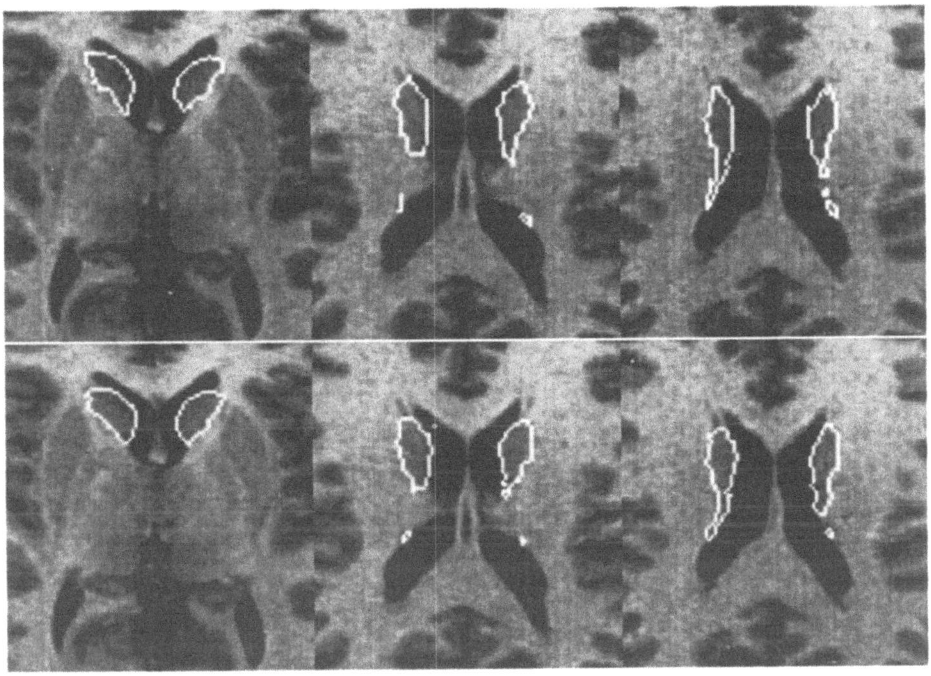

Figure 3. Typical segmentation of real MRI data.
Transverse slices through a single brain at the level of the ventricles and basal ganglia. The manually identified contours for the caudate are shown in the top row. The corresponding automatic segmentation is shown in the bottom row. Notice that there is good agreement between the manually and automatically-defined contours.

greater than in the left, after non-linear segmentation. Since the right caudate is thinner and occupies slightly less volume than the left ($5285mm^3$ vs $5482mm^3$, on average of the manually segmented data), misregistration between the automatic and manually segmented structures cause more important errors for the right side. While the value of Δ increased to more than 97% for the simulations using the brain phantom, Δ increases to only 90% with real data. There are two reasons for this lack of improvement. The main cause may be due to intra-observer variability in the definition of the caudate on different subjects. For example, the anatomist found it quite difficult to identify the anterior-ventral aspect of the caudate near the region of the *nucleus accumbens*, sometimes including it in the structure definition and sometimes not. This region represents 1-2% of the volume of the head of the caudate. Also, since the structures were painted manually, some ventricular voxels were included on the medial wall of the body of the caudate and may account for an additional 3-4% of the difference. The second reason may be due to bias incurred by using the manually segmented caudate from the single individual. Errors in manual segmentation of this structure are carried through the non-linear transformation, yielding errors in the segmentation. This will be true for in general for any atlas defined on the basis of a single brain.

TABLE 3. Segmentation results for real MRI data.

	linear		non-linear	
structure	δ	Δ	δ	Δ
caudate (L):	7.3%	77.5%	3.7%	89.5%
caudate (R):	6.6%	73.5%	6.9%	86.6%
average:	**7.0%**	**75.5%**	**5.3%**	**88.0%**

4. Discussion

The segmentation method is subject to the following caveats. Even though the goal of the non-linear registration algorithm was to account for as much anatomical variability as possible, it is impossible to eliminate all variability, because of non-equivalent topology in some brain regions. Therefore it is not clear that there can exist a completely objective *correct* answer that can be used to evaluate the algorithm. In particular, while the major cortical convolutions are present in all brains, the secondary and tertiary gyri are much more variable. Hence, at some scale, the notion of absolute correspondence breaks down. This problem may ultimately require a completely different strategy in which the cortical mantle is "unfolded" and inter-subject matching is achieved my maximizing the overlap of these unfolded sheets with a 2-D algorithm. This work has been the subject of the PhD project of another student at the MNI [16]. Eventually, these two methods must be combined into one.

The recovery of the deformation is based on the assumption that there exists a one-to-one mapping between homologous points in different brains. Since neighbouring points within a particular brain structure of the source data set should be mapped to neighbouring points in the target, the non-linear deformation is constrained to be continuous throughout the domain of the brain. However, there exist neighbouring points in unconnected structures (such as on opposite sides of a sulcus, or on either side of the longitudinal fissure) that do not need to be mapped to neighbouring points in the target. Therefore, it may be desirable to allow discontinuities in the transformation at internal brain surfaces, e.g surfaces that separate the cerebellum from the occipital lobe or that separate the temporal lobe from the inferior frontal lobe. While this may complicate the recovery of the global transformation, the definition of piece-wise continuous regions where the current algorithm can be applied is certainly possible. However, the main problem will then reside in the inversion of the dis-continuous spatial deformation transformation required for segmentation.

Finally, the target model used here is based on a single subject. Even though normal, the particular brain chosen may represent an extreme of normal variability. We have built an MRI atlas from 305 normal subjects [11] in stereotaxic space that can be used to represent the most average brain. We are currently in the process of constructing a probabilistic atlas of neuro-anatomical structures. In this atlas, each voxel of the atlas represents the probability of finding a given structure at that voxel position [9].

5. Summary

A fully automatic segmentation strategy has been presented to address the goal set forth at the beginning of this paper. The reversal of the standard segmentation strategy, from matching geometric contours directly on image data, to one of registration followed by delineation has been show to be valid. The structures identified by this method are comparable to those segmented manually. The experimental results on both simulated and real MRI data have shown that the method is accurate and robust.

The segmentation method developed here has direct application in the quantitative study of neuro-anatomical variability in the normal population, the determination of inter-group differences, the detection of subtle abnormalities, and the tracking of normal development. It also has direct utility in studies of functional neuro-anatomy with macroscopic imaging techniques such as PET and fMRI. For PET, averaging across subjects in a standardized space is used to detect subtle cognitive activation foci measured from cerebral blood flow (CBF) volumes. The removal of the non-linear individual neuro-anatomical differences will potentially remove the blurring due to morphometric variability and potentially increase the signal-to-noise ratio, and enhance the detectability of even smaller signals. For fMRI, the automatic segmentation procedure can be used to objectively define an anatomical region associated with some cognitive process. The analysis of the fMRI data within these regions may provide clues as to the existence and quantification of neuro-functional variability between subjects.

References

1. L. Arata, A. P. Dhawan, J. Broderick, and M. Gaskil. Model-based analysis of MR images of the brain. *IEEE EMBS*, 13(3):1331–1332, 1991.
2. R. Bajcsy and S. Kovacic. Multiresolution elastic matching. *Comp Vis Graph Imag Proc*, 46:1–21, 1989.
3. J. C. Bezdek, L. O. Hall, and L. P. Clarke. Review of MR image segmentation techniques using pattern recognition. *Med Phys*, 20(4):1033–48, Jul-Aug 1993. Review.
4. C. Bohm, T. Greitz, and L. Thurfjell. The role of anatomic information in quantifying functional neuroimaging data. *J Neural Transm Suppl*, 37(37):67–78, 1992.
5. F. L. Bookstein. Principal warps: Thin-plate splines and the decomposition of deformations. *IEEE PAMI*-11(6):567–585, 1989.
6. D. L. Collins. *3-D model-based segmentation of individual brain structures from magnetic resonance imaging data.* PhD thesis, McGill University, Montreal, Canada, December 1994.
7. D. L. Collins, P. Neelin, T. M. Peters, and A. C. Evans. Automatic 3D inter-subject registration of MR volumetric data in standardized talairach space. *J Comput Assist Tomog*, 18(2):192–205, 1994.
8. D. L. Collins, T. M. Peters, and A. C. Evans. Multiresolution image registration and brain structure segmentation. In *3D Advanced Image Processing in Medicine*, pages 105–109. Proc IEEE, Nov. 1992.
9. D. L. Collins, T. M. Peters, and A. C. Evans. An automated 3D non-linear image deformation procedure for determination of gross morphometric variability in human brain. In *VBC94*. SPIE, 1994.
10. A. P. Dhawan and L. Arata. Knowledge-based multi-modality three-dimensional image analysis of the brain. *Amer J Physiol Imag*, 7(3-4):210–9, 1992.
11. A. C. Evans, D. L. Collins, and B. Milner. An MRI-based stereotactic atlas from 250 young normal subjects. *Soc Neurosci Abstr* , 18:408, 1992.
12. A. C. Evans, S. Marrett, P. Neelin, D. L. Collins, K. Worsley, W. Dai, S. Milot, E. Meyer, and D. Bub. Anatomical mapping of functional activation in stereotactic coordinate space. *NeuroImage*, 1(1):43–53, 1992.
13. P. T. Fox, J. S. Perlmutter, and M. E. Raichle. A stereotactic method of anatomical localization for positron emission tomography. *J Comput Assist Tomog*, 9(1):141–153, 1985.
14. M. Kamber, R. Shinghal, D. Collins, G. Francis, and A. Evans. Model-based 3D segmentation of multiple sclerosis lesions from magnetic resonance brain images. *IEEE Trans Med Imag*, 1994.
15. J. Koenderink and A. van Doorn. Representation of local geometry in the visual system. *Biol Cybern*, 55:367–375, 1987.
16. D. MacDonald, D. Avis, and A. C. Evans. Multiple surface identification and matching in magnetic resonance images. In *VBC94*. SPIE, 1994.
17. S. Pizer, T. Culip, and R. Frederickson. Toward interactive object definition in 3D scalar images. In K. Höhne, H. Fuchs, and S. Pizer, editors, *3D Imaging in Medicine: Algorithms, Systems, Applications*, pages 83–106, Travemünde, Germany, June 25-29 1990. NATO ASI Series.
18. J. Talairach and P. Tournoux. *Co-planar stereotactic atlas of the human brain: 3-Dimensional proportional system: an approach to cerebral imaging.* Georg Thieme Verlag, Stuttgart, New York, 1988.
19. B. tar Haar Romeny, L. M. Florack, J. J. Koenderink, and M. A. Viergever. Scale space: its natural operators and differential invariants. In A. C. F. Colchester and D. J. Hawkes, editors, *IPMI91*, page 239, Wye, UK, 1991.
20. H. Zachmann. Interpretation of cranial MR-images using a digital atlas of the human head. *IEEE Trans Med Imag*, pages 99–110, 1991.
21. J. Zhengping and P. H. Mowforth. Mapping between MR brain images and a voxel model. *Med Inf (Lond)*, 16(2):183–93, Apr-Jun 1991.

DEFORMABLE TOPOLOGICAL MODELS FOR SEGMENTATION OF 3D MEDICAL IMAGES

Jean-François MANGIN[1,2], Florence TUPIN[1,2], Vincent FROUIN[1], Isabelle BLOCH[2], Rozenn ROUGETET[1], Jean RÉGIS[3] and Jaime LÓPEZ-KRAHE[2]

Service Hospitalier Frédéric Joliot, CEA, Orsay, France[1]

Département Images, ENST, Paris, France[2]

Service de Neurochirurgie, La Timone, Marseille, France[3]

Abstract–We propose to make low level segmentation of 3D medical images rely on deformable topological models. An initial label image made up of a set of deformable regions is endowed with a priori known topological properties of the desired result. Then, multi-scale topology preserving deformations are applied to this image in order to minimize a global energy whose form stems from the Gibbs Random Field domain. This new segmentation framework leads to important improvements of previous works. First, topological regularization is included in the segmentation process which results in better robustness of posterior processing like skeletonization. Second, the topological arrangement allows discrimination between objects of similar intensities. Third, interface between two deformable regions behaves like a deformable surface but surface regularization relies on Ising model rather than on differential terms. This results in a better behaviour with highly convoluted objects like the cortex. Finally, this approach allows the management of interactions between several deformable objects.

1. INTRODUCTION

Embedding of a priori knowledge in "low level" image processing appears often necessary for increasing robustness of "high level" post-processing. In fact, the development of model based treatments turns out to be easier in the field of 3D medical images than in other fields like computer vision or satellite image processing. Indeed, since apart from potential pathologies the underlying anatomical models are fixed, the development of generic algorithms dealing with any pattern of a database or any geographic map is not required. Moreover, volumetric images provide a dense 3D description of the whole object rather than several 2D views involving complex stereovision problems. Lastly, the voxel intensity value is generally related in a simpler way to the underlying physical property than when complex illumination models have to be taken into account. In this paper, we propose a new framework for the management of model-based low level processing. This framework intends to provide ways of designing high level pattern recognition systems driving low level segmentation according to their need.

During last decade, models stemming from Physics like deformable sur-

Y. Bizais et al. (eds.), Information Processing in Medical Imaging, 153–164.

faces [3] or Gibbs Random Fields (GRF) [4] have gained increasing popularity in the field of image processing. This success calls for the development of methods combining the leading points of both approaches. Recent works in 3D discrete topology have given rise to efficient tools to construct topology preserving deformations [5]. Hence, the idea of dealing with deformable regions rather than deformable surfaces can be addressed practically, which provides a way of performing such a combination. We propose to make low level segmentation rely on a model made up of deformable regions initially endowed with a priori known topological properties. This model is simply a label image, each label defining a region. Then, segmentation amounts to applying topology preserving deformations in order to reach an energetic goal, or more generally a sequence of energetic subgoals designed from a higher level of processing.

This framework can be viewed as a flexible alternative to the well-known elastic matching paradigm [1]. Indeed, we think that elastic matching is doomed to fail in regions of high anatomical variability because of problems of initialization, over-regularization (elasticity constraints), and ill-adaptation to structural variability. In return, our framework can include a wide variety of a priori knowledge and could be driven by higher levels of processing in charge of parameter adaptation and of creation or deletion of additional templates (which could allow the management of pathologies). In our opinion, this framework could be used for giving to models stemming from artificial intelligence the possibility to backtrack until the low level processing stage.

2. DEFORMABLE TOPOLOGICAL MODELS

2.1. Energy defining the segmentation goal

Surface regularization is naturally embedded in our framework using GRF forms of energy, which allows greater flexibility than Tikhonov forms of regularization [9] as regard convoluted objects like the cortex. Thus, the energy E_g whose global minimum corresponds to the goal of the segmentation has the form of a clique potential sum classically composed by two terms: a data driven term which is a simple model of intensity processes and a regularization term expressed via Ising or Potts models [4] (the Potts model is simply the extension to n classes of Ising model).

$$E_g = \sum_{C_M} V^a_{l(M)}(i(M)) + \sum_{C_{M_1M_2}} V^P(l(M_1), l(M_2)) \;, \qquad (1)$$

where $l(M)$ is the current label of point M, $i(M)$ is the intensity of M in the data image, C_M is the order 1 clique attached to M and $C_{M_1M_2}$ is the order 2 clique attached to 6-neighbours M_1, M_2. V_l^a are basin-shaped functions roughly modeling for each region statistical knowledge on intensities and V^P is a matrix of adjacency costs related to Potts model (see Fig.4). Analogy with deformable contour approaches is straightforward. The interface between two deformable regions behaves like a deformable surface regularized by Ising model rather than by a differential term. Data contribution to the global energy does not rely on potentials attracting the deformable surface towards edge points previously extracted by an edge detector [3] but on intensity-based potentials acting on both delimited regions. In this way, one interface can be selected among several ones using differences between object intensities or textures.

2.2. Topological constraints

Topological constraints are imposed region by region. The topology of a given region is defined relatively to the background made up by the union of all other regions. Connectivity choice may appear as a problem because of the well-known difficulties related to Jordan theorem. Indeed, since we are potentially dealing with more than two regions, the classical choice of different connectivities for object and background seems no more possible (considering for this choice 6-connectivity and 26-connectivity as the only available possibilities). In fact, since one region is always considered relatively to the union of the others, this classical solution remains possible: whatever the connectivity chosen for a given region, the dual connectivity has to be used for its background. Nevertheless, further investigations should be done on this subject in order to check that this cannot results in deadlocks for topology preserving deformations. For the application involving more than two regions presented in this paper, such difficulties have been avoid thanks to a particular property of the topological model used to represent the head. Since this model is made up of nested regions (see Fig.3), 6-connectivity and 26-connectivity have been alternatively chosen from the outer most region until the inner most one. Hence, each interface separates two regions endowed with dual connectivities. Three types of topological constraints can be imposed during deformations:

1. **Homotopy preservation:** the region is subject to homotopic deformations which preserve region connected components, cavities and "tunnels" (see [6] for a precise definition of homotopy relying on the notion of digital fundamental groups). Homotopic deformations are constructed as sequences of simple point additions or deletions. Simple points are efficiently characterized by the computation

of two local connected component numbers [5]. Intuitively, simple points are points whose neighbourhood contains exactly one connected component of object and background. Multi-scale homotopic deformations can be constructed thanks to the link between discrete topology and topology of the underlying continuous objects. This allows reduction of computation time and better robustness to noise.

2. **Connectivity preservation:** this weaker constraint is imposed in a similar way by sequential addition of object adjacent points or deletion of points whose neighbourhood contains exactly one connected component of the object. Thus cavities or "tunnels" can be created or deleted but the object can not split in several connected components. This constraint is used for object of complex topology like cerebro-spinal fluid (CSF) or skull.

3. **Adjacency interdiction:** this constraint is naturally embedded in interface regularization potentials through a simple modification of Potts model [8]. Prohibited adjacencies correspond to infinite values of the function V^P attached to order two cliques (see Fig.4).

It should be noted that these three types of constraints are far to be independent. Furthermore, they are often redundant. For instance, when a region A is surrounded by a region B whose homotopy is preserved, if the only authorized adjacency for A is with B, then A homotopy will be preserved too. Therefore, since computation cost of homotopy preservation is high, a judicious choice of the topological constraint set is important.

3. ENERGY MINIMIZATION

3.1. Energy non-convexity

Energy minimization based methods usually suffer from one major impediment: the non-convexity of the function to be minimized. Apart from mainly heuristic solutions, two general approaches are used to overcome this difficulty: either providing a good initialization to a fast deterministic algorithm yielding the nearest local minimum or employing a stochastic algorithm. Simulated annealing has been a great success during last decade when dealing with functions involving a large number of parameters. Nevertheless, since practical implementations are only approximations of the theoretical SA, they do not perform well for any energetic landscape. A very favourable behaviour of SA has been proved in the context of GRF [4], which has led to important results in the image processing field. Unfortunately, this important property is no more verified in the context of deformable regions, which does not belong to the

GRF framework (label images which do not respect the topological properties imposed from the initial image would have a null probability which is incompatible with Gibbs distributions). Indeed, since the introduction of global topological constraints and of infinite potentials preventing some adjacencies forbids most of label image configurations, the number of possible paths between the initial image and the global minimum (or the local minimum of interest) is very reduced. Hence, when the initial image is too far from this global minimum, the paths to follow often contain high barriers which results in a bad behaviour of SA implementations. Such barriers are mainly induced by interferences between regions attracted by similar intensities. Indeed, a misplaced region overlapping the data domain aimed at a concurrent region often has to cross a domain of repulsive intensities before reaching the correct data domain. Owing to these difficulties, SA cannot be applied in the usual way in the context of deformable regions. Nevertheless, SA will turn out to be a powerful tool during the design of problem-dedicated minimization algorithms.

3.2. Problem-dedicated minimization algorithms

In favourable cases, a simple initialization is sufficient to obtain good results with deterministic algorithms like Iterated Conditional Modes (ICM) [2]. For more complex situations, problem-dedicated initialization algorithms have to be designed in order to feed ICM. Such an initialization algorithm has to position each interface in such a way that no interference between concurrent regions remains. A powerful way of designing such algorithms consists in the construction of a sequence of energetic intermediate goals corresponding to a sequence of locations in the configuration space which last element is the minimum of interest of E_g. Each energetic intermediate goal is constructed through modifications of the conventional energy E_g intending to flatten a barrier which prevents to reach the next location in the sequence. This approach shows strong analogies with a well-known toy where the player has to drive a marble between places of a plane maze by modifying the plane inclination relatively to the gravity field. Energy modifications are performed either through deletion of some potential families constituting E_g (see Fig.4) or through addition of new potentials introducing for instance global knowledge on object shapes (see Fig.5). These additional potentials tend to act like pressure forces used in active contour approaches [3]. The whole deformation process is performed via a multi-scale implementation. Therefore, the initialization process is restricted to the higher level of the scale pyramid. Thus the use of SA for each successive minimization is possible keeping reasonable computation times, which results in a greater robustness to local minima. SA is initialized at an intermediate temperature and thus is not able to cross high barriers which legitimates this sequence of successive SA initializations (SA is trapped in a restricted configuration area delimited by high barriers). After this initialization process, segmentation is refined by successive deterministic minimizations of E_g with increasing resolution (see Fig.1).

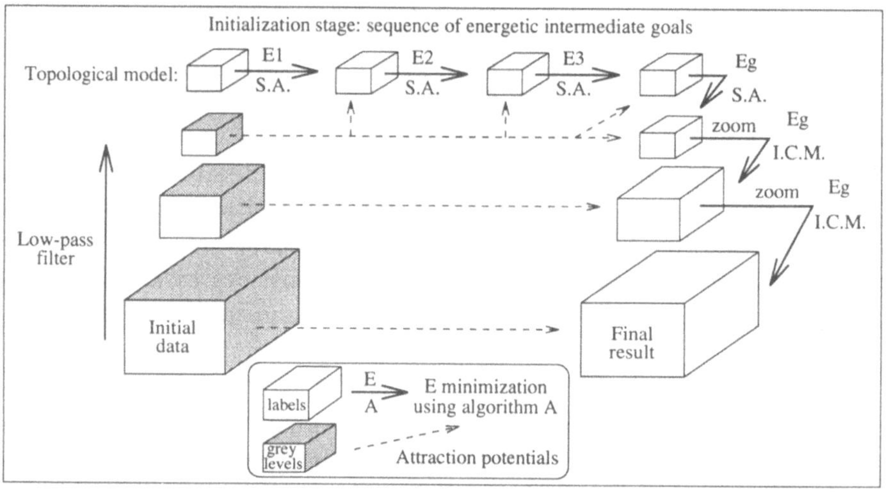

Figure 1: Multi-scale implementation of the segmentation process

4. PRACTICAL RESULTS

Two different applications of deformable topological models are proposed. The first one is an illustration of contributions of this approach in the preprocessing steps of a high level pattern recognition system. The second one, which uses a more complex model, describes the implementation of a dedicated minimization algorithm based on the principle described above.

4.1. Segmentation of the union {CSF, gray matter}

The deformable region principle has been included in a sequence of processes allowing the construction of a high level description of the cortex topography from a T1-weighted magnetic resonance (MR) image [6]. The aim of this work is the automatic identification of the main cortical folds. Successive steps of the method are the following:

1. Brain segmentation using 3D mathematical morphology.
2. Segmentation of the union of gray matter and CSF using a single homotopically deformable region (HDR) embedded in the background.

The initial region is the empty parallelepipedic bounding box of the brain. Hence the HDR is endowed with the homotopy of an empty sphere. Several multi-scale homotopic geodesic dilations allow a first deformation of the initial region in order to make it match the brain hull. Then homotopic deformations are applied in order to make the region internal surface reach the gray matter/ white matter interface (see Fig. 2). Parameters of attraction potentials are estimated from an initial rough classification using the K-means algorithm. Parameters measuring the relative influence of attraction potentials and regularization (Ising model) are chosen according to local considerations on the interface geometry [6]. Deterministic minimization is performed using an ICM-like algorithm.

3. Homotopic skeletonization of the previous region (see Fig. 2).
4. Segmentation of the skeleton in simple surfaces mainly corresponding to cortical folds [5].
5. Construction of an attributed relational graph whose nodes are these simple surfaces.

In this application, contribution of the deformable topological model approach can be considered from two point of views. First, compared with conventional deformable surface methods, it allows segmentation of a highly convoluted contour including surface regularization. Second, compared with conventional classification methods (like GRF), the result topology is imposed (with other approaches, post-processing could fill cavities but there is no simple way to detect "tunnels") which increases the robustness of the following skeletonization. The whole method described above, applied to fifteen brains, has turned out to be especially robust as regard various acquisition parameters. A graph database has been constituted from which a structural model of the cortical topography has been inferred [7]. We experiment today with the final goal, the automatic recognition of the cortical folds.

4.2. Tissue classification in MR images of the head

Segmentation of 3D T1-weighted MR images of the head has been performed with a simple model made up of nested regions, namely from the outer most to the inner most region: *void, skin, skull, CSF, gray matter, white matter and ventricles* (this last one made up of two connected components) (see Fig.3). Attraction potential parameters are estimated through a rough procedure relying on a sample of regions of interest (see Fig.4). Different topological constraints are imposed according to the

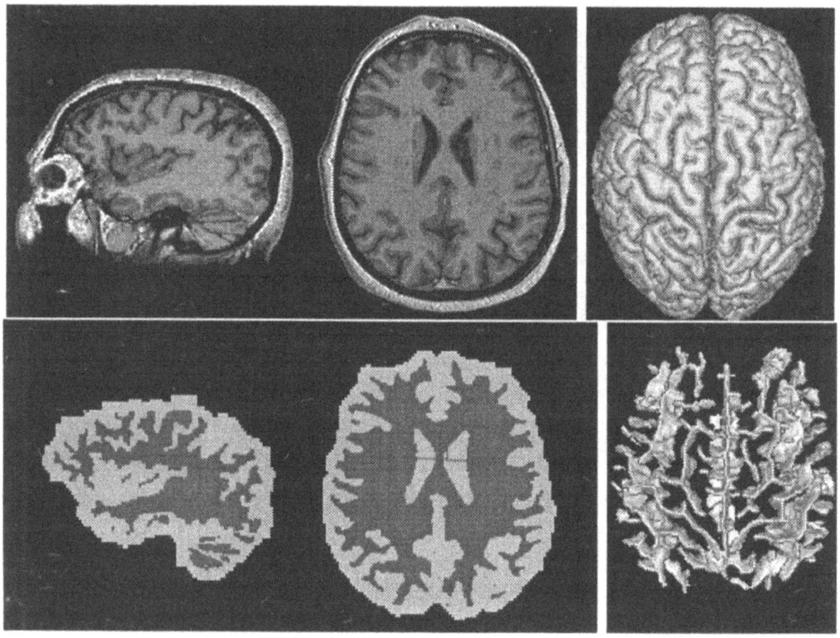

Figure 2: *Up left:* Two orthogonal slices of the T1-weighted MR image (124 axial slices 256×256 , pixel size: 1mm, slice thickness: 1.3mm, scanner: Signa GE); *Up right:* 3D rendering of the brain surface; *Down left:* Segmentation of the union {gray matter, CSF} using the HDR method (it should be noted that since we are only interested in the shape of the segmented object, the bad localization of the skull/CSF interface provided by morphological steps is not a problem); *Down right:* 3D rendering of a subpart of the segmented object skeleton.

topology of real objects (see Tab.1). The dedicated initialization process makes use of three types of pression-like additional potentials (see Fig.5). Three subgoals $E1, E2, E3$ are defined to drive three problematic region interfaces in respectively aimed areas (see Tab.1). A simple analogy with nested inflating or deflating balloons help to understand the process.

$E1$: Attraction potentials of *skin, skull and CSF* are superseded by low positive external fields and a negative external field is added to *void.* Thus *void* drives *skull* and *CSF* inside the head which allows a good positioning of the *void/skin* interface. It should be noted that keeping attractive potentials of *gray, white* and *ventricles* active prevents *skin* penetrating too deep inside the brain. This illustrates a leading guideline during energetic subgoal design: when flattening an obstructing barrier in order to reach a specific configuration area, other barriers delimiting this area have to be preserved.

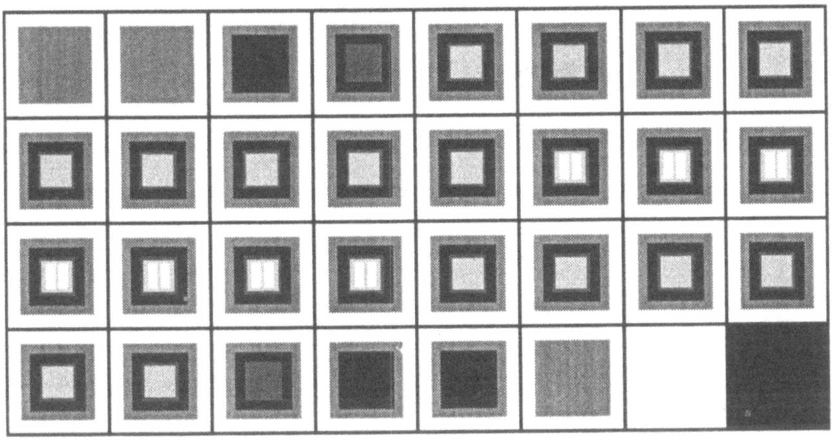

Figure 3: Topological model of the head made up of nested regions (void, skin, skull, CSF, gray and white matter, ventricles) apart from the lowest part modeling the neck.

Table 1: Sequence of Energetic intermediate goals (H denotes homotopy preservation and C connectivity preservation)

		Order 1 cliques								Order 2 cliques				Shape global terms							
	Top.	Attraction				Ext. field				Potts				Volume				Surface			
	E	1	2	3	g	1	2	3	g	1	2	3	g	1	2	3	g	1	2	3	g
v	H	×	×	×	×	×	×			×	×	×	×								
s	H			×	×	×	×			×	×	×	×								
s	C			×	×	×	×			×	×	×	×								
C	C			×	×	×	×			×	×	×	×							×	
g	H	×	×	×	×					×	×	×	×			×				×	
w	H	×	×	×	×					×	×	×	×			×					
v	H	×	×	×	×					×	×	×	×			×					

*E*2: The second energy stems from a slight modification of *E*1 (increased external field for ***skin***) intending to penalize ***skin*** in order to expel it from inside the brain. ***Skin*** is then reduced to a thin layer which results in the positioning of the ***skin/skull*** interface in the aimed area. Decoupling first and second steps allows greater robustness. Indeed, since barriers are flattened one by one, side effects of energy modifications are better controlled.

*E*3: The third energy relies on introduction of simple knowledge on the brain shape. A first global term acting on the number of points of the union {***gray, white, ventricles***} imposes a range of acceptable brain volumes, a second global term acting on the number of voxel faces belonging to the ***CSF/gray*** interface imposes a range of acceptable brain surfaces. Indeed, since the initialization process occurs at a low resolution level, brain folds do not contain CSF. Hence, this

Attraction potentials:
basin-shaped functions V_l^a are attached to order 1 cliques.

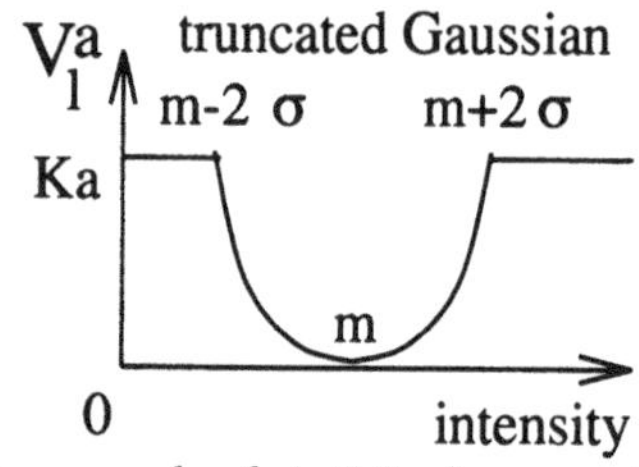

An example of statistical parameters*

	v	s	s	C	g	w	v
m	2	80	10	40	85	110	40
σ	10	20	10	10	5	5	10

Modified Potts model:
(regularization and adjacency interdiction)
$V^P(l_1, l_2)$ is attached to order 2 cliques.

$$V_P = \begin{pmatrix} v & s & s & C & g & w & v \\ \hline 0 & K_r & \infty & \infty & \infty & \infty & \infty \\ K_r & 0 & K_r & \infty & \infty & \infty & \infty \\ \infty & K_r & 0 & K_r & \infty & \infty & \infty \\ \infty & \infty & K_r & 0 & K_r & \infty & \infty \\ \infty & \infty & \infty & K_r & 0 & K_r & \infty \\ \infty & \infty & \infty & \infty & K_r & 0 & K_r \\ \infty & \infty & \infty & \infty & \infty & K_r & 0 \end{pmatrix}$$

$\star$: the statistical parameters (means and standard deviations) are roughly estimated from regions of interest.

Figure 4: Potentials of the conventional energy

External field: this corresponds simply to non zero constant potentials attached to order one cliques for some labels (a positive constant favours this region expansion, while a negative one has the reverse effect). This kind of potentials is related to external magnetic fields in the GRF domain.

Volume information: global potentials of the form $K_v(n - N)$ intend to introduce volume information in the global energy, where n is the current number of points belonging to a union of regions, N is an ideal volume to be reached and K_v is a positive constant which balances this contribution.

Surface information: global potentials of the form $K_s(s - S)$ intend to introduce surface information in the global energy, where s is the current number of voxel faces belonging to a specific interface between two regions, S is an ideal surface to be reached and K_s is a positive constant which balances this contribution.

Figure 5: Toolbox of pression-like potentials for initialization

shape parameters are relatively stable between individuals and have been estimated from several previously segmented brains. This step intends to expel the brain from the skin area which results in a good positioning of the *CSF/gray* interface.

This initialization process is followed by a deterministic minimization of the conventional energy *Eg* from which all pression-like terms have been deleted.

In order to check the relevance of the method, the whole segmentation process has been applied first with a simulated image containing six convoluted areas whose respective intensities had same statistics than ***void, skin, skull, CSF, gray*** and ***white matters*** in a sample MR image. Moreover this simulated image was containing topogical incoherences relatively to the model (connections between gray matter and skull). The segmentation process has nicely retrieved the non-noisy image and has corrected these topological incoherences (see Fig.6). Then, the segmentation process has been applied to two different high resolution MR images with the same sequence of steps (the whole process takes about one hour on a SPARC station 20). In both cases, segmentation results appear especially good in the head part where the topogical model is relevant, namely slices located above eyes (see Fig.7). In the lowest part of the head, result quality decreases progressively with the apparition of large objects not taken into account in the model, which could be foreseeable.

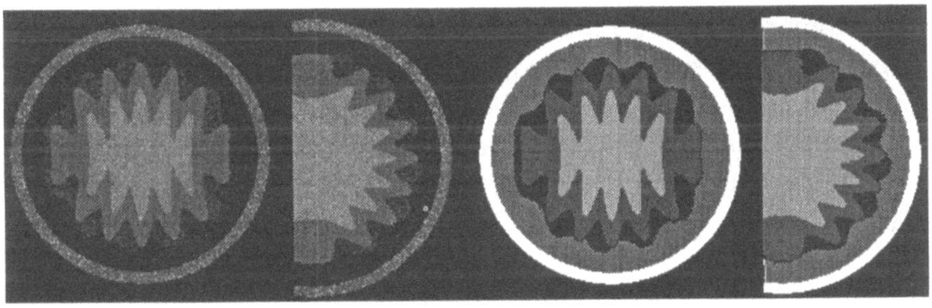

Figure 6: Segmentation result with the simulated image (left: two orthogonal slices of initial data, right: segmentation result)

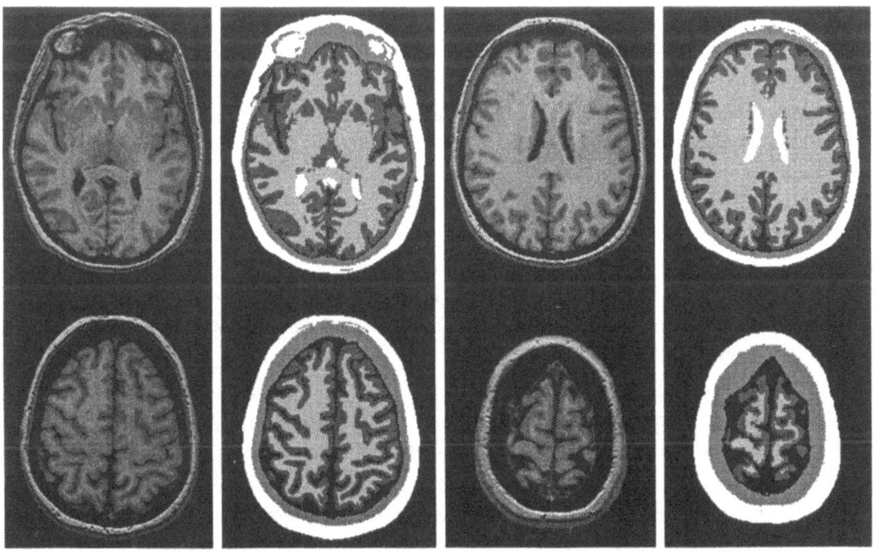

Figure 7: Segmentation result for a few slices located above the eyes

5. CONCLUSION

A new framework has been proposed in order to include topological regularization in a segmentation method dedicated to 3D medical images. This approach allows robust detection of convoluted interfaces, robust skeletonization and discrimination of tissues showing similar intensities, which has been illustrated by two different applications with 3D MR images. Further work intends to apply this framework to the segmentation of various nuclei embedded inside the brain. In order to overcome complex interferences with the lowest part of the head, this work will rely on a prior robust segmentation of the brain using 3D mathematical morphology. The toolbox of initialization dedicated potentials will be enriched in order to use other kinds of global knowledge (object location in a reference frame, distances between object centers of gravity, finer shape information...). The challenge will be to make a higher level system drive the initialization process, in order to make backtrack possible. Then problems induced by potential presence of local pathologies could be addressed.

References

[1] R. Bajcsy and S. Kovacic. "Multiresolution elastic matching", *Comput. Vision, Graph. Image Processing*, vol. 46, pp. 1-21, 1989.

[2] J. Besag. "Spatial interaction and statistical analysis of lattice systems", *A. Royal Stat. Soc. Serie B*, vol. 36, pp. 721–741, 1976.

[3] I. Cohen, L. Cohen, and N. Ayache. "Using deformable surfaces to segment 3-D images and infer differential structures", *CVGIP: Image Understanding*, vol. 56(2), pp. 242–263, 1992.

[4] S. Geman and D. Geman. "Stochastic relaxation, Gibbs distributions and the bayesian restoration of images", *IEEE Trans. Pattern Anal. Machine Intell.*, vol. 6(6), pp. 721–741, 1984.

[5] G. Malandain, G. Bertrand, and N. Ayache. "Topological segmentation of discrete surfaces", *International Journal of Computer Vision*, vol. 10(2), pp. 158–183, 1993.

[6] J.-F. Mangin, V. Frouin, I. Bloch, J. Régis, and J. Lopez-Krahe. "Automatic construction of an attributed relational graph representing the cortex topography using homotopic transformations", in *SPIE Mathematical Methods in Medical Imaging III*, vol. 2299, pp. 110–121, San Diego, 1994.

[7] J.-F. Mangin, J. Régis, I. Bloch, V. Frouin, Y. Samson, and J. Lopez-Krahe. "A MRF based random graph modelling the human cortical topography" in *First Computer Vision, Virtual Reality and Robotics in Medicine*, Nice, France, 1995.

[8] M. Sigelle and R. Ronfard. "Relaxation of previously classified images by a Markov field technique and its relationship with statistical physics" in *7th SCIA*, vol. 2, pp. 387–394, Aalborg, Denmark, 1991.

[9] A. N. Tikhonov and V. Y. Arsenin. *Solution of ill-posed problems.* Winston, New York, 1977.

DETECTION AND CHARACTERIZATION OF UNSHARP BLOBS BY CURVE EVOLUTION

Guido Gerig, Gábor Székely, Georg Israel and Martin Berger
Communication Technology Laboratory, Image Science,
ETH-Zentrum, CH-8092 Zurich, Switzerland
e-mail: gerig@vision.ee.ethz.ch

Abstract This paper presents a novel method for the multi-scale segmentation and shape characterization of unsharp, blobby structures in gray-valued image data. The iso-level curves of unsharp blobs depict a set of concentric curves, representing the shape by the geometric properties of the level curves and by the radial intensity function. Applying *Euclidean shortening flow* progressively smoothes the level-curves and lets them converge to circles before they disappear at singularity points. We interleave object detection with shape computation by analyzing the continuous extremum paths created by singularities in scale-space. Assuming radially symmetric image structures, such singularity traces attributed with evolution time and gray level fully represent the shape information and allow reconstruction and quantitative evaluation. More complex geometry of simple, closed curves can be considered by analyzing the set of level curves associated with each trace.

The project is driven by the medical problem of detecting white matter lesions in medical images. Here, a complete characterization of the gray-level structure represents a crucial descriptive feature for monitoring the progress of disease. The 2-D scheme is finally extended to the evolution of level surfaces in volume data, addressing the remaining open issues.

1 Introduction

The segmentation and shape description of unsharp structures still represents a challenging issue for image analysis. Most of the commonly applied segmentation algorithms identify objects by detecting regions either determined by homogeneity constraints or by clearly defined contours, favoring region growing or edge-detection/grouping. None of these assumptions hold for unsharp structures. Blobs are characterized by a smoothly changing radial intensity function and therefore do

Y. Bizais et al. (eds.), Information Processing in Medical Imaging, 165–176.

not reveal sharp contours or homogeneous brightness.

Pizer et al. [1] propose a concept called "object shape before boundary shape", with the intention to capture a rough object description at the appropriate scale before focussing back to the boundary. The method converts gray-level image data into a "medialness scale-space" and captures the object skeleton by tracing extremality features. This concept attempts shape characterization and segmentation as a one-step procedure. Lifshitz et al. [2] present multi-resolution image segmentation by characterizing paths of extremal regions in successively blurred versions of the initial image. They convert the image into a tree data structure expressing the motion and merging of extremal paths, with the aim to develop an interactive segmentation/visualization tool by manipulating the graph structure.

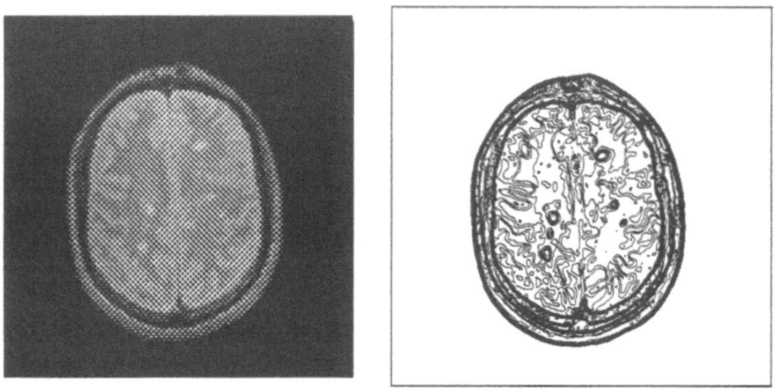

Figure 1: Gaussian smoothed version of an MRI slice through the brain of an MS patient (left), and the level curves of the image (right).

The analysis of level curves of the image intensity function offers a possility to identify and characterize blobby structures. Figure 1 shows a (Gaussian smoothed) MRI slice through the brain of a multiple sclerosis (MS) patient and the level curves of this image. The blobs corresponding to lesions can easily be identified as collections of *concentric, simple closed curves.* This observation leads to a characterization of unsharp blobs by an *onion-skin model.* The segmentation can be formulated as a search for concentric curves organized as the layers of the onion-skin. Shapes would be characterized by the common centers and the radial intensity function.

The theory of *curve evolution* provides a nice framework for the development of a simple, generic algorithm. Instead of analyzing the structure of the level sets of the *static* image, we *let all the level curves evolve simultaneously* according to geometric heat flow [3]. Our new algorithm performs annihilation of structures and simultaneously a detection of singularities. Similar to [2, 4] a structure segmentation and description will be derived from the symbolic representation of singularity paths.

2 Mean curvature flow

Theory of curve evolution The theory of curve evolution [5, 6] investigates the evolution of a curve $\vec{C}_0(p)$ according to the following equation

$$\frac{\partial \vec{C}(p,t)}{\partial t} = \alpha \vec{\mathcal{T}} + \beta \vec{\mathcal{N}}, \qquad \vec{C}(p,0) = \vec{C}_0(p) \tag{1}$$

where $\vec{\mathcal{N}}$ is the (inward) unit normal and $\vec{\mathcal{T}}$ the unit tangent of the curve $\vec{C}$. It can be shown that the tangential component of the evolution only effects the parameterization of the evolving curve. So, if the normal component does not depend on the actual parameterization, the evolution can also be written in the simplified form

$$\frac{\partial \vec{C}(p,t)}{\partial t} = \beta \vec{\mathcal{N}}, \qquad \vec{C}(p,0) = \vec{C}_0(p) \tag{2}$$

By selecting β as the Euclidean curvature of the curve κ we get the equation

$$\frac{\partial \vec{C}(p,t)}{\partial t} = \kappa \vec{\mathcal{N}} \tag{3}$$

which is often called mean curvature flow or Euclidean shortening flow. It has been shown [7] that, evolving under the condition of the above equation, any convex curve deforms into a circle and shrinks to a point as $t \to t^*$. Further, a planar embedded smooth, non-convex curve converges smoothly to a convex one and subsequently evolves according to the rules for convex curves [8].

Evolution of image surfaces under the conditions of the general diffusion equation is a well investigated topic in the recent image analysis research [9, 10]. A special form of the nonlinear diffusion equation

$$\frac{\partial \mathcal{L}}{\partial t} = \|\nabla \mathcal{L}\| \, div \left(\frac{\nabla \mathcal{L}}{\|\nabla \mathcal{L}\|} \right) \tag{4}$$

plays a central role in the analysis of level curve motion [9, 3]. It can be shown that if the image function $\mathcal{L}$ evolves according to the differential equation 4 its level curves $\vec{C}_l$ of the image undergo mean curvature flow as formulated in equation 3. This result provides an elegant way to perform mean curvature flow calculation on the level set of an image function by just applying nonlinear diffusion.

3-D extension: Evolution of level surfaces The basic ideas of the previous section can be in principle applied to the evolution of the level surfaces of 3D volumetric images, expecting that the proposed procedure can be straightforwardly extended to the detection of unsharp 3D blob structures. It has been shown [11] that any *convex surface* successively deforms to a spherical shape and contracts smoothly to a point by evolving in the normal direction with speed equal to the mean curvature. However, the 2D results regarding the behavior of non-convex level curves do not generalize to 3D. Under mean curvature flow, non-convex surfaces can develop singularities during evolution or even change their topology [12].

3 Blob characterization by extremal paths

Local minima or maxima of the image intensity function generated by the nonlinear evolution process correspond to level curves which are contracted to a single point, whereas the attributed intensities define the gray value of the corresponding level curves. The relationship between evolution time associated with singularities and a rough estimate of the original size of the level curves will be presented in the following section. We will further demonstrate how the radial intensity function is coded in the singularity path.

3.1 Circular object models

Considering the special case of circles the partial differential equation 3 simplifies to an ordinary differential equation, describing the radial change of the circle

$$\frac{dr(t)}{dt} = -\kappa(t) \tag{5}$$

where $r(t)$ is the actual radius of the circle, and $\kappa = \frac{1}{r}$ is the constant curvature. This results in $\frac{dr(t)}{dt} = -\frac{1}{r(t)}$, $r(0) = r_0$ where r_0 is the original radius of the circle. The solution of this initial value problem is

$$r(t) = \sqrt{r_0^2 - 2t} \tag{6}$$

which means that the singular condition $r(t^*) = 0$ is reached at $t^* = r_0^2/2$. The original radius can be calculated back from t^* by $r_0 = \sqrt{2t^*}$.

3.2 Unsharp circular blobs

The following discussion is based on the simplified assumption that the objects would have a circular shape. An unsharp, circular blob can be regarded as a set of circular level curves with a common center $\vec{x}_c$. Applying mean curvature motion causes each level curve with radius r_{0i} to shrink until it vanishes at scale $t_i^* = \frac{r_{0i}^2}{2}$. The set of level curves is concentric, which results in the common center $\vec{x}_c$ for all level curves.

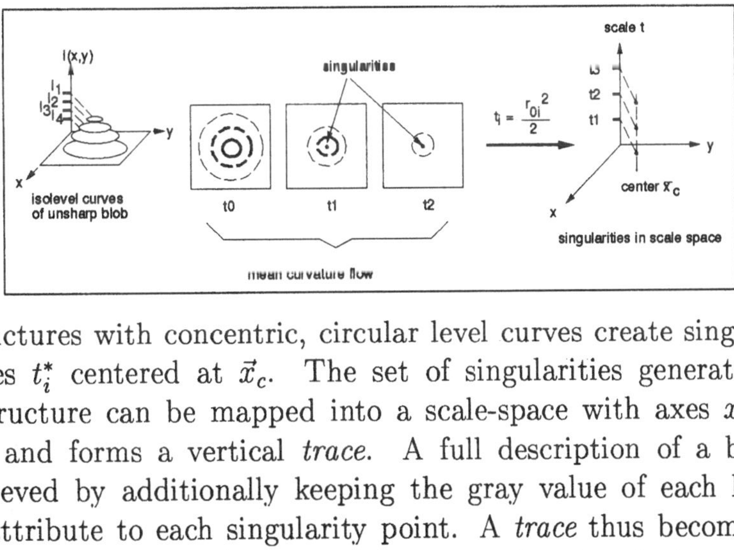

Structures with concentric, circular level curves create singularities at scales t_i^* centered at $\vec{x}_c$. The set of singularities generated by a blob structure can be mapped into a scale-space with axes x, y and scale t and forms a vertical *trace*. A full description of a blob can be achieved by additionally keeping the gray value of each level set as an attribute to each singularity point. A *trace* thus becomes a set $\{(\vec{x_i}, t_i^*, I_i),\ i = 1, \cdots, n\}$. We observe that a) a singularity at $(\vec{x_i}, t_i^*)$ completely defines the original circular curve $C(r_{0i})$, b) a singularity at $(\vec{x_i^*}, t_i)$ and the corresponding gray level I_i completely define the circular isodensity curve $C(r_{0i}, I_i)$, and c), a set of singularities, called a *trace* in scale-space, with gray level attributes $\{(\vec{x_i}, t_i^*, I_i),\ i = 1, \cdots, n\}$ completely defines the stack of original concentric circular isodensity curves and hence the *shape of the circular blob*. Therefore it is possible to reconstruct a circular blob from its trace in scale space (see illustration below).

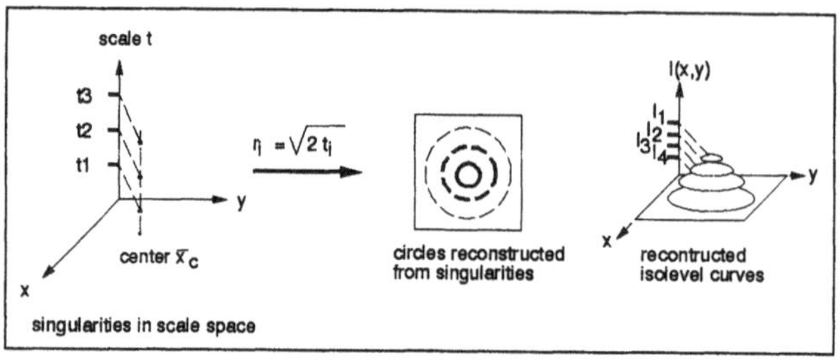

3.3 Representation of blobs with different radial intensity functions

The previous discussion shows that the set of isolevel contours and the corresponding gray levels can be treated separately. Blobs of a certain maximum size but with different gray level profiles will generate exactly the same *trace* in scale space, since the *trace* only describes the location of the singularities. The actual radial shape of blobs, e.g. its contrast, polarity or even a non-convex shape is described by the gray-value attributes of each *trace* (Figure 2). The three different circular blobs (**c**,**d**,**e**) all generate the same trace in scale space (**b**) since their range of radii is equivalent. The gray values of the level curves corresponding to each singularity, however, differ significantly. The difference between the high contrast and low contrast Gaussian blob is clearly visible by the different intensity ranges of plots (**f**) and (**g**). The plots display the parameter t as the vertical and the corresponding radial intensities I as the horizontal axis. The blob with a dimple (**e**) shows increasing intensities for small and decreasing level curves for large level curves (**h**).

3.4 Implementation of the diffusion equation

The anisotropic diffusion equation 4 can be rewritten into cartesian coordinates

$$\mathcal{L}_t = \frac{\mathcal{L}_x^2 \mathcal{L}_{yy} - 2\mathcal{L}_x \mathcal{L}_y \mathcal{L}_{xy} + \mathcal{L}_y^2 \mathcal{L}_{xx}}{\mathcal{L}_x^2 + \mathcal{L}_y^2} \tag{7}$$

where the indices t, x, y denote partial differentiation according to the independent variables. The equation shows that discontinuity and consequently instability problems are to be expected at places where $||\nabla \mathcal{L}||$ is small.

In our implementation we set a lower limit to the $\mathcal{L}_x^2, \mathcal{L}_y^2$ terms of the above differential equation, achieving similar effects as proposed by

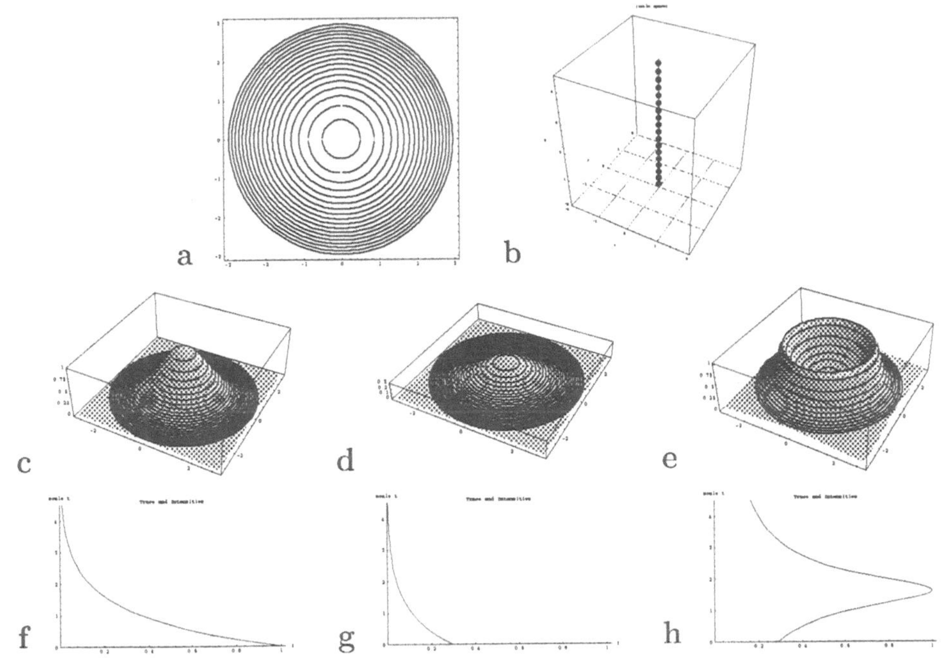

Figure 2: Singularity representation of blobs of different shape in scale space. The three shapes (**c,d,e**) generate the same set of contours, as plotted in (**a**), and the same set of singularities in scale space (**b**), since the radius range amounts to $\{0, \cdots, 3\}$. The gray values of the level curves that create singularities are different. Plots (**f,g,h**) display the intensity functions attributed to the traces for the corresponding figures in the upper row.

the theory of weak solutions based on the viscosity solution theory [13]. This stabilizing viscosity effect is governed by the parameter ε, leading to the equation

$$\mathcal{L}_t = \frac{(\mathcal{L}_x^2 + \varepsilon^2)\mathcal{L}_{yy} - 2\mathcal{L}_x\mathcal{L}_y\mathcal{L}_{xy} + (\mathcal{L}_y^2 + \varepsilon^2)\mathcal{L}_{xx}}{(\mathcal{L}_x^2) + (\mathcal{L}_y^2) + \varepsilon^2} \tag{8}$$

In case $||\nabla\mathcal{L}|| \to 0$ the partial differential operator converges to a scaled version of the Laplace operator $\mathcal{L}_{xx} + \mathcal{L}_{yy} = \Delta\mathcal{L}$. For the discretization of the equation 8 we used finite difference schemes (centered differences). The forward difference has been used for the approximation in the time domain, leading to $\mathcal{L}(\vec{x}, t + \Delta t) \approx \mathcal{L}(\vec{x}, t) + \Delta t \frac{\partial\mathcal{L}(\vec{x},t)}{\partial t}$. For the reasons of numerical stability Δt has been selected in the range $1/8 \leq \Delta t \leq 1/4$. We are currently working on the implicit implementation of equation 8 since it offers improved stability and the choice of larger integration steps Δt [9].

3.5 The blob detection procedure

The numerical integration scheme defined by equation 3.4 provides discrete samples of the evolving image $\mathcal{L}(x, y)$ at selected discrete times t_n, which are defined on the discrete image grid $\{x_i, y_j\}$. These $\mathcal{L}_n(i, j, n)$ are then searched for local extrema in the spatial domain, representing candidates for level curve singularities. Singularities are followed across scale-space to form continuous singularity paths which are stored as vector lists.

While the spatial and time variables of the problem are discretized by the numerical solution, the image intensity itself is theoretically not quantized and handled as a real variable. Searching for extrema of this "continuous" variable raises some numerical problems. Special problems are caused by completely flat regions which can become indistinguishable from very shallow extremal regions. Such flat regions grow out of saddle points of the original image function and do not correspond to point singularities of disappearing level curves.

A simple solution to these problems is offered by applying a filtering scheme to the complete set of singularity traces. The filter heuristics applied so far are measures which are directly related to *shape characteristics* (contrast, size, radial symmetry). They can be easily chosen based on a prior shape model. The **contrast** of a blob structure can be determined from the intensity attributes of each trace, e.g. $I_{max} - I_{min}$. A simple **size measurement** is generated from the largest radius of a structure and is calculated from the largest scale t of each trace (equation 6). The location of singularities remains stable for concentric level curves but moves considerably for arbitrary shapes. The standard deviation of the singularity location along each trace is calculated as a measure for separating structures with **radial symmetry** from highly asymmetric shapes.

Having obtained the symbolic attributed graph description of singularity traces one can still think of applying more complex filtering schemes. Lifshitz et al. [2], e.g., generate a complete graph structure by linking extremum paths. As a first step towards more general descriptions, traces are linked across flat regions to form super-structures (see Fig. 4c).

4 Experiments and Results

Synthetic 2D images Mean curvature flow and detection of singularities was applied to a test image presenting a collection of three Gaussian blobs (Figure 3). The iterative series demonstrates that the three blobs first shrink to form flat regions (**b**). The evolution creates three distinct traces of singularities in scale space. In the next iterations there is no singularity detected until the flat region gets a smooth peak (**c**). The structure as a whole then changes its shape to a circular blob and forms the 4th trace.

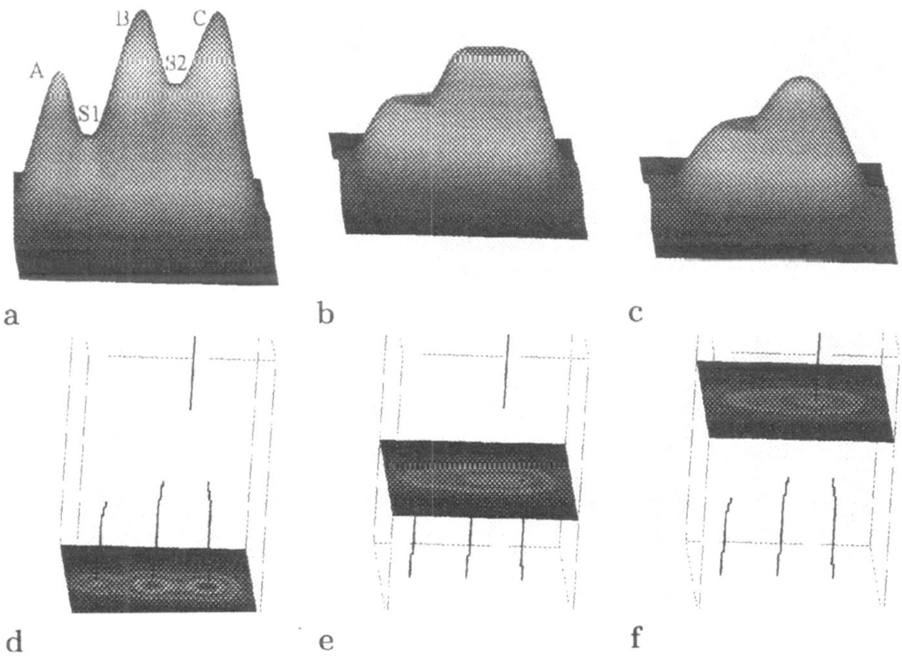

Figure 3: Test image with three Gaussian blobs (A,B and C) generating two saddle points (S1 and S2). The evolution time, or the scales, respectively, are 0, 57 and 100 (**a,b,c**). The second row (**d,e,f**) displays the corresponding traces formed by the singularities in scale space with overlay of the current image at evolution time. Note that the traces are not connected and form 4 isolated strings.

Application to 2D MR images The new blob detection procedure is driven by the problem of detecting and characterizing unsharp blob structures. A typical application can be found in detecting white matter lesions in magnetic resonance data. While MRI is the imaging technique of first choice in diseases of the white matter of the central nervous system due to its high sensitivity, the signal changes observed are not specific. Several inflammatory diseases of the CNS present similar patterns in MR intensity. Lesions appear in the MR images as blob-like structures that are usually brighter than the surrounding white matter tissue (see Figure 1). The inspection of MRI data by clinicians includes *detection, size measurement, assignment of anatomical location and shape characterization* . Guttman also [14] performed measurements of *intra-lesion heterogeneity* as opposed to the *inter-lesion diversity.* Other studies with Gadolinium-DTPA enhanced MRI showed ring- or crescent-shaped enhancement of the edge of a lesion.

Mean curvature flow and tracking of singularities was applied to an axial MR slice of a multiple sclerosis patient (Figure 4). The complete set of traces **a** illustrates the low specifity of blob characterization by singularity paths. Image **b** clearly demonstrates that the number of traces is significantly reduced after filtering with simple shape criteria. We used criteria for minimum contrast and size and for high radial symmetry. Image **c** illustrates the hierarchy created by linking single traces to larger entities.

5 Conclusions

This paper proposes a new framework for the segmentation and description of blob structures based on nonlinear filtering and detection of extremum paths. The procedure aims at combining the usually distinct processing steps of segmentation and shape description. The characterization of the shape of unsharp structures will be divided into a description of the radial intensity function and of the level curve geometry (not discussed in this paper), thus potentially offering simple descriptors for a large variety of structures. Figure 4 clearly demonstrates that a filtering of singularity traces even with simple heuristics results in a significant reduction of blob candidates. A more accurate assessment of the geometric shape, e.g. elliptic structures, would require feedback to the original level curves and is a topic for continuing research.

A simple extension to 3-D is not possible, since nonconvex surfaces do not simply converge to spherical structures as would be desirable. However, we plan to tackle this problem by developing a strategy

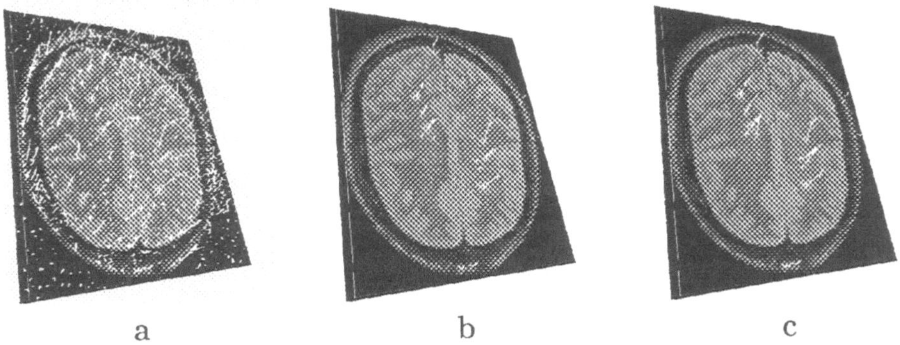

Figure 4: Application of blob detection to an axial MR slice of a patient presenting MS lesions. The complete set of singularities is displayed in **a**. Image **b** shows the result after filtering the traces (minimum contrast, minimum size, radial symmetry). Image **c** illustrates the structure to substructure relationship provided by the linked graph structure.

for containment relation by linking single singularity traces to graphs (Fig. 4c). In 3-D, complex shape would be first broken up into simple shapes and then again combined to super-structures at the symbolic description level.

The segmentation of white matter lesions in MS cases represents the driving force for future research. The segmentation and shape characterization of the unsharp lesions require volume description *without explicit edge finding*. The specifity of lesion classification has to be improved by the generation of morphometric measures (radial intensity and outline shape) and by the analysis of time evolution.

Acknowledgements The MRA data were provided Dr. F. Jolesz and Dr. R. Kikinis, Brigham and Women's Hospital, Boston, USA.

References

[1] S.M. Pizer, C.A. Burbeck, J.M. Coggins, D.S. Fritsch, and B.S. Morse. Object shape before boundary shape: scale-spoace medial axes. In *Shape in Picture*, NATO Advanced Research Workshop, 1994. to appear J. of Math. Imaging and Vision.

[2] Lawrence M. Lifshitz and Stephen M. Pizer. A Multiresolution Hierarchical Approach to Image Segmentation Based on Intensity Extrema. *IEEE Trans. Pattern Analysis and Machine Intelligence*, 12(6):529–540, June 1990.

[3] S. Osher and S. Sethian. Fronts propagating with curvature dependent speed: algorithms based on the hamilton-jacobi formalism. *J. Computational Physics*, 79:12–49, 1988.

[4] Tony Lindeberg. Scale-space behaviour of local extrema and blobs. *Journal of Mathematical Imaging and Vision*, 1(1):65–99, March 1992.

[5] Benjamin B. Kimia, Allen Tannenbaum, and Steven W. Zucker. On the evolution of curves via a function of curvature I: the classical case. *Journal of Mathematical Analysis and Applications*, 163:438–458, 1992.

[6] G. Sapiro, R. Kimmel, D. Shaked, B. B. Kimia, and A. M. Bruckstein. Implementing continuous-scale morphology via curve evolution. *Pattern Recognition*, 26(9), 1993.

[7] M Gage and R. S. Hamilton. The heat equation shrinking convex plane curves. *J. Differential Geometry*, 23:69–96, 1986.

[8] M. Grayson. The heat equation shrinks embedded plane curves to round points. *Journal of Differential geometry*, 26:285–314, 1987.

[9] Luis Alvarez, Pierre-Louis Lions, and Jean-Michel Morel. Image selective smoothing and edge detection by nonlinear diffusion. ii. *SIAM J. Num. Anal.*, 29(3):845–866, June 1992.

[10] Bart M. ter Haar Romeny (Ed.). *Geometry Driven Diffusion in Computer Vision.* Computational Imaging and Vision. Kluwer Academic Publishers, NL, 1994.

[11] G. Huisken. Flow by mean curvature of convex surfaces into spheres. *Journal of Differential Geometry*, 20:237–266, 1984.

[12] Peter.J. Olver, Guillermo Sapiro, and Allen Tannenbaum. Invariant Geometric Evolutions of Surfaces and Volumetric Smoothing. Technical Report LIDS-P-2238, Dept. of Mathematics, Minneapolis / MIT, 1994.

[13] L. C. Evans and J. Spruck. Motion of level sets by mean curvature I. *J. Differential Geometry*, 33:635–681, 1991.

[14] Charles R.G. Guttmann, Sungkee S. Ahn, Liangge Hsu, Ron Kikinis, and Ferenc A. Jolesz. The Evolution of Multiple Sclerosis Lesions on Serial MRI. submitted, September 1994.

TRACKING MYOCARDIAL DEFORMATION USING SPATIALLY-CONSTRAINED VELOCITIES

FRANCOIS G. MEYER, R. TODD CONSTABLE
ALBERT J. SINUSAS AND JAMES S. DUNCAN
Image Processing and Analysis Group, Dept. of Diagnostic Radiology, Yale University, New Haven, CT 06520, USA

Abstract. This work proposes a unified framework to track the deformation of the myocardium using velocity fields and boundary information. The deformation of the myocardium is characterized by a deforming mesh. A general framework for locally regularizing the velocity field has been developed. The tracking is modeled as an estimation problem which makes it possible to properly take into account uncertainties in the velocity and boundary measurements. The results of experiments conducted on phantom data and on *in vivo* data demonstrate the strength of the approach.

Keywords: non-rigid motion, phase contrast MRI, tracking, left ventricular motion

1. Introduction

The ability to analyze and quantize the internal deformation of the myocardium from a sequence of images is of fundamental importance for the diagnosis of heart disease. A deformed body can be described by its boundary, and recent efforts have been aimed at understanding the complex motion of contours and surfaces [5, 13]. However, as opposed to a rigid object, the knowledge of internal deformations is necessary to uniquely describe the state of a deformed material. A suitable measure of deformation is provided by the strain tensor, or the rate of deformation tensor [15]. With the availability of dense 3-D velocity maps, estimating the rate of deformation of the myocardium becomes possible. Recently the authors in [4, 12] have demonstrated that velocity maps in three orthogonal directions can be measured at multiple instants throughout the cardiac cycle. This new tech-

Y. Bizais et al. (eds.), Information Processing in Medical Imaging, 177–188.

nique, termed *phase contrast cine magnetic resonance imaging*, combines phase contrast imaging methods with cardiac cine techniques.

One limitation of the description provided by velocity maps is that it does not establish the correspondence between material points over time. An immediate understanding and analysis of the deformation of the same region of the myocardium is not available from this information alone. In principle the integration of the spatial velocity yields an estimate of the position of each material point, or particle at the next instant. In practice, serious numerical difficulties can arise. The estimate of the new position of a given particle at time $t + dt$ may not correspond to the true position at this instant. Therefore the instantaneous spatial velocity at this point will not coincide with the true velocity of the particle. If we iterate this scheme again, errors in particle positions will grow rapidly in time. Another related problem stems from the fact that any noise in the velocity will be transmitted to the position estimates. These errors may also accumulate over time. The authors in [4, 12] have encountered these difficulties when tracking small isolated regions of interest in the myocardium using phase contrast MRI velocities. They have tried to correct them by forward and reverse integration in time. Establishing trajectories of the same material points from a sequence of images is an intricate problem [1]. Except for feature points or special landmarks, there is no unique way to establish the correspondence of material particles between two successive frames. Methods exploiting geometric cues such as curvature have tackled the difficult issue of non rigid correspondence for curves [5] and surfaces [13]. Tissue tagging with MRI makes it possible to embed in the image a large number of material landmarks. These non invasive landmarks move along with the tissue as if they were implanted in the tissue. However, one severe limitation of this imaging technique is the decay of the tagging grid over the heart cycle. Establishing correspondence of these points between different images makes it possible to recover a sparse set of trajectories [7]. Another limitation is the limited transmural resolution of the tags. Most effort in the field of non-rigid motion analysis has been expended to understand the deformation of curves and surfaces [1]. However, we believe that such a description remains incomplete. We advocate therefore a continuum approach that uses velocity fields to describe the internal twisting and stretching of a deforming body.

The major contribution of this work is to explore and propose a unified framework that exploits velocity fields and boundary information to track non-rigid motion. The idea of exploiting velocity and shape information has been previously proposed in [10] in the context of the tracking of a rigid object, and we intend to merge this with the work described in [5] to estimate non-rigid motion. The approach makes it possible to track and analyze the temporal variation of stretching and twisting of any region inside the myocardium. This method should provide a powerful tool for the diagnosis of local abnormalities in the thickening of the myocardium. We emphasize the case where velocity measurements are obtained by phase-contrast MRI, although we note that velocities can be directly estimated from the spatiotemporal changes of the intensity function [14]. The tracking is modeled as an estimation problem which makes it possible to properly take into account uncertainties in the velocity and boundary measurements. Stochastic optimal linear estimation techniques yield a solution of the estimation problem. Experiments have been conducted with real data. Encouraging results demonstrate the strength of our approach.

The paper is organized as follows. In the next section we explain how the physical assumption of local homogeneity of the rate of deformation yields an affine model of the velocity field. It is finally shown that the continuous deformation of the myocardium can be described by the positions and velocities of the vertices of a deforming mesh. The measurement algorithms that give the position and the velocity of each node of the mesh are given in Section 3. The dynamic model that describes the evolution of the deforming mesh, and the associated Kalman filter are presented in Section 4. Results of experiments conducted on real data are given in Section 5.

2. Representation of the myocardium

We represent each slice of myocardium that is imaged as a continuous medium. To construct a satisfactory description we subdivide the domain into elementary cells. We use triangles in 2-D and tetrahedra in 3-D. The deformation in each cell will characterize the continuum deformation of the body in this neighborhood. We characterize now the velocity field inside each cell.

2.1. DEFORMATION AND VELOCITY MODEL

Before describing the work any further we need to introduce a few concepts from continuum mechanics [15] . Let us consider a deformable body $\mathcal{B}$ in motion ($\mathcal{B}$ is an open set of $\mathbb{R}^3$). A motion of the body is defined as a mapping from $\mathcal{B}$ to $\mathbb{R}^3$

$$\begin{array}{rcl} \mathcal{B} \times \mathbb{R}+ & \longrightarrow & \mathbb{R}^3 \\ (\mathbf{X}, t) & \longrightarrow & \mathbf{x}(\mathbf{X}, T) \end{array}$$

Informally we can think of $\mathbf{X}$ as being the label of a given particle, and $\mathbf{x}$ the position of this particle at time t. We can define the velocity of the particle $\mathbf{X}$ at time t as

$$\mathbf{v}(\mathbf{X},t) = \frac{\partial \mathbf{x}}{\partial t}(\mathbf{X},t)$$

In the continuum mechanics literature, $\mathbf{v}(\mathbf{X},t)$ is called the material, or Lagrangian, velocity. Further we define $\mathbf{u}(\mathbf{x},t) = (u(x,y,t), v(x,y,t))$ the Eulerian, or spatial velocity, as being the velocity of the particle that occupies the position $\mathbf{x}$ at time t. We have $\mathbf{u}(\mathbf{x}(\mathbf{X},t),t) = \mathbf{v}(\mathbf{X},t)$.

The deformation of a body depends on the relative motion of neighboring particles, and can be characterized in a number of ways [15]. For our purposes, we will describe the relationship between velocities at neighboring points with the rate of deformation tensor (or rate of strain tensor)

$$\mathbf{D} = \begin{bmatrix} u_x & \frac{1}{2}\{u_y + v_x\} \\ \frac{1}{2}\{u_y + v_x\} & v_y \end{bmatrix}$$

$\mathbf{D}$ corresponds to the symmetric part of the velocity gradient tensor $\mathbf{L}$. The antisymmetric part is called the vorticity tensor $\mathbf{W}$. In our analysis of the deformable motion we assume that the rate of deformation tensor and the vorticity tensor are locally homogeneous. However we note that $\mathbf{D}$ and $\mathbf{W}$ will change from one neighborhood of the myocardium to another. Similar assumptions in the biomechanical literature have assumed that the deformation gradient tensor in the myocardium was locally homogeneous [8]. We can perform a local analysis of the Eulerian velocity field. Let $\mathbf{x}$ be a point in the deformed body. For any point $\mathbf{x}+\delta\mathbf{x}$ in a small neighborhood around $\mathbf{x}$ we have

$$\mathbf{u}(\mathbf{x}+\delta\mathbf{x},t) = \mathbf{u}(\mathbf{x},t) + \mathbf{L}\delta\mathbf{x} \tag{1}$$

The physical assumption of local homogeneity of rate of deformation results in the locally linear (or affine) model (1). This model appears to be a good tradeoff between the exhaustive description of the entire set of noisy observed velocities, and a crude model of the velocity, which would assume a constant motion inside each cell. Furthermore this model allows us to estimate the rate of deformation, an essential parameter in the analysis of the myocardial deformation.

Let Δ be the mesh associated with the triangulation of the myocardium. Let N be the number of nodes of the partition. We consider now the space P^0_Δ of continuous piecewise polynomial functions in two arguments and of degree one. A local basis of P^0_Δ is composed of the collection $(\xi_k)_{k=1\ldots N}$, where ξ_k is the linear function which takes the value 1 at the vertex k and vanishes on all other nodes (see Figure 1a). The basis $(\xi_k)_{k=1\ldots N}$ is not orthogonal but the product $\int_\Delta \xi_k\xi_j$ is equal to zero each time that the nodes k and j are not in the same triangle. We can now express our model of the entire velocity field inside the myocardium. Inside each triangle the velocity is described by the linear model (1). Across the boundaries the velocity is continuous. Each component of the velocity is thus a function of P^0_Δ, and it can be expanded in the local basis $(\xi_k)_{k=1\ldots N}$. We have

$$(u,v)(t) = \left(\sum_{k=1}^{N} u_k\xi_k, \sum_{k=1}^{N} v_k\xi_k\right)(t) \tag{2}$$

The advantage of this local basis is that the entire velocity field is completely described by the value of the velocity at each vertex of the triangulation. This important result sets up a one-to-one correspondence between the dense motion of the material points inside the medium, and the motion of the nodes of the mesh. We note that we are not working with a fixed grid, but rather with a deforming mesh that follows the myocardium. Therefore the position of the support of the function ξ_k will change at each instant. At any given instant t the state of the myocardium is uniquely described by the position $(x_i, y_i)(t)$, and the velocity $(u_i, v_i)(t)$ of each vortex of the partition. Our aim is to generate the best estimates of the sequence of $(x_i, y_i)(t)$ and $(u_i, v_i)(t)$, $i = 1, \ldots, N$ over the heart cycle, given observations of the velocity and of the position.

3. Position and velocity measurements

3.1. VELOCITY MEASUREMENTS

Velocity fields obtained from optical-flow method tend to be noisy. Velocity fields obtained from phase-contrast measurements are visually smoother but yield significant outliers on regions near the boundary of the myocardium, as shown in figure 1b. A general framework for locally regularizing the velocity field has been developed. The global vector field is approximated locally by continuous linear piecewise vector fields. Furthermore the approximation will be continuous across the common boundary of two cells. This model yields a local description of the velocities with a small number of crucial parameters.

Given an observation of the velocity field $\mathbf{u} = (u, v)$, our goal is to estimate the best approximation, $\mathbf{p} = (p, q)$ of $\mathbf{u}$ in $P^0_\Delta \times P^0_\Delta$ that minimizes the error

$$< \mathbf{p} - \mathbf{u}, \mathbf{p} - \mathbf{u} >= \int_\Delta (\mathbf{p} - \mathbf{u})^2 = \int_\Delta (p-u)^2 + \int_\Delta (q-v)^2 \tag{3}$$

The two terms $\int_\Delta (p-u)^2$ and $\int_\Delta (q-v)^2$ are independent and can be minimized independently. Therefore we need only to solve the problem

$$\min_{p \in P^0_\Delta} \int_\Delta (p-u)^2 \tag{4}$$

where u is a one scalar function. Let $\boldsymbol{\alpha}$ be the coordinates of the expansion of p in the local basis $(\xi_k)_{k=1...N}$. The problem (4) can be expressed as

$$\min_{\alpha_i \in \mathbb{R}^N} \int_\Delta \left(\sum_{i=1}^{N} \alpha_i \xi_i - u \right)^2 \tag{5}$$

(5) is a classic least-squares problem. Taking the partial derivative with respect to α_i, the solution $\boldsymbol{\alpha}$ of (5) is given by the normal equations

$$\mathbf{A}\boldsymbol{\alpha} = \mathbf{r} \tag{6}$$

where $A_{ij} = < \xi_i, \xi_j >, i, j = 1, \ldots, N$, and $r_i = < \xi_i, u > \; i = 1, \ldots, N$. As mentioned before, (ξ_i) is a local basis, and most of the inner products A_{ij} vanish to zero. There exists a very effective way to compute the matrix $\mathbf{A}$ and the vector $\mathbf{r}$ [2]. Again we note that the approximation method naturally extends to the 3-D case, although here, instead of solving two independent approximation problems, three problems need to be solved. Figure 1b shows the phase contrast velocity field in a small region around the myocardium. We note the turbulent flow in the blood pool, and the noisy vectors outside in the pericardial space. Figure 1c shows the result of the approximation algorithm. While keeping the same pattern and structure, the flow field is smoother. Noisy vectors have been discarded. This smooth velocity field can now serve as the input of our tracking algorithm.

The quality of the triangulation has a major influence on the accuracy of the approximation. On the one hand, triangles should be small to yield an accurate approximation of the velocity, and an accurate approximation of the contour. On the other hand, the number of points inside each triangle needs to be sufficient to estimate the parameters of the local affine model. We did not investigate here the problem of finding the best triangulation with respect to these two conflicting goals. We have developed an automatic 2-D triangulation scheme that exploits the shape particular to the left ventricle [11].

3.2. POSITION MEASUREMENTS

At each instant the contours of the epicardium and of the endocardium have been obtained using the deformable boundary/region growing approach presented in [3]. We update the position of the vertices of the triangulation that correspond to points on the boundaries of the myocardium. Our analysis exploits the prediction of the mesh position generated by the filter. Provided the dynamic model of the myocardial deformation is accurate, the position of a given material point at the next instant should lie in a small neighborhood around the predicted position. We project the prediction, $(x, y)(t + \delta t|t)$ onto the boundary $\partial\mathcal{B}(t + \delta t)$. The projection is defined

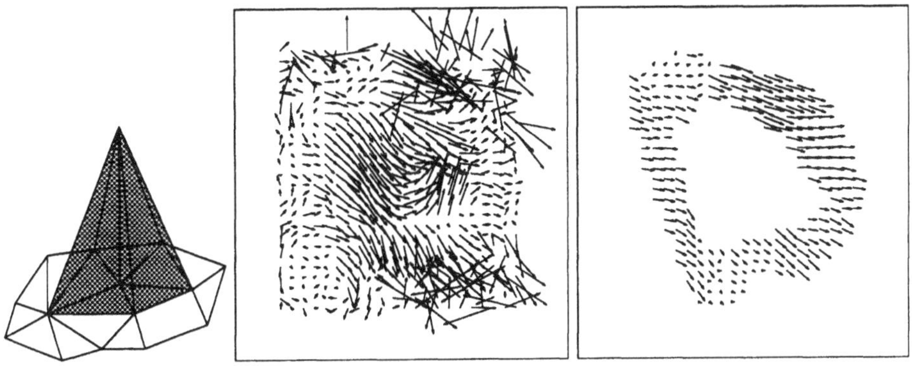

Figure 1. (a) The hat function ξ_k. (b) phase contrast MRI velocity field around the myocardium. (c) Continuous piecewise linear approximation of the velocity field shown in (b).

as the closest point from $(x,y)(t+\delta t|t)$ on $\partial\mathcal{B}(t+\delta t)$. As shown in the experiments, our algorithm does a very good job at predicting the position of the mesh at the next instant.

4. Tracking the partition

We assume that the vertices of the triangulation can be tracked independently. The assumption is realistic since we are tracking a deformable object, and we have independent position and velocity measurements for each node of the mesh. In order to obtain more tractable equations we simplify the Kalman filter and decouple the components x and y. We give in the following the equations for one of the components (x for instance).

We model the derivative of the Lagrangian velocity of each vertex as a sequence of zero-mean Gaussian noise. Let $\mathbf{x} = [x, \dot{x}]^T$ be the state vector that describes the x component of a given vertex of the mesh. We consider a discrete formulation of the problem where the discrete-time index k corresponds to $t + k\delta t$. The state vector evolves according to the following linear dynamic system

$$\mathbf{x}(k+1) = \mathbf{\Phi}\mathbf{x}(k) + \boldsymbol{\zeta}(k) \qquad \text{with} \qquad \mathbf{\Phi} = \begin{bmatrix} 1 & \delta t \\ 0 & 1 \end{bmatrix} \tag{7}$$

where $\boldsymbol{\zeta}(k)$ is a two dimensional, zero mean Gaussian noise vector of covariance matrix $\mathbf{Q}(k)$. The above model provides only an approximation of the trajectory of the vertex over a short time interval. However, the noise $\boldsymbol{\zeta}(k)$ takes into account errors in the model and makes it possible to accommodate significant changes in velocity, as demonstrated in the experiments. From the position and the velocity measurements we can construct the measurement vector $\mathbf{z}(k)$. The measurement system takes the form

$$\mathbf{z}(k) = \mathbf{x}(k) + \boldsymbol{\eta}(k) \tag{8}$$

where $\boldsymbol{\eta}$ is a two dimensional, zero mean Gaussian noise vector of covariance matrix $\mathbf{R}(k)$. Given the observations : $\mathbf{z}(1), \mathbf{z}(2), \ldots, \mathbf{z}(k)$, we wish to obtain an estimate of the state vector $\mathbf{x}$ at time k, $\mathbf{x}(k|k)$, which minimizes the mean-norm-squared-error. The optimal filtered estimate $\hat{\mathbf{x}}(k+1|k+1)$ can be efficiently computed with a recursive algorithm, the Kalman-Bucy filter [6]. We have

$$\hat{\mathbf{x}}(k+1|k+1) = \hat{\mathbf{x}}(k+1|k) + \mathbf{K}(k+1)\left[\mathbf{z}(k+1) - \hat{\mathbf{x}}(k+1|k)\right] \tag{9}$$

$$\hat{\mathbf{x}}(k+1|k) = \mathbf{\Phi}\mathbf{x}(k|k) \tag{10}$$

The gain matrix $\mathbf{K}(k+1)$ is given by

$$\mathbf{K}(k+1) = \mathbf{P}(k+1|k)^t\left[\mathbf{P}(k+1|k)^t + \mathbf{R}(k+1)\right]^{-1} \tag{11}$$

$\mathbf{P}(k+1|k)$ is the covariance matrix of $\hat{\mathbf{x}}(k+1|k)$. $\mathbf{P}(k+1|k+1)$ is the covariance matrix of $\hat{\mathbf{x}}(k+1|k+1)$. We have

$$\mathbf{P}(k+1|k) = \mathbf{\Phi}\mathbf{P}(k|k)\mathbf{\Phi}^t + \mathbf{Q}(k) \tag{12}$$

$$\mathbf{P}(k+1|k+1) = \left[\mathbf{I} - \mathbf{K}(k+1)\right]\mathbf{P}(k+1|k) \tag{13}$$

where $\mathbf{I}$ is the 2×2 identity matrix. Equation (9) consists in updating the prediction $\hat{\mathbf{x}}(k+1|k)$ with the new measurement $\mathbf{z}(k+1)$. The gain matrix $\mathbf{K}(k+1)$ controls the relative importance of the measurement with respect to the prediction. We note in (11) that if $\mathbf{R}(k+1)$, the covariance of the measurement, has very large components with respect to $\mathbf{P}(k+1|k)$, the covariance of the prediction, then $\mathbf{K}(k+1)$ is very small. Then $\hat{\mathbf{x}}(k+1|k+1)$ given by (9) essentially ignores the new measurement $\mathbf{z}(k+1)$, and is almost equal to the prediction. On the other hand if $\mathbf{R}(k+1)$ has very small components with respect to $\mathbf{P}(k+1|k)$, then $\mathbf{K}(k+1)$ is almost equal to $\mathbf{I}$, and $\hat{\mathbf{x}}(k+1|k+1)$ is almost equal to $\mathbf{z}(k+1)$. As explained by Gelb [6], the gain matrix is "proportional" to the uncertainty in the estimate, and "inversely proportional" to the measurements noise.

5. Experiments

We present here two experiments conducted with phase contrast MRI that best illustrate the approach. The protocol for the acquisition is fully described in [4]. In order to judge the accuracy of the trajectories obtained with our method we first present an experiment conducted with a phantom.

5.1. PHANTOM DATA

The phantom consisted of a gel-filled disk undergoing rotation and translation in the plane. The disk had four small markers on its outer edge which allowed to recover the rigid motion of the disk. The phantom was imaged in a quadrature head coil using cine PC gradient echo images sequence.

We have superimposed the result of the triangulation on a magnitude image in figure 2a. The trajectories of the marker were manually hand-traced. The trajectories of the nodes, and the trajectories of the markers are shown in figure 2b. The contours of the phantom at time 4 and at time 12 are also shown. The estimated trajectories are in close agreement with the trajectories of the markers. Our approach does not model the motion of the mesh as a periodic function. However, as shown in figure 2b almost all the trajectories are closed. This good result could clearly be improved by imposing a periodicity constraint as proposed in [9]. Figure 3a shows the piecewise linear approximation of the 2-D velocity field at time 9. We note that the approximation is smooth and accurately captures the complex pattern of the complete flow.

The quality of the prediction generated by the Kalman filter is illustrated in figure 3b. We have displayed the predicted positions of the nodes at time 10 and the real contours at time 9 and 10. For the clarity of the figure we have only displayed the inner contour. We note the accuracy of the prediction. This experiment validates the tracking algorithm in the case of complex rigid motion.

5.2. *IN VIVO* DATA

To further validate our approach with non-rigid motion the tracking algorithm has been tested with *in vivo* data. A dog was positioned in the quadrature head coil for MR imaging. The tracking started at time 13. Figure 5 shows the deformed mesh at time 1 (end-diastole), 5, 9 (end-systole), and 13. The piecewise linear velocity field is shown in figure 6 at the corresponding instants. Figure 4 shows the trajectory of the nodes attached to the endocardium. To help visualize the non-rigid motion of the endocardium, contours have been plotted at time 2, 6, 10 and 14. We first note that the tracking is stable and not perturbed by the flow inside the LV. Also, as mentioned previously the periodicity of the motion is not a priori modeled in the estimation problem. This explains why a few trajectories are not closed. However, given the significant gross motion of the heart (5 pixels from end-diastole to end-systole), the estimated trajectories are reasonable. These encouraging results should clearly be improved when the analysis will be performed in 3-D.

6. Conclusion

We have proposed a new unified framework to track the non-rigid motion of the myocardium using velocity maps and boundary information. The deformation of the myocardium is characterized by a deforming mesh. A general framework for locally regularizing the velocity field has been developed. The tracking is modeled as an estimation problem which makes it possible to properly take into account uncertainties in the velocity and boundary measurements. Results of experiments conducted on phantom data and on *in vivo* data demonstrate the strength of the approach. We are currently working on the 3-D extension of this work.

Acknowledgments

This work is supported in part by NIH Grant R01HL44803 from the National Heart, Lung and Blood Institute and by United States Air Force Grant F49620-93-1-0575. The authors gratefully acknowledge helpful discussions with Amit Chakraborty and Pawel Skudlarski.

References

1. J.K. Aggarwal, Q. Cai, W. Liao, and B. Sabata. Articulated and elastic non-rigid motion: A review. In *Proc. IEEE Workshop on Motion of Non-Rigid and Articulated Objects, Austin, Texas, November 11-12*, pages 2–14, 1994.
2. E.B. Becker, G.F. Carey, and J.T. Oden. *Finite Elements, An Introduction, Volume I.* Prentice-Hall, 1981.
3. A. Chakraborty, L.H. Staib, and J.S. Duncan. Deformable boundary finding influenced by region homogeneity. In *Proc. of CVPR'94, Seattle, Washington*, pages 624–627, 1994.
4. R.T. Constable, K.M. Rath, A.J. Sinusas, and J.C. Gore. Development and evaluation of tracking algorithms for cardiac wall motion analysis using phase velocity MR imaging. *Magnetic Resonance in Medecine*, Vol. 32:33–42, 1994.
5. J.S. Duncan, R.L. Owen, L.H. Staib, and P. Anandan. Measurement of non-rigid motion using contour shape descriptors. In *Proc. Conference Computer Vision and Pattern Recognition, Hawaii*, pages 318–324, June 1991.
6. Arthur Gelb. *Applied Optimal Estimation.* MIT Press, 1974.
7. M.A. Guttman, J.L. Prince, and E.L. McVeigh. Tag and contour detection in tagged MR images of the left ventricule. *IEEE Trans. on Medical Imaging*, Vol. 13, No.1:74–88, March 1994.
8. A.D. McCulloch and J.H. Omens. Non-homogeneous analysis of three-dimensional transmural finite deformation in canine ventricular myocardium. *J. Biomechanics*, Vol. 24, No 7:pp 539–548, 1991.
9. J.C. McEachen, A. Nehorai, and J. Duncan. A recursive filter for temporal analysis of cardiac motion. In *Proc. IEEE Workshop on Biomedical Image Analysis, Seattle, Washington*, pages 124–133, 1994.
10. F.G. Meyer and P. Bouthemy. Region based tracking using affine motion models in long image sequences. *CVGIP: Image Understanding*, Vol. 60 No. 2:119–140, Sept. 1994.
11. F.G. Meyer, R. T. Constable, A. J. Sinusas, and J. S. Duncan. Tracking myocardial deformation using spatially-constrained velocities. Submitted to *IEEE on Medical*

Imaging, also available as Technical Report 951-1, 1995.

12. N.J. Pelc, R.J. Herfkens, and L.R. Pelc. 3-D analysis of myocardial motion and deformation with phase contrast cine MRI. In *Proc. of SMRM, 11th Annual Meeting, Berlin*, page 18, 1992.
13. P. Shi, A. Amini, G. Robinson, A. Sinusas, R.T. Constable, and J. Duncan. Shape-based 4-D left ventricular myocardial function analysis. In *Proc. IEEE Workshop on Biomedical Image Analysis, Seattle, Washington*, pages 88–97, 1994.
14. S.M. Song, R.M. Leahy, D.P. Boyd, B.H. Brundage, and S. Napel. Determining cardiac velocity fields and intraventicular pressure distribution from a sequence of ultrafast CT cardiac images. *IEEE Trans. on Medical Imaging*, Vol. 13, No.2:386–397, June 1994.
15. A.J.M. Spencer. *Continuum Mechanics*. Longman, 1980.

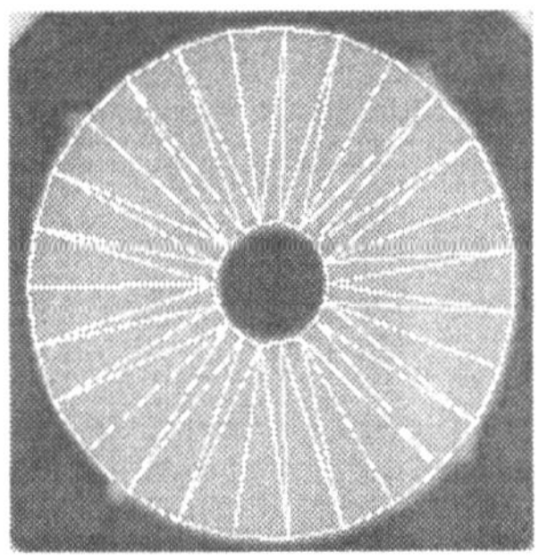

Figure 2. Phantom at instant 0, with superimposed mesh.

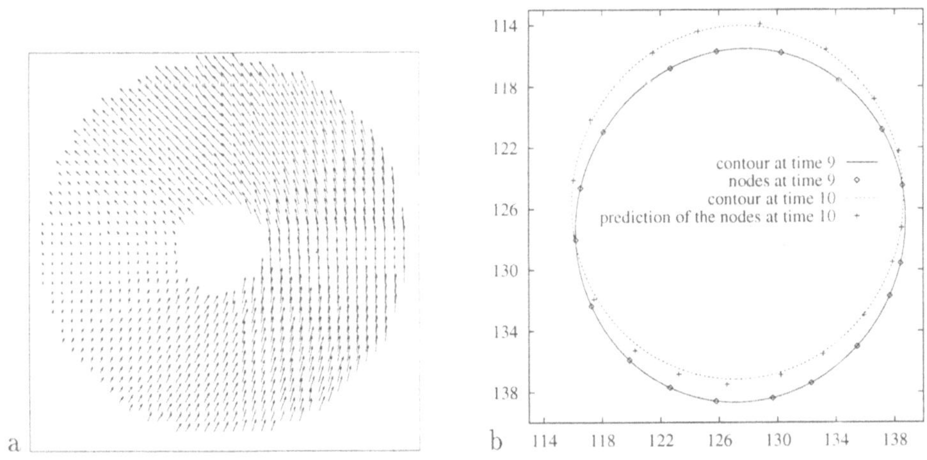

Figure 3. (a) Piecewise linear flow field at time 9. (b) Prediction of the nodes on the inner contour at time 9. The actual contours at time 9 and 10 are shown to demonstrate the accuracy of the prediction.

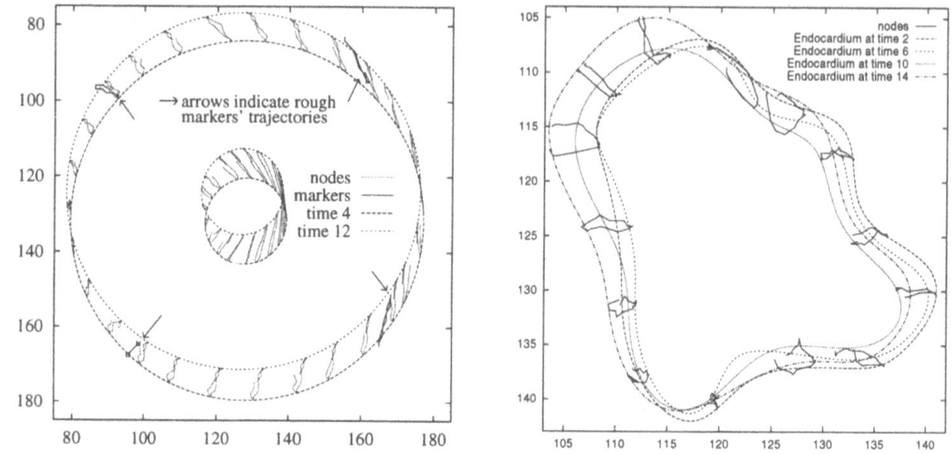

Figure 4. (a) Trajectories of the tracked nodes and trajectories of the hand-traced markers. The contours of the phantom are superimposed at time 4 and at time 12 only. (b) Trajectories of the nodes attached to the endocardium, with contours of the endocardium at time 2, 6, 10 and 14.

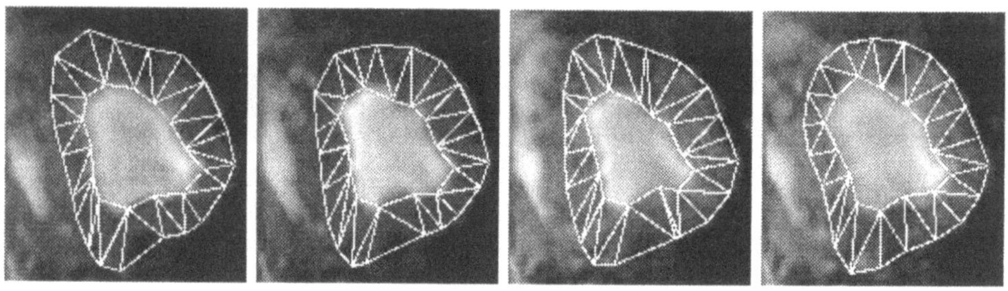

Figure 5. From left to right: results of the tracking at time 1 (end-diastole), 5, 9 (end-systole), and 13.

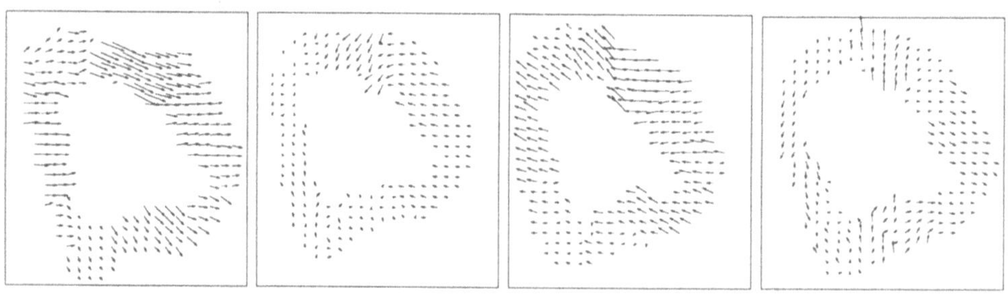

Figure 6. Continuous piecewise linear approximation of the velocity field at time 1, 5, 9, and 13.

INTEGRATION OF BOUNDARY FINDING AND REGION-BASED SEGMENTATION USING GAME THEORY

AMIT CHAKRABORTY AND JAMES S. DUNCAN
Image Processing and Analysis Group,
Depts of Elect. Engg. and Diag. Radiol., Yale University, USA

Abstract. Robust segmentation of structures from an image is essential for a variety of applications in biomedical image analysis. Here we propose a method that integrates region based segmentation and gradient based boundary finding using game theory in an effort to form a unified approach that is robust to noise and poor initialization. The novelty of the method is that this is a bi-directional framework whereby the two seperate modules improve their results through mutual information sharing.

Keywords: *game theory, boundary finding, region based segmentation, Maximum A posteriori probability*

1. Introduction

Precise segmentation of underlying objects in an image is very important especially for biomedical image analysis where it constitutes an important pre-processing step to such tasks as the registration of images obtained from two modalities, quantitative analysis of anatomical structures, the derivation of priors for image reconstruction in another modality and cardiac motion tracking. Two basic approaches to image segmentation are as follows. On one hand, we have region-based methods, which rely on the homogeneity of spatially localized features and other pixel statistics. The other group of work consists of gradient-based edge detection and boundary finding methods. However, the success of either of the methods is often limited by poor image quality. Integration of these methods can potentially result in an improved output as pointed out in [9]. However, integration always result in a larger complicated system, analyzing which is a daunting task. Probably the easiest way to do that is to divide the system into functional modules and analyze them separately. Hence, the reverse problem i.e. the development of such large and complex systems (which is our goal) often consists of intelligently assembling smaller well-understood modules,

Y. Bizais et al. (eds.), Information Processing in Medical Imaging, 189–200.

leading to what can correctly be termed *integrated* systems. Here *integration* refers not only to succinctly describing the mathematical relations between these modules but also describing algorithms that can perform the integration.

Thus in effect, we have in the general case, a team of modules (two for the present segmentation problem) working in a coordinated fashion in a possibly hostile and uncertain environment towards common or at least related goals. It is important to note that the members do not necessarily acquire the same information and thus may need to operate in a decentralized mode of decision making. If we presume the presence of a common probability space, thereby allowing the modules to share a similar view of the world, this could lead to probably the most common method of optimizing multi-objective functions, namely the single objective function approach.

Since the single objective approach seems to often be the method of choice in the vision literature [7, 1], we note that there are occasions when it is inadequate or at least not the best approach. One of the important concerns could be the robustness of such models and the *optimum* solutions that it produces under small variations in the underlying assumptions. This could especially be important in situations where the models perceive the outside world differently i.e. if they have different probability descriptions of the environment. This would then lead to different expected cost functions for the different modules, making the use of a single objective approach at best sub-optimal. One of the biggest advantages of using game theory is that it can dealing with this problem by treating it as a stochastic non-zero sum game with the modules taking on the roles of decision makers having different subjective probability measures.

Another advantage of using the game-theoretic approach is that the modules need not know the cost functions of the others. This keeps them decoupled and hence, computationally light. The flow of information is restricted to passing only the results of the decisions between the modules. Thus we could combine for example incommensurate objectives through this approach which might not be possible using the conventional single objective approach.

Coming back to the integration problem, both region based-methods, and gradient-based methods have their advantages and their disadvantages. While the region-based methods can handle better noise and loss of high frequency information, gradient-based methods are less error prone to changes in the grey scale values and can deal in a better way with shape changes. Our aim in this paper is to combine the strengths of both the approaches in an integrated method.

Among the available methods, AI based techniques have been used where production rules are invoked for conflict removal [9]. Other efforts have used probability-based approaches (see e.g. [5, 1]) where often the aim is to maximize the *a posteriori* probability of the region classified image given the raw image data by optimization methods like simulated annealing. Integration here is achieved in the local or dense field sense where the edges are used as line processes and the optimization is achieved both over the location of the line processes as well as the pixel classification.

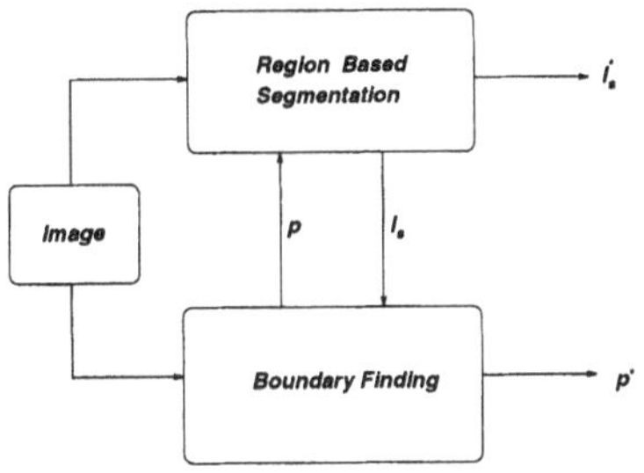

Figure 1. Flow diagram for game-theoretic integration of region based segmentation and boundary finding. The outputs of each of the models, I_s for the region module and p for the boundary module are fed back to each other after every decision making step. The game stops when none of the modules can improve their positions unilaterally. The final solution pair, (I_s^*, p^*) constitutes the Nash Equilibrium

The main contribution of this paper is thus to present an integrated bi-directional approach for segmentation by combining boundary finding and region based segmentation using a game-theoretic framework. This leads to a system where the two modules would operate in parallel so that at each step the outputs of each model get updated using information from the outputs of both the models from the previous iteration. Thus as the game progresses, both the modules improve their positions through mutual information sharing. The final boundary obtained is a compromise between the region segmented image and the gradient image, the edges are better localized. On the other hand, the region based segmentation is forced to be consistent with the presence of real objects in the image, making small islands of mis-segmented pixels highly unlikely. Further, we also present an theorem that concerns with the existence of the Nash Equilibrium. It is related to a theorem in [6, 2] but is more relevant to the present segmentation problem.

2. Game Theory

In this section we briefly introduce the concepts of parallel decision making and state a theorem for the stability and existence of a Nash Equilibrium for the types of objective functions that we would be using for our segmentation problem. The game is being played by a set of decision makers (or players) which for our case will correspond to the two segmentation modules that we try to integrate. Let $N = 1, 2$ be the player set, P^1 and P^2 be the strategy spaces of the decision makers D1 and D2 respectively (i.e. $p^1 \in P^1$ and $p^2 \in P^2$ constitute their strategies) and let

$$F^1 : P^1 \times P^2 \longrightarrow R \;;\; F^2 : P^1 \times P^2 \longrightarrow R$$

be the cost function for the decision makers 1 and 2 respectively. Then the normal form of the game is described by $\{F^1, F^2\}$ where it is assumed that P^1 and P^2, the strategy spaces are also Hilbert spaces and the decision makers strive to minimize their cost functions at each turn. This constitutes a rational form of decision making for the individual players.

The game continues unless the decision makers cannot unilaterally improve their positions (i.e. further minimize their cost functions). This natural stopping point is termed as the Nash Equilibrium. Formally, a pair of strategies ($\bar{p}^1 \in P^1$, $\bar{p}^2 \in P^2$) constitutes a Nash Equilibrium solution if

$$\forall p^1, p^2 \quad F^1(\bar{p}^1, \bar{p}^2) \leq F^1(p^1, \bar{p}^2) \;;\; F^2(\bar{p}^1, \bar{p}^2) \leq F^2(\bar{p}^1, p^2) \tag{1}$$

The Nash Equilibrium is obtained by a set of rational choices made by the players. If we consider k to be the time index, then this procedure of parallel decision making can be viewed as

$$p^1_{k+1} = \arg\min_{p^1 \in P^1} F^1(p^1, p^2_k) \;;\; p^2_{k+1} = \arg\min_{p^2 \in P^2} F^2(p^1_k, p^2) \tag{2}$$

where $p^1_0 \in P^1$ and $p^2_0 \in P^2$ is the initial point. A Nash Equilibrium is stable if it can be obtained as the limit of the above for any initial point. It is locally stable if convergence is valid for all initial conditions in some ϵ-neighborhood of the equilibrium solution.

For our segmentation problem, we shall be concerned with cost functions of the following structure

$$F^1(p^1, p^2) = f_1(p^1) + \alpha f_{21}(p^1, p^2) \;;\; F^2(p^1, p^2) = f_2(p^2) + \beta f_{12}(p^1, p^2) \tag{3}$$

where α and β are some scaling constants.
We also consider the following constraints on F^1 and F^2.
(i) F^i is bounded in $p^i \in P^i$.
(ii) F^i is continuously second order differentiable in $p^i \in P^i$.
(iii) $\exists$ a closed neighborhood $u^i \subseteq P^i$ such that F^i is strongly convex in u^i.
For the above assumptions, we have the following:
Theorem: For the given above structure of $F^1(p^1, p^2)$ and $F^2(p^1, p^2)$ there exists a locally stable Nash Equilibrium solution, ie. for any $p^1 \in U^1 \subseteq P^1$ and $p^2 \in U^2 \subseteq P^2$, the sequence of rational choices generated by the parallel

decision making process converges and the limit point is a Nash equilibrium solution if α and β satisfy the following condition:

$$\left\|\left[\left(\alpha^{-1}\frac{\partial^2}{\partial p^1 \partial p^1} f_1(p^1) + \frac{\partial^2}{\partial p^1 \partial p^1} f_{21}(p^1,p^2)\right)^{-1}\left(\frac{\partial^2}{\partial p^1 \partial p^2} f_{21}(p^1,p^2)\right)\right] * \left[\left(\beta^{-1}\frac{\partial^2}{\partial p^2 \partial p^2} f_2(p^2) + \frac{\partial^2}{\partial p^2 \partial p^2} f_{12}(p^1,p^2)\right)^{-1}\left(\frac{\partial^2}{\partial p^2 \partial p^1} f_{12}(p^1,p^2)\right)\right]\right\| < 1$$

What this theorem tells us is that there exists a Nash Equilibrium and also it states a condition under which that exists provided the modules described by their corresponding objective functions are well behaved. But, assuring the above condition is non trivial. However, as can easily be verified, for simple objectives this does not severely constrain the problem. For complex systems as in the present case, it is almost impossible to analytically verify the above, but most reasonable choices of α and β results in convergence.

3. The modules

The modules that we consider here are the two types of segmentation procedure (see Figure 1), one region based and the other gradient based boundary finding. Even though the method works for other more general image models, for simplicity of analysis, we shall assume that the image can be modeled by

$$y(i,j) = x(i,j) + n(i,j); \quad 1 \leq i \leq M; \quad 1 \leq j \leq N \tag{4}$$

where $n(i,j)$ corresponds to additive white Gaussian noise (AWGN). Now, as for $x(i,j)$, we make the further assumption that it can be modeled as a collection of homogeneous regions of uniform or slowly varying intensities.

3.1. REGION BASED SEGMENTATION MODULE

Here we try to describe the region based segmentation model. For this one we have $Y = I$ corresponds to the image data and X corresponds to the segmented image consisting of uniform regions. The task of this module is to come up with a segmentation X given the raw image Y and our prior knowledge of the process X. Thus the aim is $\arg\max_X P(X|Y)$. We model the region process X by a Markov random field. For recovering piecewise smooth functions as hypothesized in this problem, we shall use the weak membrane model [1]. Hence, maximizing the *a posteriori* probability in the above equation becomes equivalent to minimizing the energy function

$$E = \sum_{i,j}[y_{i,j} - x_{i,j}]^2 + \lambda^2\left(\sum_{i,j}(x_{i,j} - x_{i-1,j})^2 + \sum_{i,j}(x_{i,j} - x_{i,j+1})^2\right) + P \tag{5}$$

In the above equation, the first term is the data fidelity term, the second enforces smoothness and the last term is the penalty term which is non-zero if we want to recover the line processes. But, again for the simplicity of analysis we shall ignore that term and instead assume that the image intensities change smoothly over the regions. Such a model works reasonably for most of the situations. However, in some cases it becomes necessary to incorporate the line processes.

3.2. BOUNDARY FINDING MODULE

In this module, the aim is to find a class of objects with smooth boundaries that are continuously deformable. A variety of deformable contour methods have been proposed. We use the approach due to Staib and Duncan [10] which uses a Fourier parameterization which we have adopted in this work as our primary approach. It expresses a curve in terms of an orthonormal basis. Thus

$$v(t) = \begin{pmatrix} x(t) \\ y(t) \end{pmatrix} = \begin{bmatrix} a_0 \\ c_0 \end{bmatrix} + \sum_{k=1}^{\infty} \begin{bmatrix} a_k & b_k \\ c_k & d_k \end{bmatrix} \begin{bmatrix} cos(kt) \\ sin(kt) \end{bmatrix} \tag{6}$$

where, $v(t)$ is the contour vector consisting of the x and y coordinates and $p = (a_0, c_0, a_1, b_1, c_1, d_1, ...)$ represent the parameterized boundary. The estimation of these parameters to find the boundary is posed as an optimization problem, where the objective function measures the strength of the boundary given the set of parameters from the image. (see [10] for more details)

4. Interacting Modules

Here, in this section, we shall describe the interaction between the two modules. This would represent a modification of the cost function and we shall show that the format of the terms will represent that in the parallel decision theoretic model that we had described earlier.

4.1. REGION BASED SEGMENTATION INFLUENCED BY BOUNDARY FINDING

We have already described the region based method. Initially the requirement was to develop a region model given the raw image and the prior knowledge which was modeled using a MRF which in fact was no more than a smoothness constraint resulting in what is termed as a weak membrane model. Now, besides the input image and the smoothness constraint, we also have as a prior the output of the boundary finder in terms of a contour p. Thus the aim is $\arg\max_X P(X|Y,p)$, where as already defined, Y is the raw image data and X is the segmented image. Since we are concerned with maximization w.r.t. X, our aim is

$$\arg\max_X lnP(X|Y,p) \equiv \arg\max_X lnP(X|Y) + \arg\max_X lnP(p|X) \quad (7)$$

The first term in the above equation is derived from the image data and the previous output of this module, whereas the second term represents the influence of the previous output of the boundary finding module.

As we can recognize, the first term in the above equation (7) is the same as that in the original stand alone model. The second term, however corresponds to the information derived from the boundary finding module.

Even though this procedure can be extended to images with more than two regions, for the time being we shall continue with the assumption that there are only two homogeneous regions, one of which represents the target object. Thus the contour vector p represents the boundary of that object. Now, let A_p correspond to the points that lie inside the contour and $\bar{A}_p$ those points that lie on the background. Thus, $A_p \bigcup \bar{A}_p = \{(i,j);\ 1 \le i \le M,\ 1 \le j \le N\}$ i.e. the whole image consists of those two parts only.

A very simple model to represent this information is thus given by,

$$lnP(p|X) \propto \sum_{(i,j)\in A_p} (x_{i,j} - u_{i,j})^2 + \sum_{(i,j)\in \bar{A}_p} (x_{i,j} - v_{i,j})^2 \quad (8)$$

where $u_{i,j}$ represents the information regarding the intensity of those points that lies inside the contour and $v_{i,j}$ of those that lie outside. For most simple images with uniform regions, they could actually represent just the expected value of the pixels at those points, something that can easily be computed *a priori* [8]. Thus, finally the region based module including the interacting term is mathematically equivalent to the minimization of the following:

$$\begin{aligned}
\min_X E &= \min_X \Bigg[\sum_{i,j} \left([y_{i,j} - x_{i,j}]^2 + \lambda^2[(x_{i,j} - x_{i-1,j})^2 + (x_{i,j} - x_{i,j+1})^2] \right) \\
&+ P + \alpha \left\{ \sum_{(i,j)\in A_p} (x_{i,j} - u_{i,j})^2 + \sum_{(i,j)\in \bar{A}_p} (x_{i,j} - v_{i,j})^2 \right\} \Bigg] \quad (9) \\
&= \min_X [f_1(X) + \alpha f_{21}(X,p)]
\end{aligned}$$

The thing to note here is the fact that we were able to pose the problem in the format of the parallel decision making module as shown in the above, with the first term of the objective function $f_1(X)$ representing the contribution of the region based module by itself and the other term $f_{21}(X,p)$ representing the interaction term which uses the latest available output p of the boundary module.

4.2. BOUNDARY FINDING INFLUENCED BY REGION BASED SEGMENTATION

Unlike the stand alone boundary finding model, here, besides the gradient image, we will also use as input, the output from the region based method. Once again, the entire problem will be posed in a Bayesian framework as we shall shortly see. Thus, we want to maximize $P(p|I_g, I_s)$, where as described in the previous section, p is the vector of parameters used to parameterize the contour and I_g represents the gradient image and I_s (obtained from X of the region module) is the region segmented image. It can be shown that our aim is [4]

$$\max_p P(p|I_g, I_s) \;\propto\; \max_p P(I_g|p)P(I_s|p)P(p) \tag{10}$$

$$\equiv\; \max_p \left[lnP(p) + lnP(I_g|p) + lnP(I_s|p)\right] \tag{11}$$

The first two terms in (11) as we can recognize, are identical to the terms that were present in the stand alone boundary finder and have the same significance as before.

The last term, however incorporates the region based information into the boundary finding framework. It can be shown, (see [4] for details) that this term can be represented as

$$lnP(I_s|p) \propto \iint_{A_p} I_s(x, y)dA \tag{12}$$

The intuition, here is that if by any means we could identify the target region by positive values, and the other by negative values, the above area integral is maximum when the boundary correctly encircles the target region. For the simple two region problem, after the segmentation has been done, we could linearly translate the intensity values to satisfy the above requirement. Here, that would just mean subtracting out the mean of the means of the two characteristic regions. In other words for this case, $I_s(i, j) = X(i, j) - m$, where (i, j) represents the pixel location and m is the mean of the means of the two regions.

Omitting the details, it can be derived that we finally have:

$$\begin{aligned}
&\max_p P(I_g|I_s, p)P(I_s|p)P(p) \\
&= \max_p \left(\left[lnP(p) + \int_{C_p} K_1 I_g(x(p,t), y(p,t))dt \right] \right. \\
&+ \beta \left. \left[\int_{C_p} \{ N_s(x(p,t), y(p,t)) \frac{\partial x}{\partial t} + M_s(x(p,t), y(p,t)) \frac{\partial y}{\partial t} \} \, dt \right] \right) \qquad (13) \\
&\qquad = \max_p [f_2(p) + \beta f_{12}(I_s, p)]
\end{aligned}$$

Here, K_1 is a constant and so is β. However, the point to note is that the construct of the objective function is once again consistent with the game theoretic framework where $f_2(p)$ is the contribution of the gradient based boundary finder and $f_{12}(I_s, p)$ is the interaction term that uses the latest available output from the region module.

Thus, in this section we have presented the segmentation problem in a game-theoretic framework where the players correspond to the two different segmentation procedures. At each turn, each module tries to find a new output that best satisfies it's cost function based on the previous set of outputs. It then updates it's former output to this new output and broadcasts it to the other module as in Figure 1. The game finishes when both the modules have converged to their optimal strategies (see [3] for further details). That this process of decision making is acceptable and that the algorithm indeed converges to the Nash equilibrium has been shown in [6, 2].

5. Results

Experiments were carried out both with synthetic and real biomedical images to verify the performance of game theoretic integration against the individual models without any information sharing. The equations (9) and (13) jointly describe the outputs of the game-theoretic integration where equation (9) gives the region output and equation (13) the boundary output under the integrated framework. When the coupling coefficients are set to zero, we have the stand alone models.

First, we use a synthetic image that has one target object in the center surrounded by a background as shown in Figure 2 where the SNR is 0.67. The outputs of the two modules with and without game theoretic integration are shown in the above figure. Clearly, both the modules improve their outputs under game theoretic integration.

In the next figure we try to analyze the effects of integration for both the modules under increasing amounts of noise. In Figure 3 (a) we compare the output of stand alone region based segmentation with the same output obtained using game-theoretic integration as in equation (9). In Figure 3 (b) we compare the output of stand alone gradient-based boundary finder with the same output obtained using game-theoretic integration as in equation (13). We use the same template as in the last figure. The contrast difference between the background and the target is 200.0 and the x-axis in the Figures 3 (a) and 3 (b) correspond to the standard deviation σ for the white noise added. Both the experiments vary from $\sigma = 25$ to $\sigma = 400$. In other words, the SNR varies from 8.0 to 0.5. As for the y-axis, in Figure 3 (a), it represents the average per-pixel intensity error, and in Figure 3 (b), it represents a scaled version of the average Euclidean distance between

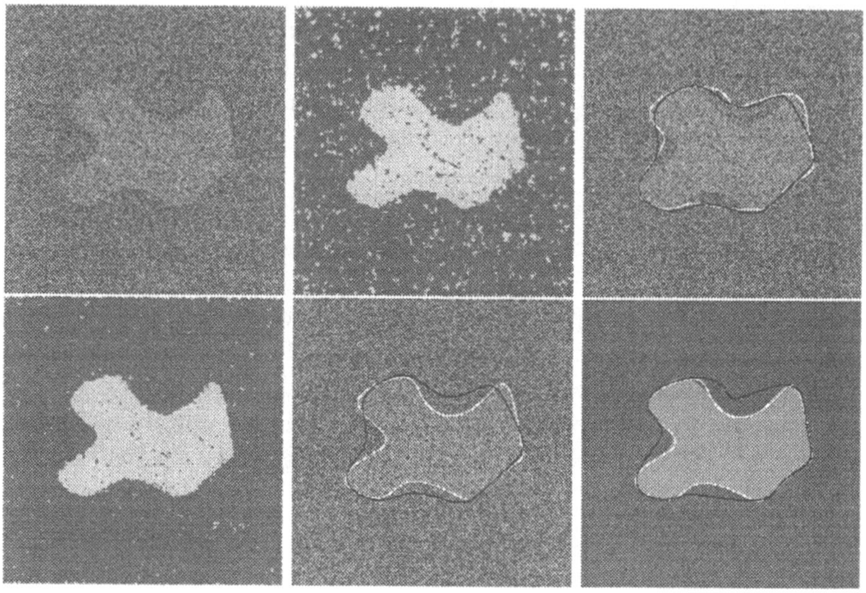

Figure 2. Example with a synthetic image to show the effects of game-theoretic integration. (a)Top,Left: Original image with $SNR = 0.67$. (b)Top,Middle: Output of region based segmentation without using game-theoretic integration. (c)Top,Right: Output of boundary finding without using game-theory; The dark contour represents the initialization, and the brighter one is the final output. (d)Bottom,Left: Output of region based segmentation using game-theoretic integration. (e)Bottom,Middle: Output of boundary finding using game-theory; The dark contour represents the initialization, and the brighter one is the final output. (f)Bottom,Right: The final boundary outputs generated by the two methods overlayed on top of the template. The dark contour represents the output of the gradient-based boundary-finder without using game-theory, and the brighter one is the final output using game-theoretic integration.

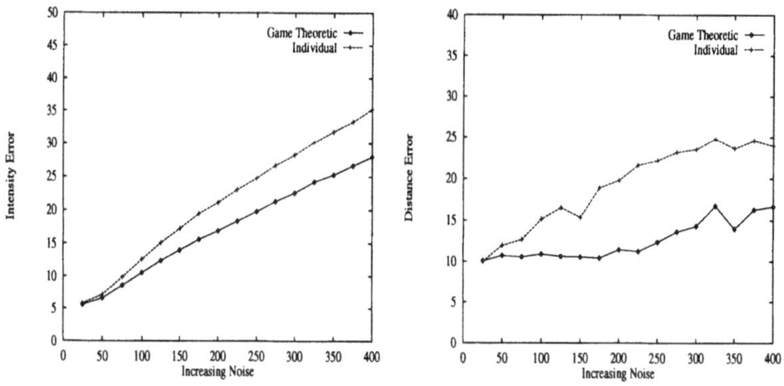

Figure 3. Performance of the two modules under increasing noise conditions without and with using game theoretic integration (a)Left: Comparison of region module with and without game-theoretic information sharing. (b)Right: Comparison of boundary finding with and without game-theoretic information sharing. Clearly, the error is larger without information sharing that what we get using game-theoretic integration for both the cases.

the true contour and the final one obtained (see [4] for details). Clearly, the results are superior for both the modules when there is game-theoretic information fusion.

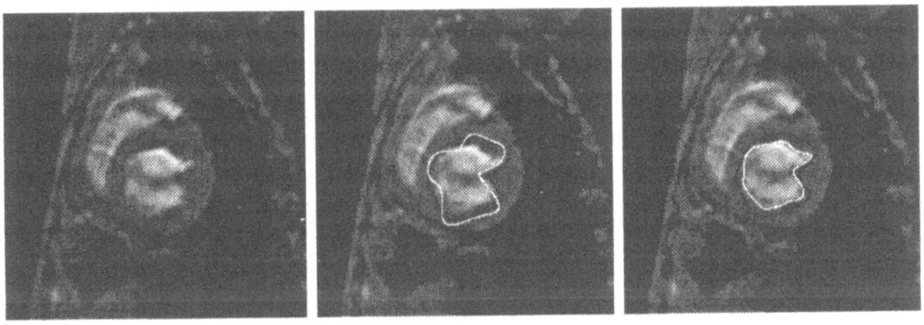

Figure 4. Example with the short axis MR image of a canine heart. (a)Left: Original image; (b)Middle: Output of gradient based boundary finding without using game theoretic integration. The dark contour represents the initialization, and the brighter one is the final output. (b)Right: Output of gradient based boundary finding using game theoretic integration. The dark contour represents the initialization, and the brighter one is the final output.

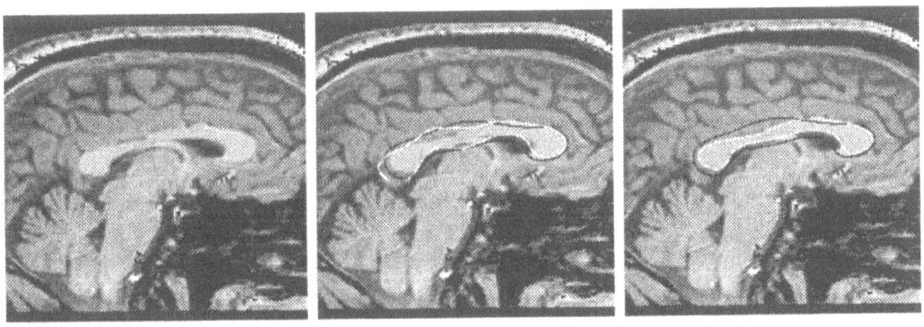

Figure 5. Example with the MR image of the brain. (a)Left: Original image; (b)Middle: Output of gradient based boundary finding without using game theoretic integration. The dark contour represents the initialization, and the brighter one is the final output. (b)Right: Output of gradient based boundary finding using game theoretic integration. The dark contour represents the initialization, and the brighter one is the final output.

In Figures 4 and 5 we show the results of game theoretic integration for some typical clinical images. While in the heart image, noisy and fuzzy edges seem to be the problem, in the brain image the problem seems to be the poor initialization. However, the game theoretic outputs in both the cases are satisfactory. This clearly demonstrates the robustness of the game theoretic method to both noise and poor initialization.

6. Conclusions

We have presented in this paper a new technique for integrating region based segmentation and gradient based boundary finding using the game theoretic framework. We have posed this as a non-zero sum two-person non-cooperative game where, the cost functions of both the modules are based on the Bayesian theory of maximization of the *a posteriori* probability. The main advantage of using the game-theoretic integration is that it can bring together the incommensurate region and boundary methods that operate in different probability spaces into a common information sharing framework of rational decision making. This leads to an improved output for both the modules. As the examples show, the integrated approach is much more robust to both noise as well as poor initialization. Application of this method on real medical images results in noticeable improvement as has been shown.

Acknowledgement: This work was supported in part by NIH-NHLBI grant R01-HL-44803. We also thank Dr. A. Sinusas and Dr. L.H. Staib for providing us with the images.

References

1. A. Blake and A. Zisserman. *Visual Reconstruction.* MIT Press, Cambridge, MA, 1987.
2. H.I. Bozma. *Decentralized Integration in Modular Systems Using a Game-Theoretic Framework.* PhD thesis, Yale University, 1992.
3. H.I. Bozma and J.S. Duncan. A game-theoretic approach to integration of modules. *IEEE Transactions on Pattern Analysis and Machine Intelligence*, 16:1074–1086, 1994.
4. A. Chakraborty, L.H. Staib, and J.S. Duncan. An integrated approach to boundary finding in medical images. *Proc. IEEE Workshop on Biomedical Image Analysis*, pages 13–22, 1994.
5. D. Geman and S. Geman. Stochastic relaxation, gibbs distribution and bayesian restoration of images. *IEEE Transactions on Pattern Analysis and Machine Intelligence*, 6:721–741, 1984.
6. S. Li and T. Basar. Distributive algorithms for the computation of noncooperative equilibria. *Automatica*, 23:523–533, 1987.
7. R.C. Luo and M.G. Kay. Multisensor integration and fusion in intelligent systems. *IEEE Transactions on System, Man and Cybernetics*, 19, 1989.
8. B.S. Manjunath and R. Chellappa. Unsupervised texture segmentation using markov random field models. *IEEE Transactions on Pattern Analysis and Machine Intelligence*, 13:478–482, 1991.
9. T. Pavlidis and Y. Liow. Integrating region growing and edge detection. *IEEE Transactions on Pattern Analysis and Machine Intelligence*, 12:225–233, 1990.
10. L.H. Staib and J.S. Duncan. Boundary finding with parametrically deformable models. *IEEE Transactions on Pattern Analysis and Machine Intelligence*, 14:161–175, 1992.

AN ADAPTIVE BAYESIAN APPROACH TO THREE-DIMENSIONAL MR BRAIN SEGMENTATION

Michelle X.H. YAN[1], **Joel S. KARP**[2]

Department of Psychiatry[1] *and Radiology*[2]
University of Pennsylvania, Philadelphia, PA. USA
Email: yan@einstein.psycha.upenn.edu

Abstract. We present an algorithm for segmenting 3D MR brain image into K different tissue types, which include gray, white matter and CSF, and maybe other abnormal tissue types. MR images considered can be either single- or multi-valued. Each single-valued image is modeled as a collection of regions with slowly varying intensity, described by B-spline functions, plus a white Gaussian noise. Tissue type is modeled by a Markov random field with the second order neighborhood in a 3D lattice. The proposed algorithm maximizes the *a posteriori* probability jointly over tissue types and mean intensities. Its implementation is iterative and adaptive. Each iteration consists of two maximization steps: estimation of mean intensities for each type by the least squares fitting of B-spline function, and tissue types using the ICM algorithm. The adaptation of the algorithm is achieved by increasing the number of control points of the B-spline functions, its purpose is to make the algorithm to slowly adapt to the local intensity variation of each region. Due to the use of both adaptation and B-spline function, the algorithm is robust to the "shading" effect. It has the potential to routinely process clinical MR images with minimal user involvement. Its performance is tested using patient data.

1. INTRODUCTION

Numerous techniques have been developed and reported in the literature for segmenting MR brain images. Many of them are based on clustering techniques which group voxels with similar features into the same tissue type. Methods that rely exclusively on partitioning the feature space such as the histogram or scattergram [1] [2] [3] work well only for MR images with excellent soft tissue contrast, high signal-to-noise ratio (SNR) and negligible "shading" effect. Unfortunately, most of clinically available MR images do not yet meet such a quality standard. It has been shown that

Y. Bizais et al. (eds.), Information Processing in Medical Imaging, 201–213.

by incorporating spatial interactions between neighboring voxels into the segmentation procedure, the degradation of segmentation outcome, due to poor SNR and poor feature contrast, is significantly reduced [4] [5] [6]. But the quality of segmentation is still limited by the failure of effectively compensating for the "shading" effect, which is due to an inhomogeneous RF field across the imaging volume. In contrast, edge based segmentation techniques, which explore local intensity contrast to partition an image into regions through the Marr-Hildreth operator [7] [8] or graph theory [9], are insensitive to the "shading" effect. But they may generate regions with mixed tissue types which are difficult to detect and correct for without user intervention.

The adaptive K-mean clustering algorithm [10] is an appealing approach to MR image segmentation in the sense that it allows to incorporate spatial interactions between adjacent voxels, and more importantly, to model mean intensity of each tissue type as spatially varying function, which can follow the local variations of the feature characteristics of the image. Based on the work in [10], an algorithm [11] was developed to segment three-dimensional (3D) single- or multi-valued brain MR images by alternating between the estimation of tissue types and mean intensities. To further improve the segmentation results, in this work, a cubic B-spline function is introduced to accurately model slowly varying mean intensity functions of each tissue type through the least squares fitting technique. Thus, under our image model, the estimated tissue types and mean intensities maximize the *a posteriori* probability of MR images. Furthermore, the use of spline helps to reduce the bias against small and isolated regions, for example, some of CSF regions. The algorithm is implemented adaptively by gradually increasing the number of control points (less than grid points of MR images) of the B-spline function, and provides an optimum estimate for each fixed number of control points. This permits local mean intensities, starting from almost constant values, slowly adapt to local intensity variations of each region of MR images. Tissue types are estimated based on from global to local characteristics of MR images, as the process proceeds.

2. IMAGE MODEL

Let $y = \{(y_{1,i}, y_{2,i}, \cdots, y_{M,i})^T\}$ be a M-valued 3D image, where $y_{m,i}$ is the intensity value of the m-th single image y_m at the i-th voxel location. The task of brain MR segmentation is to label each one of the voxels as one of K different tissue types, which could be gray matter (GM), white matter (WM), cerebrospinal fluid (CSF), and possibly other abnormal tissues. Denote this tissue label image by $x = \{x_i\}$, where $x_i = k \in \mathcal{K} = \{1, 2, \ldots, K\}$ is the tissue type at the i-th voxel. Thus the goal of segmentation can be stated as, under a certain image model, to seek a "good" label image x given data y.

We model the input M-valued 3D MR image y as a mosaic of regions of uniform or slowly varying mean intensities corrupted by additive white Gaussian noise. Specifically, the following assumptions are made:

1. Every M-valued voxel $(y_{1,i}, y_{2,i}, \cdots, y_{M,i})^T$ belongs to one of the K tissue types;
2. For each single image y_m, the intensity values are conditionally independent given the label image x and the conditional probability density $p(y_{m,i}|x_i)$ is Gaussian with unknown mean and constant variance σ_m^2.
3. Each mean is modeled as a slowly varying function of the voxel location i, denoted as $\mu = \{(\mu_{1,i}^k, \cdots, \mu_{M,i}^k)^T\}$, where $\mu_{m,i}^k$ is the mean value of the m-th single image at the voxel location i with tissue type k.
4. At each voxel i, the M intensity values are conditionally independent given the label x_i.

Denote $\sigma^2 = (\sigma_1^2, \sigma_2^2, \cdots, \sigma_M^2)^T$, where σ_m^2 is the noise variance in y_m and function only of m. In our experiments, the standard deviation of the background noise of MR images is generally much smaller than the mean intensities of all tissue types, hence the variation of σ^2 due to different tissue types is negligible [12]. Thus the M-valued MR image y is completely characterized by the label image x, mean intensity μ and noise variance σ^2. Since σ^2 can be easily estimated from the background voxels of the input image y, it will be considered as known for the purpose of segmentation.

Markov random field (MRF) [13] is used to introduce spatial interaction between neighboring label voxels, where the probability density of the tissue label image x, $p(x)$, is a Gibbs distribution, expressed as:

$$p(x) = \frac{1}{Z} \exp\left\{ -\sum_{C} V_C(x) \right\},$$

where Z is a normalizing constant and the summation is over all cliques C (a clique is defined as a set of voxels that are neighbors of each other). In our work, the 3D second-order neighborhood system is employed with the two-voxel clique potential, defined as follows:

$$V_C(x) = \begin{cases} 0, & \text{if } x_s = x_q, \\ \beta, & \text{if } x_s \neq x_q, \text{ connected by plane,} \\ \beta/\sqrt{2}, & \text{if } x_s \neq x_q, \text{ connected by line,} \end{cases} \tag{1}$$

where the parameter β is positive. This expression reflects our believe that neighboring voxels more likely belong to the same tissue type than different types. Using the above image model and the Bayes' rule, the logarithm of the *a posteriori* probability, $p(x|y;\mu)$, is proportional to the following:

$$Q(x|y;\mu) \equiv -\sum_m \left[\frac{1}{2\,\sigma_m^2}\sum_i \left(y_{m,i} - \mu_{m,i}^{x_i}\right)^2\right] - \sum_C V_C(x). \tag{2}$$

Under this Bayesian model, the task of segmentation is to estimate mean intensity μ and label image x, which is achieved by maximizing the *a posteriori* probability $p(x|y;\mu)$, or equivalently, the function $Q(x|y;\mu)$ in Eq.(2).

3. BRAIN MR SEGMENTATION

In this section, three approaches to MR image segmentation are described: maximum likelihood (ML) and maximum *a posteriori* probability (MAP) estimation, where mean intensity is a constant for each tissue type; and an adaptive MAP approach with spatially varying mean intensity, represented by a B-spline function.

3.1. Constant mean: ML estimation *via* K-means clustering

A special case of our model in Eq. (2) is that the prior is constant, and the mean intensity functions, $\mu_{m,i}^k = \mu_m^k$, is constant for each tissue type. Drop the term $\sum_C V_C(x)$ and replace $\mu_{m,i}^k$ by μ_m^k in Eq. (2), it yields the log-likelihood function of $(x;\mu)$ given y:

$$L(y|x;\mu) \equiv -\sum_m \frac{1}{2\,\sigma_m^2}\left[\sum_i (y_{m,i} - \mu_m^{x_i})^2\right]. \tag{3}$$

Then the segmentation is to find both μ and x which maximize $L(y|x;\mu)$.

This problem can be solved using the K-means clustering algorithm [14], which groups voxels in an image into K clusters through the minimization of the total inter-cluster variance. This procedure, starting with an initial mean intensity estimate μ^0, maximizes the likelihood function Eq. (3) by alternating between two maximization steps:

$$\begin{aligned} x^{n+1} &= \arg\{\max_x L(y|x;\mu^n)\}, \\ \mu^{n+1} &= \arg\{\max_\mu L(y|x^{n+1};\mu)\}, \quad n \Leftarrow n+1. \end{aligned} \tag{4}$$

In order to reduce the sensitivity of the clustering outcome to the choice of an initial guess μ^0 and to avoid the final segmentation being trapped at a poor local optimum, the algorithm builds the clusters gradually as described below:

$$
\begin{aligned}
&\text{For each } l = 1, 2, \ldots, K-1: \\
\text{S-step:}\quad & (\mu_m^k)^{(0)} = (\mu_m^k)^*, \quad k = 1, \ldots, l, \\
& (\mu_m^{l+1})^{(0)} = (1+\Delta)(\mu_m^k)^*; \\
\text{M-step:}\quad & \{x^*, \mu^*\} = \arg\left\{\max_{x,\mu}\left[L(y|x;\mu)\right]\right\}: \qquad (5) \\
& \text{M1-step: } x_i^{(n+1)} = \arg\left\{\min_{x_i \in \mathcal{K}}\left[\sum_m \left(y_{m,i} - (\mu_m^{x_i})^{(n)}\right)^2\right]\right\}, \\
& \text{M2-step: } (\mu_m^k)^{(n+1)} = \frac{\sum y_{m,i}\delta\left(x_i^{(n+1)} = k\right)}{\sum \delta\left(x_i^{(n+1)} = k\right)}, \quad n \Leftarrow n+1,
\end{aligned}
$$

where x^* and μ^* denote an optimum estimate of label image and mean values. The function $\delta(x_i^{(n+1)}=k)$ is an indicator that is equal to one if $x_i^{(n+1)}=k$; zero, otherwise. This algorithm is designed one level at a time, consisting of the splitting step (S-step) and the maximization step (M-step). The S-step splits one mean value to create an additional cluster by introducing a positive perturbation constant Δ, which should be small relative to the mean differences. The mean to be split is one of the optimal estimates of mean values from the previous level. $\Delta=0.08$ was chosen for all cases presented. After each S-step, the M-step is carried out by the iterations that alternates between the M1- and M2-steps. The M1-step updates the label image x by maximizing the Eq. (3) for the given mean values. The M2-step updates the mean intensity values also through the maximization of the Eq. (3), which leads to the averages of image values of all voxels that belong to the same tissue type, based on the updated labels from the M1-step. At each level l in the increasing order from 1 to $K-1$, the estimated label image x^* and mean intensity μ^* is a maximum of the likelihood function $L(y|x;\mu)$. Iterations proceed until the label image is separated into K types of tissue, and the likelihood function $L(y|x;\mu)$ is maximized. It is not clear which cluster should be chosen for splitting in order to produce the best clustering result. In our application, the procedure begins with an initial estimate for $(\mu_m^1)^*$, chosen as the intensity mean of the m-th single brain MR image, and splits the cluster with the smaller mean value when $l=2$.

The K-means clustering algorithm under this model leads to a maximum likelihood (ML) estimation. The quality of ML segmentation is, however, limited by the assumption that each tissue type is characterized by a constant mean intensity and the failure to take into account the spatial relationship between neighboring voxels.

3.2. Constant mean: MAP estimation *via* generalized K-means

Let us consider another special case of our model, where the label image is modeled by a MRF defined in Eq. (1), and the mean intensity is still an unknown constant for each tissue type, again denoted as μ_m^k. The function $Q(x|y;\mu)$ in Eq.(2) then become

$$Q(x|y;\mu) = -\sum_m \left[\frac{1}{2\,\sigma_m^2}\sum_i (y_{m,i} - \mu_m^{x_i})^2\right] - \sum_C V_C(x). \qquad (6)$$

Applying the K-means algorithm to this problem yields the generalized K-means clustering algorithm, which provides a MAP estimate. This procedure is very similar to the K-means clustering procedure Eq.(5), except that the M1-step is done here at every voxel in the image using ICM [15]. It searches for tissue labels x which maximize the function $Q(x|y;\mu)$ in Eq.(6) for given mean intensity μ. The M2-step is exact the same as in the K-means clustering algorithm, *i.e* averaging image values of all voxels that are labeled as the same tissue type for each single MR image. So again, both x and μ are estimated based only on the global characteristics of input MR images. Although with a suitable choice of the β value, the method produces the better results to some extent, but they seem over-smoothed particularly across true boundaries in the image, and are sensitive to slow variations of MR image intensity, because of the constant mean assumption.

3.3. Slowly varying mean: adaptive MAP estimation

It has been shown that the use of spatially varying mean intensity model is important to account for the "shading" effect for better MR image segmentation [11]. But how to estimate the smooth function is another practical and theoretical aspect of this problem. One approach to this problem is the use of polynomial forms to describe each mean intensity function. However, the simple shape represented by polynomials is not able to accurately follow local intensity variations of MR images. In [11], we used local average to provide an estimation for mean intensity of MR images. But it seems theoretically unappealing that we begin by searching for an optimum solution for both label image x and intensity functions μ, but end up with a non-optimum one. In this work, we use a cubic B-spline function to model spatially varying mean, the estimated label x and mean μ is a maximum of the function $Q(x|y,\mu)$ for given data y.

B-spline representation of mean intensity: First, we consider the problem of estimating the local intensity function for given label image x. A spline representation can approximate a smooth function very well with relatively few parameters, and it provides a closed form and analytic description of the function. Let $(z_{1,i}, z_{2,i}, z_{3,i})$ denote the coordinates of the

i-th voxel of the 3D image. Given a 3D array of control points $\{V_{j,l,q}(k,m)\}$, where $V_{j,l,q}(k,m)$ is the value of control point at an index (j,l,q) for the tissue type k and the m-th single image, a cubic B-spline representation for the mean intensity function, $\mu_m^k(z_{1,i},z_{2,i},z_{3,i})$ can be described as

$$\mu_{m,i}^k \equiv \mu_m^k(z_{1,i},z_{2,i},z_{3,i}) = \sum_{j,l,q} V_{j,l,q}(k,m)B_{j,l\,q}(z_{1,i},z_{2,i},z_{3,i}), \qquad (7)$$

where $B_{j,l,q}(z_{1,i},z_{2,i},z_{3,i}) = B_j(z_{1,i})B_l(z_{2,i})B_q(z_{3,i})$ is the tensor product of cubic B-spline basis functions [16]. The B-spline representation is uniquely determined by the control points. In practice, we often have a large number of finely sampled data points, for example, $256 \times 256 \times 100$ points from each single MR image. The number of control points necessary to represent a smooth function accurately may be much smaller. Therefore, the control points $\{V_{j,l,q}(k,m)\}$ for each tissue type k and each single MR image m can be obtained by the least squares fitting of a B-spline expression in Eq.(7) to the entire m-th single MR image. More specifically, given the label image $x = \{x_i\}$ and the total number of control points used, the estimated control points $\{V_{j,l,q}^*(k,m)\}$ minimize the following form:

$$\sum_{x_i=k}(y_{m,i}-\mu_{m,i}^k)^2 = \sum_{x_i=k}\left(y_{m,i} - \sum_{j,l,q} V_{j,l,n}(k,m)B_{j,l,q}(z_{1,i},z_{2,i},z_{3,i}).\right)^2$$

over the entire m-th MR image for the voxels whose labels are equal to k. Therefore, given label x and MR image y, the estimated mean intensity $\mu^* = \{((\mu_{1,i}^k)^*,\cdots,(\mu_{M,i}^k)^*)^T\}$ is also a minimum of the following function:

$$\sum_m \left[\frac{1}{2\,\sigma_m^2}\sum_i \left(y_{m,i} - \mu_{m,i}^{x_i}\right)^2\right]. \qquad (8)$$

In addition, by changing the total number of control points $\{V_{j,l,n}(k,m)\}$ used in the least squares fitting, we will be able to control the local variation or smoothness of estimated mean intensity represented by B-spline function.

ICM for label image estimation: Second, we consider the problem of estimating the label image x for given mean intensity functions μ. This is done by the ICM algorithm [15], which maximizes the conditional density at each voxel i given MR image y, intensity function μ and the current labels at other voxels, *i.e.*:

$$\min_{x_i\in\mathcal{K}}\left[\sum_m \left(y_{m,i}-\mu_{m,i}^{x_i}\right)^2 + \sum_{x_i\in C} V_C(x)\right].$$

Now, with each fixed total number and locations of control points, one can find an optimum solution to the MR segmentation problem by maximizing the function $Q(x|y;\mu)$ over x and μ, which requires the alternation between the estimation of label image x and intensity functions μ for given initial estimates x^0 and μ^0.

Adaptive MAP implementation: The purpose of the adaptive implementation is to allow the final decision on segmentation outcomes to be based on both global features (the entire image) as well as local feature contrast. The adaptation is achieved by reducing the spacing between two adjacent control points, *i.e.*, increasing the total number of control points. As the number of control points gradually increases, the resulting estimate of intensity functions changes from almost constant values to smooth functions which accurately approximate the local variations of MR image intensity. Thus, the segmentation results can slowly approach to local estimates governed by local features of MR images. We begin with an initial estimate from the ML segmentation, provided by the procedure Eq.(5), and the control point spacing $S = (S_{z_1}, S_{z_2}, S_{z_3})$ equal to half of the whole image in each dimension. The spacing and locations of control points are kept fixed until the function $Q(x|y;\mu)$ reaches a local optimum. The whole process is then repeated with a reduced control point spacing, chosen to be half of the previous one in each direction. The algorithm stops when the largest spacing among S is not greater than 8 voxels. This adaptive procedure is summaried as follows:

Initial x^0, μ^0: ML; for $\{\ S = S_0\ ; S \leq 8\ ; S \Leftarrow S/2\ \}$

$$\text{M-step: } \{x^*, \mu^*\} = \arg\left\{\max_{x,\mu}\left[Q(x|y;\mu)\right]\right\}: \tag{9}$$

$$\text{M1-step: } x_i^{(n+1)} = \arg\left\{\min_{x_i \in \mathcal{K}}\left[\sum_m \left(y_{m,i} - (\mu_{m,i}^{x_i})^{(n)}\right)^2 + \sum_{x_i \in C} V_C(x^{(n)})\right]\right\};$$

$$\text{M2-step: } (\mu_{m,i}^k)^{(n+1)} = \sum_{j,l,q} V_{j,l,q}^{(n+1)}(k,m) B_{j,l,q}(z_{1,i}, z_{2,i}, z_{3,i}), \quad n \Leftarrow n+1;$$

$$V_{j,l,q}^{(n+1)}(k,m) \Leftarrow \min_{\{V_{j,l,q}\}} \sum_{x_i^{(n+1)}=k} \left(y_{m,i} - \sum_{j,l,q} V_{j,l,q}(k,m) B_{j,l,q}(z_{1,i}, z_{2,i}, z_{3,i})\right)^2.$$

Similar to the K-means clustering procedure in Eq.(5), the adaptive MAP algorithm is also designed one level at a time, that is, for each fixed spacing and locations of control points in B-spline representation, a local optimum of the function $Q(x|y;\mu)$ in Eq.(2) is obtained by alternating the M1- and M2-steps. The use of adaptation will help the algorithm converge to a good local optimum. In the early stage of the iterative algorithm, the segmentation is inaccurate. So a large spacing between control points is necessary for mean intensity estimates to be mostly based on the whole image. As the algorithm progresses, the result of segmentation is getting better. A small spacing should be used to obtain more accurate local intensity estimation, so that the label image would be able to follow local features of each region in MR images. The combination of B-spline representation and adaptation is especially desirable for overcoming the "shading" effect that most of MR images suffer from. Because of this combination, there is no need to further split the same tissue type, therefore the number of tissue types interested is the choice of K. This is because estimated mean intensities of each type will closely follow the local intensity variation of the same type of MR images.

4. EXPERIMENTAL RESULTS AND SUMMARY

As an initial step to evaluate its performance, we have applied the adaptive MAP algorithm to 3D MR brain images acquired using two protocols: dual-echo (proton density and T_2-weighted) and spoiled GRASS (SPGR) pulse sequence. All data presented are collected using GE 1.5 tesla scanners. The voxel sizes are $0.8594 \times 0.8594 \times 5$ and $0.9375 \times 0.9375 \times 1$ (mm^3) for dual-echo and SPGR MR images, respectively. There is no gap between slices in either case. Since we are interested in GM, WM and CSF volumetric measurements, the number of tissue types K is equal to 3. The parameter β from the Gibbs prior is chosen to be 0.35 for all cases presented below. For dual-echo MR images, the two-dimensional second-order spatial interaction is only imposed to in-plane voxels, due to the large slice thickness.

We used an algorithm [17], developed for detection of PET and MR brain contours for registration, as a pre-segmentation procedure to estimate and extract a 3D MR brain volume before the segmentation is performed. Since this algorithm for detecting MR brain contours is not fully successful yet for slices below eyeballs, brain volume considered in this paper included only the slices above eyeballs. Figure 1 and 2 illustrate the segmentation results of the two protocols using the three algorithms: ML, MAP and adaptive MAP. In Figure 1, as expected, there is some pepper and salt noise in the ML segmentation, which is less smooth than the MAP segmentation. The same statement applies to the segmentation of SPGR. Figures 1(e) and 2(c) are the segmentations using the adaptive MAP algorithm. These results show that there is a clear benefit from the use of B-spline representation and adaptation in comparison to the ML or MAP segmentation in both protocols. They are with clear and smooth structure boundaries, and seem to be more accurate particularly for small structures and near structure boundaries of MR images.

To study the consistency of measurements from different tissue types, produced by our adaptive MAP algorithm, the following evaluation has been conducted, which involves SPGR scans of two normal subjects at

	Scan-1	*Scan-2*	*Scan-3*	*Scan-4*	*Scan-5*	*Scan-6*	*Mean*	*STD*
Gray	389342	386385	382450	390311	379472	383876	385306	3802
White	404149	412239	403743	404599	413279	410502	408085	4012
CSF	129621	124418	136825	128162	130286	128777	129681	3702
Total	923112	923042	923018	923072	923037	923155	923073	47

TABLE 1. Volumetric measurements of gray, white matter and CSF from *Subject One* based on the total brain volume about 923073 voxels.

their 20's and 40's. *Subject One* was scanned six times using the same scanner (unit one): three scans were taken consecutively and the other three taken one week later. *Subject Two* was scanned nine times using two different scanners. Six scans were collected using the identical procedure as for *Subject One*, the other three were taken from a different scanner (unit two) at the same day as the first three scans were taken. Our adaptive MAP algorithm was then applied to each brain volume, so there were a total of 15 independent segmentations performed, where the standard deviation σ of image noise was equal to 6.6 for all scans from unit one and 5.8 for unit two. In order to compare the volumetric measurements of each tissue type, an image registration algorithm [17] was used to align all the original scans to one arbitrarily selected scan for each subject. As a result, one intersected brain volume of all scans from each subject could be computed using the estimated transformation parameters. Table 1 lists the total voxel number of each tissue type and the intersected brain volume from each scan, the mean and standard deviation (STD) of each tissue type for *Subject One*, calculated from the six scans. Similar consistency results were also obtained for *Subject Two*, where the scans were from two different scanners. The measurements from the two subjects indicate our algorithm produces quite consistent results. From Table 1, the ratio between STD and mean for GM and WM is about 1%, and is higher for CSF. The main reason for this, we believe, is that many of CSF regions are small and isolated relative to regions of GM and WM, therefore, the relative percentage of boundary voxels is higher. These boundary voxels are more susceptible to the partial volume effect, which are not considered here, and result in a higher variance.

In summary, we have developed an adaptive MAP segmentation algorithm for 3D single- or multi-valued brain MR images, which directly incorporates some specific characteristics of MR images and requires no user interaction. The algorithm is implemented adaptively and maximizes the *a posteriori* probability over the label image and mean intensity function. The selection of final cluster numbers is not an important issue in our algorithm, and is chosen to be the number of tissue types of interest. The algorithm produces quite consistent volumetric measurements for GM, WM and CSF, and is insensitive to the "shading" effect of MR images. Our results thus far are encouraging, we will further validate the proposed algorithm.

Acknowledgments The authors would like to thank Dr. Zhenyu Wu for his many valuable suggestions and Drs. Raquel Gur and David Mozley for providing us with MR images presented in this paper. This work was supported by National Institute of Mental Health Grant MH48539.

References

[1] M. Kohn *et al.*, "Analysis of brain and cerebrospinal fluid volumes with MR imaging," *Radiology*, vol. 178, pp. 115–122, 1991.

[2] M. W. Vannier *et al.*, "Validation of magnetic resonance imaging (MRI) multispectral tissue classification," *in proc. 9th Conf. on Pattern recognition, Rome, Italy*, pp. 1182–1186, Nov. 1988.

[3] G. Gerig *et al.*, "Automatic segmentation of dual-echo MR head data," *in Proc. 12th Intl. Conf. on Information Processing in Medical Imaging*, 1991.

[4] D. A. Ortendahl and J. W. Carlson, "Segmentation of magnetic resonance images using fuzzy clustering," *in Proc. 10th Info. Process. in Med. Imaging Conf.*, 1988.

[5] S. P. Raya, "Low-level segmentation of 3-d magnetic resonance brain images - A rule-based system," *IEEE Trans. on Medical Imaging*, vol. 9, pp. 327–337, 1990.

[6] H. S. Choi, D. R. Haynor, and Y. Kim, "Partial volume tissue classification of multichannel magnetic resonance images - A mixel model," *IEEE Trans. on Medical Imaging*, vol. 10, pp. 395–407, 1989.

[7] H. H. Ehricke, "Problems and approaches for tissue segmentation in 3D-MR imaging," *SPIE: Medical Imaging IV*, vol. 1233, pp. 128–137, 1990.

[8] M. Bomans *et al.*, "3-D segmentation of MR images of the head for 3-D display," *IEEE Trans. on Medical Imaging*, vol. 9, pp. 177–183, 1990.

[9] Z. Wu and R. Leahy, "An optimal graph theoretic approach to data clustering: Theory and its application to image segmentation," *IEEE Trans. on Pattern Analysis and Machine Intelligence*, vol. 15, no. 11, 1993.

[10] T. N. Pappas, "An adaptive clustering algorithm for image segmentation," *IEEE Trans. on Signal Processing*, vol. 40, pp. 901–914, 1992.

[11] M. X. Yan and J. S. Karp, "Segmentation of 3D brain MR using an adaptive K-means clustering algorithm," *Proceedings of IEEE Medical Imaging Conference, Norfolk, Virgina*, November 1994.

[12] R. M. Henkelman, "Measurement of signal intensity in the presence of noise in MR images," *Med. Phys.*, vol. 12, March/April 1985.

[13] S. Geman and D. Geman, "Stochastic relaxation, Gibbs distributions, and the Bayesian restoration of images," *IEEE Trans. on Pattern Analysis and Machine Intelligence*, vol. 6, pp. 721–741, 1984.

[14] R. M. Gray and Y. Linde, "Vector quantization and predictive quantizer for Gauss-Markov sources," *IEEE Trans. on Communications*, vol. 30, pp. 381–389, 1982.

[15] J. E. Besag, "On the statistical analysis of dirty pictures," *Journal of Royal Statistical Society, Series B*, vol. 48, pp. 259–302, 1986.

[16] B. A. Barsky, "A description and evaluation of various 3D models," *IEEE Trans. on Computer Graphics and Applications*, vol. 4, pp. 38–52, 1984.

[17] M. X. Yan and J. S. Karp, "Image registration of MR and PET based on surface matching and principal axes fitting," *Proceedings of IEEE Medical Imaging Conference, Norfolk, Virgina*, November 1994.

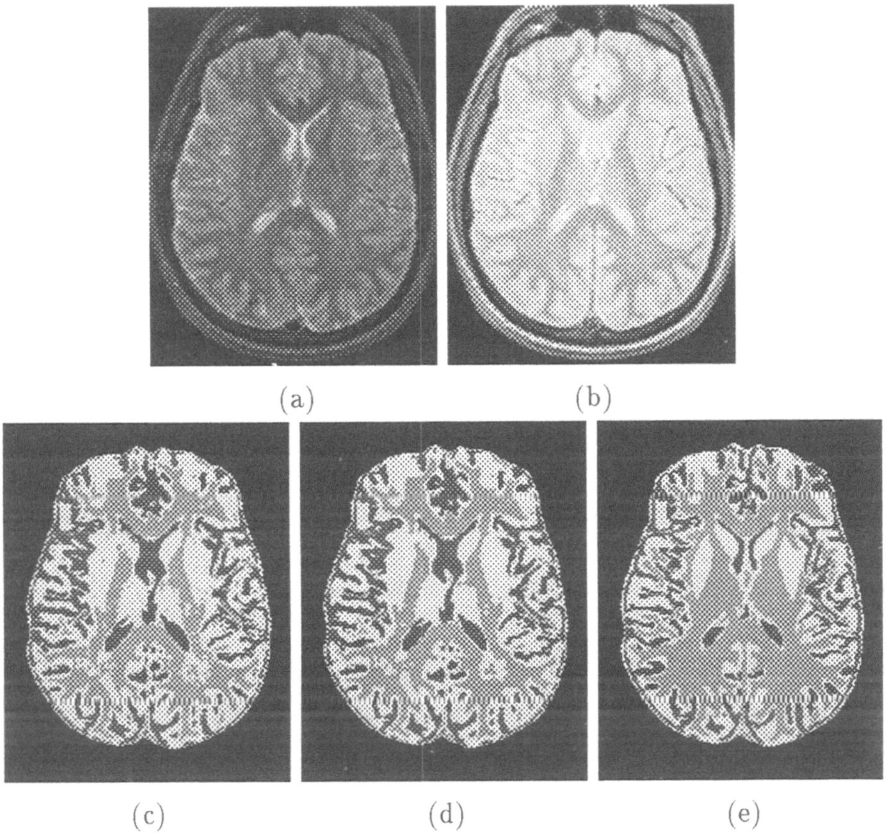

Figure 1. Segmentation of dual-echo MR data. Original transaxial slice of T_2-weighted (a) and proton density (b) MR image. (c): ML segmentation. (d): MAP segmentation. (e): Adaptive MAP segmentation.

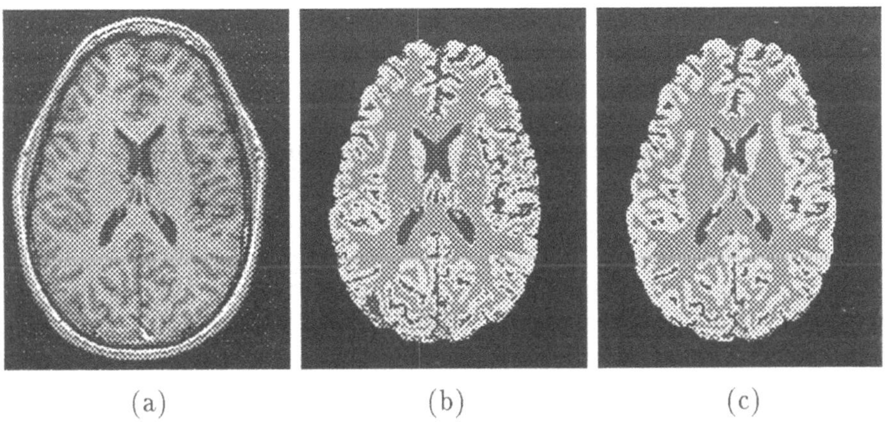

Figure 2. Segmentation of spoiled GRASS pulse sequence MR data. (a): Original transaxial slice. (b): MAP segmentation. (c): Adaptive MAP segmentation.

AUTOMATIC IMAGE PARTITIONING FOR GENERIC OBJECT SEGMENTATION IN MEDICAL IMAGES

F.MAES, D.VANDERMEULEN, P.SUETENS AND G.MARCHAL
Laboratory for Medical Imaging Research† *, ESAT-Radiology*
K. U. Leuven, Kardinaal Mercierlaan 94, 3001 Heverlee, and
U. Z. Gasthuisberg, Radiology, Herestraat 49, 3000 Leuven, Belgium.

Abstract.
In this paper a generic strategy is presented to facilitate the segmentation of anatomical objects in medical images by providing powerful and intelligent assistance to the clinical user. Partitioning of the image in segmentation primitives, such that all pixels within the same primitive are likely to belong to the same object, reduces the segmentation problem to selecting those primitive regions that belong to the object of interest. This dramatically increases segmentation efficiency, because no individual pixels need to be considered. The power of this approach is that such a partitioning can be computed independently of the object to be segmented, such that this strategy is very general and can be applied for the segmentation of a variety of anatomical shapes in different imaging modalities. This is illustrated for MR images of the brain, for CT images of the abdomen and for angiographic images of the coronary arteries.

1. Introduction

Segmentation of the anatomical objects of interest in medical images is a necessary requirement for their quantification and three-dimensional (3D) visualization. Manual delineation of objects in 3D images is very time-consuming and fast and accurate computer-aided segmentation methods are needed to improve the physician's productivity in clinical practice [9]. Because clinical users are in many ways superior to any form of current au-

†F. Maes is Aspirant of the Belgian National Fund of Scientific Research (NFWO).

Y. Bizais et al. (eds.), Information Processing in Medical Imaging, 215–226.

tomation, semi-automatic methods that improve interaction by providing intelligent assistance are often superior to fully automatic methods, which may be computationally expensive and may require a lot of manual correction. Because of the growing number of applications and the increasing requirement for accurate representation of a diversity of complex shaped objects, the segmentation algorithms need to be adaptable in a flexible way to a variety of anatomical (and artificial) shapes and should be robust with respect to varying imaging conditions. Such a multi-purpose approach should make it possible to offer the clinical user a single tool that enables him to quickly segment objects with different appearance in the image, without limiting restrictions regarding image contrast or imaging modality.

In this paper we propose such a generic strategy, which aims at facilitating the segmentation of anatomical objects in a variety of different applications by providing powerful and intelligent assistance to the clinical user. This intelligence is based on the idea that partitioning of the image in segmentation primitives such that all pixels within the same primitive are likely to belong to the same object, reduces the segmentation problem to selecting those primitive regions that belong to the object of interest. Because all pixels that belong to the same primitive region can be treated as one single entity, the number of segmentation decisions to be made is drastically reduced, which increases segmentation robustness and efficiency.

The power of this approach is that such a partitioning is independent of the object of interest to be segmented and can be computed in advance as a pre-processing step without the need for user interaction to specify the object of interest in the image. Moreover, the method presented here to partition the image in meaningful segmentation primitives is independent of image specific parameters and can be applied without adaptation for the segmentation of a variety of objects in very different images obtained with different imaging modalities. Because of these features, the approach presented here is very well suited to be used in a generic segmentation algorithm with large applicability.

Section 2 of this paper describes how relevant segmentation primitives, whose boundaries correspond to image edges, can be computed automatically from the image data using a fast pseudo-watershed procedure, resulting in an oversegmentation with many, relatively small regions. Section 3 describes how the Minimum Description Length (MDL) criterion is used to reduce the oversegmentation by selectively merging neighbouring watershed regions, such that the total boundary length of the partitioning is reduced while the relevant object boundaries are maintained. Segmentation of the object of interest from the image partitioning is discussed in section 4. The power and the generality of the approach is illustrated in section 5 for MR images of the brain and CT images of the abdomen.

2. Construction of segmentation primitives

We want to partition the image in primitive regions such that all pixels within one region belong to the same object. Boundaries between different objects in a medical image are often characterised by a significant change in image intensity across the boundary, resulting in an image edge. Thus, for all pixels within one region to be likely to belong to the same object, there should be no evidence of edges within that region and all edges in the image should coincide with one of the boundaries of the primitive regions. However, edge detection is very sensitive to noise and image contrast and usually requires additional processing in order to obtain closed boundaries. We use the watershed of the gradient magnitude image (see [4, 15] and the references therein) to partition the image in relevant segmentation primitives. Because the boundaries of the watershed regions coincide with the ridges of the gradient magnitude surface, pixels that are separated by an image edge that corresponds to a gradient magnitude ridge will be assigned different labels by the watershed transformation. When compared with classical edge detection techniques, such as the Canny edge detector [3], the watershed procedure has no problems detecting junctions and, because of the region-based approach, the boundaries found are always closed, independent of image contrast or noise level.

A true watershed transformation, defined by an immersion process analogy, is a complex and computationally expensive procedure [15]. Instead, we use a simple and fast pseudo-watershed algorithm [13], which involves growing downhill Maximum Gradient Paths (MGP, [11]), starting from each pixel in the image, by selecting in a 3×3 (in 2D, or $3 \times 3 \times 3$ in 3D) neighbourhood of each pixel the pixel with the smallest gradient magnitude. In order to reduce the influence of image noise on the gradient computation, the image is smoothed first with a Gaussian kernel before computing the gradient magnitude image using the Sobel [6] operator. Each MGP terminates at a local minimum of the gradient magnitude and each minimum an MGP terminates in is given a distinct label. Watershed regions are obtained by assigning to each pixel in the image the label of the minimum to which its downhill MGP leads (figure 1).

Except for the smoothing parameter used to calculate the gradient magnitude image, the watershed procedure is completely parameterless and does not require any a priori knowledge regarding the objects to be segmented. First results show that this method is sensitive to very small intensity changes in the image, as long as these correspond to a ridge in the gradient magnitude image. This might not be the case for the boundaries of thin line-like (or plate-like) structures, such as for instance the small blood vessels in figure 1(c), which will therefore probably be less well segmented.

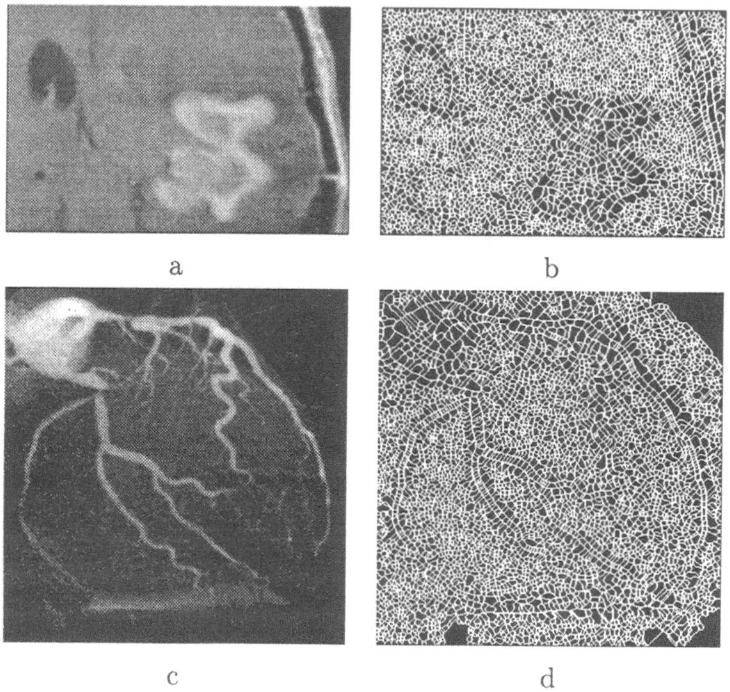

Figure 1. (a) Part of one slice of a coronal MR image of the brain, showing a brain tumor (385 × 241). (b) Result of the pseudo-watershed partitioning procedure applied on the image of figure a (1913 regions). (c) Angiographic image of the coronary arteries (405 × 423). (d) Result of the pseudo-watershed partitioning procedure applied on the image of figure c (4434 regions). In both examples, Gaussian smoothing with $\sigma = 2$ was applied before gradient computation.

3. Merging regions using the MDL criterion

The partitioning obtained with the watershed procedure consists of many, relatively small primitive regions. In order to make this partitioning more useful for segmentation purposes, we reduce the oversegmentation by selectively merging those primitive regions that are likely to belong to the same object, without merging regions that belong to different objects. Selecting the neighbouring regions to be merged is based on the observation that the image intensity is often rather homogeneous within one object, but differs significantly between objects, such that the image intensity distributions of two neighbouring primitive regions that belong to the same object are likely to be more similar than the intensity distributions of two neighbouring regions that belong to different objects. Therefore, merging regions with similar intensity distributions only will most likely preserve the relevant region boundaries in the partitioning. This similarity can be measured by the entropy of the distributions, as described below. Apart from intensity similarity, merging small regions that share a large common boundary segment compared to their size reduces the oversegmentation more than merging large regions that only share a small common boundary segment and should therefore be preferred.

These criteria can be formulated mathematically as a global optimisation problem using the Minimum Description Length (MDL) principle: the optimal image partitioning is the one for which the total number of bits required to completely describe the image data in terms of a specified descriptive language is minimal [12, 14]. In our case, such a description consists of the number of bits needed to code the image intensity of each pixel in each region and the number of bits needed to code the region boundary information. The image intensity data are coded using separate code alphabets for each region R_i by treating the original image intensity x of all pixels in R_i as a stochastic variable with distribution $p_{R_i}(x)$. According to information theory [5], the number of bits $B_I(R_i)$ required to code the image intensity of all pixels in R_i using an optimal coding scheme is given by $B_I(R_i) = n(R_i)H(R_i) = -n(R_i)\sum_x p_{R_i}(x)log_2(p_{R_i}(x))$, with $n(R_i)$ being the number of pixels in R_i and $H(R_i)$ the entropy of $p_{R_i}(x)$. We estimate $p_{R_i}(x)$ from the histogram $h_{R_i}(x)$ of the original image in the region R_i as the convolution of $h_{R_i}(x)$ with a Gaussian window, assuming that the number of pixels in each region is sufficient to obtain reliable image statistics over the range of image intensity values in that region.

The boundary of a 2D region can be coded by specifying one starting point and a chain code of boundary line elements. The overall number of bits needed to code the region boundary information in the image partitioning R using this scheme is given by $B_B(R) = N_r(R) \times b_1 + N_b(R) \times b_2$, with $N_r(R)$ being the number of regions in R, $N_b(R)$ the total boundary length of the partitioning, b_1 the number of bits required to code the starting point and b_2 the number of bits required to code each element of the boundary chain code. We take $b_1 = 2 \times 8 = 16$ for a 2D 256×256 image and $b_2 = log_2(4) = 2$, as the alphabet size of the chain code elements is 4.

The surface of a 3D binary object can be represented as a directed graph [1, 10], linking each boundary surface element with two neighbouring boundary elements. Given one boundary element as a starting point and using this graph representation, the surface is completely defined by specifying for each boundary element the orientation of 2 neighbouring boundary elements, for which only 4 different values are possible for each neighbour. Therefore, the number of bits needed to code the boundaries of 3D regions is given by the same expression as for the 2D case, using $b_1 = 3 \times 8 = 24$ for coding the starting point and $b_2 = 2 \times log_2(4) = 4$ for each boundary surface element.

The total number of bits to code the image data from the image partitioning R using this coding scheme for both 2D and 3D partitionings is given by $B(R) = B_I(R)+B_B(R) = \sum_i B_I(R_i)+B_B(R)$. Merging two neighbouring regions R_i and R_j of R results in a new partitioning R', containing the region $R_k = R_i+R_j$. While more bits will be required to code the image information of R_k than of R_i and R_j separately, the total description length might decrease because no bits are needed to code the common boundary segment of R_i and R_j. The description length reduction $\delta B(R_i, R_j)$ (either positive or negative) of merging R_i and R_j is

$$\begin{aligned}\delta B(R_i, R_j) &= B(R) - B(R') \\ &= [n(R_i) \times H(R_i) + n(R_j) \times H(R_j) + b_1 + n_b(R_i, R_j) \times b_2] \\ &\quad -(n(R_i) + n(R_j)) \times H(R_i + R_j)\end{aligned}$$

with $n_b(R_i, R_j)$ being the number of common boundary elements of R_i and R_j. If $\delta B(R_i, R_j) < 0$, the entropy increase of merging R_i and R_j is not completely compensated by the description length reduction of eliminating the boundary between these regions and these regions should not be merged. If $\delta B(R_i, R_j) > 0$, the image intensity distributions of R_i and R_j are sufficiently similar to consider these regions to belong to the same object and R_i and R_j should be merged into one region R_k in order to minimise $B(R)$. This changes the neighbour relations, requiring an update of the description length reductions of merging this new region R_k with all of its neighbours. Therefore, the order in which the regions are merged influences the outcome of the merging process, making global minimisation of the objective function $B(R)$ a computationally expensive procedure.

Instead, we use an iterative procedure in which regions are merged in a deterministic order. First, the histogram is constructed for each region, the neighbour relations of all regions are extracted from the original image partitioning and the common boundary line length and the description length reduction is computed for every pair of neighbouring regions. At each step the pair of neighbouring regions with the largest positive description length reduction is merged, the neighbour relations are updated and all values $\delta B(R_i, R_j)$ that are affected are recalculated. This is repeated as long as regions can be merged with positive description length reduction. Merging thus terminates when a partitioning is obtained that corresponds to a local minimum of the description length function $B(R)$.

Figure 2 shows the result of applying this procedure on the image partitionings of figure 1. While the oversegmentation has been drastically reduced, most region boundaries corresponding to relevant object boundaries were preserved. Generally, the merging procedure does not completely resolve the oversegmentation problem and additional processing is required in order to segment the object of interest, which is discussed in the next section.

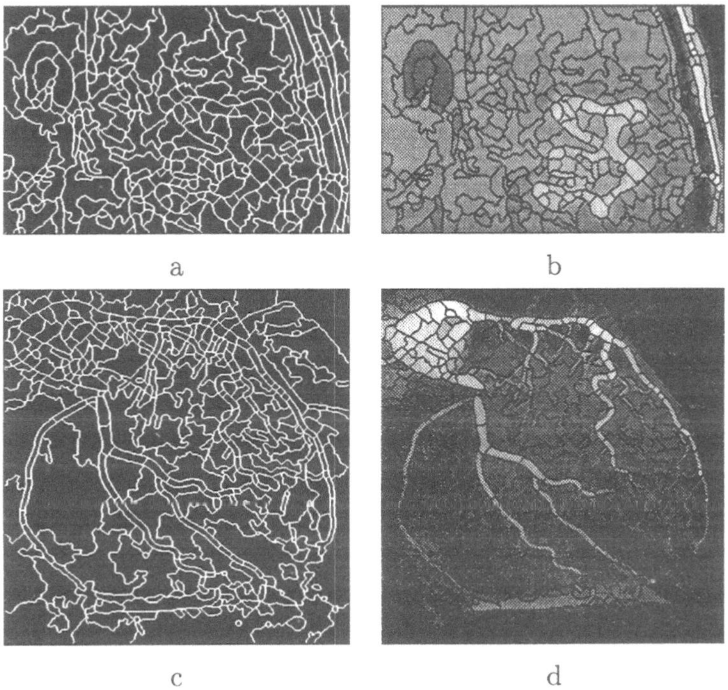

Figure 2. (a) Result of the MDL merging procedure applied on the partitioning of figure 1b (279 regions). (b) Overlay of the partitioning of figure a on the original image. (c) Result of the MDL merging procedure applied on the partitioning of figure 1d (387 regions). (d) Overlay of the partitioning of figure c on the original image.

4. Generic object segmentation

The power of partitioning the image in segmentation primitives is that the segmentation problem is reduced to identifying all primitives in the partitioning that belong to the object of interest in the image. Instead of having to consider all pixels individually, the segmentation algorithm can treat all pixels that belong to the same primitive region as a single entity. This dramatically reduces the number of decisions that needs to be made by the segmentation algorithm, thus increasing segmentation efficiency and robustness. Because of the generality of the partitioning approach, several different segmentation strategies, which have been designed for specific applications, can benefit from the approach and use the information of the image partitioning to facilitate the segmentation task.

One way to extract the object to be segmented from the image is to do this manually. The user can interactively select all regions in the image partitioning that make up the object of interest by dragging the mouse pointer over the object in the original image, while the partitioning itself

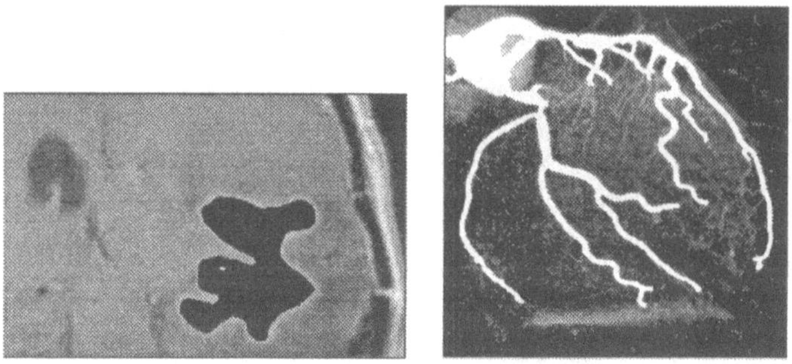

a b

Figure 3. (a) Segmentation of the brain tumor in the image of figure 1a using the partitioning of figure 2a. (b) Segmentation of the coronary arteries in the image of figure 1c using the partitioning of figure 2c. In both examples, the segmentation was done interactively using the intelligent paintbrush tool.

remains hidden to the user. The mouse pointer hereby behaves like an intelligent paintbrush [13, 2] and colors or erases each primitive region it hits, reducing the interaction to selecting one point of each region that belongs to the object to be segmented. Because of the generic properties of the image partitioning, the same tool can be applied successfully to a variety of different applications (figure 3). The paintbrush technique is very efficient and accurate when compared to manual delineation of object contours. When compared to most (semi)automatic segmentation methods, this (semi)interactive method eliminates the need for object and image dependent models and parameters completely, thus avoiding problems with robustness and repeatability. The accuracy of the method depends on the overlap of the boundaries in the image partitioning with the true object boundaries in the original image. Manual editing might be needed to correct for segmentation errors.

More automated and application-specific methods to extract the object of interest from the image partitioning can be considered, in order to further speed up the segmentation process. Because the primitive regions are computed such as to contain only pixels that are likely to belong to the same object, image statistics calculated over all pixels within one region are directly correlated with the image properties of the object to which the region belongs and are therefore more reliable and more significant than statistics calculated rather heuristically within a pixel neighbourhood or within a user specified region. Region growing methods could benefit from this directly, by extending these algorithms to merging primitive regions instead of pixels. The procedure described in section 3 to merge the regions obtained by the watershed procedure is an example of such a strategy. This method can be transformed into a true region growing method by imposing the restriction that regions can only be merged with an initially specified region, which belongs to the object to be segmented. Additional a priori knowledge regarding the object of interest can be incorporated in the formulation of the MDL criterion, for instance by using image intensity models when coding the image data [12].

Another feature that might be useful to extract the object of interest from the image partitioning is the shape of the primitive regions. In the image of figure 2(c) for instance, the regions that belong to the blood vessels have a rectangular shape and the long axis of this rectangle is indicative for the direction of the blood vessel through that region. Therefore, it is to be expected that vessel tracking algorithms exploiting this property will perform better than pixel based algorithms, for which the most likely direction of a vessel through a pixel is usually estimated by local neighbourhood evaluations.

5. Results

The image of figure 4(a) is one slice of a CT image of the abdomen, showing the liver in an advanced state of metastasis. Figure 4(b) shows the boundaries of the segmentation primitives after merging of the original watershed regions. The total computation time on an IBM RS/6000 370 workstation was about 10 seconds. Although there is no clear boundary between the metastatic tissue and the background, the partitioning procedure succeeded in discriminating between them and large parts of the slightly darker metastatic tissue were identified to form one single region.

Figure 5 illustrates the segmentation of 3D objects using the image partitioning approach. This image is one slice of a transversal 3D MR image of the brain with a voxel size of 1 mm $\times$ 1 mm $\times$ 1.2 mm. Slice-by-slice 2D image partitionings were computed for all 33 image slices in which a brain tumor was visible, using Gaussian smoothing with $\sigma = 2$ before calculating the gradient magnitude image. This takes less than 10 seconds per slice on an IBM RS/6000 370 workstation, including gradient computation, watershed transformation and region merging. It takes the user about 2 minutes to completely segment the tumor on all slices when using the intelligent paintbrush tool. A 3D watershed partitioning resulted in 26127 3D regions, which was reduced to only 111 after merging. Computation time was about 10 minutes. In this case, the tumor object corresponded precisely to exactly one region in the image partitioning, such that the tumor could be extracted entirely from the original image by selecting one point of it. Tumor volume calculated from the number of pixels in this region turned out to be 73.4 cm^3.

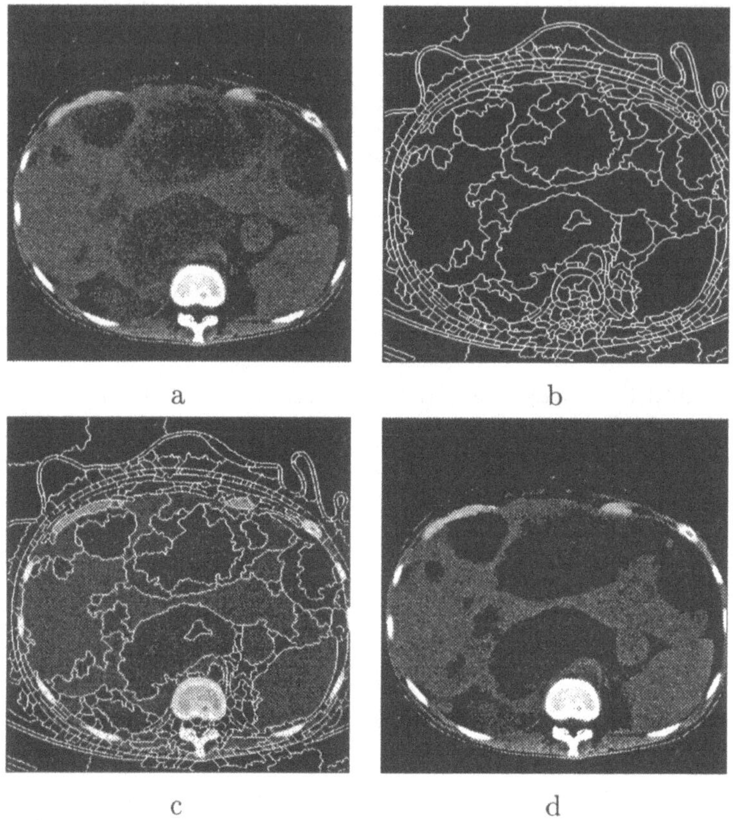

Figure 4. (a) One slice of a CT image of the abdomen (512 × 512). (b) Image partitioning of the image of figure a (595 regions). (c) Overlay of the partitioning on the original image. (d) Segmentation of metastatic liver tissue using the partitioning of figure b.

6. Discussion and conclusion

We use an automatic watershed-like transformation of the gradient magnitude image to compute a partitioning of the image in segmentation primitives. This method is somewhat similar to the method described by Griffin [11], but has the advantage that no continuous representation of the image intensity function needs to be constructed and that we don't need to first detect the saddle points of the gradient magnitude image before tracking the maximum gradient paths. Other image partitionings can be obtained that are based on another definition of ridges and crest lines. For instance, Eberly [8] uses a ridge flow model based on the height ridge definition [7], in which ridge points are computed from the image intensity itself, rather than from the gradient magnitude image. It is not clear to us how the results obtained with these methods compare to ours.

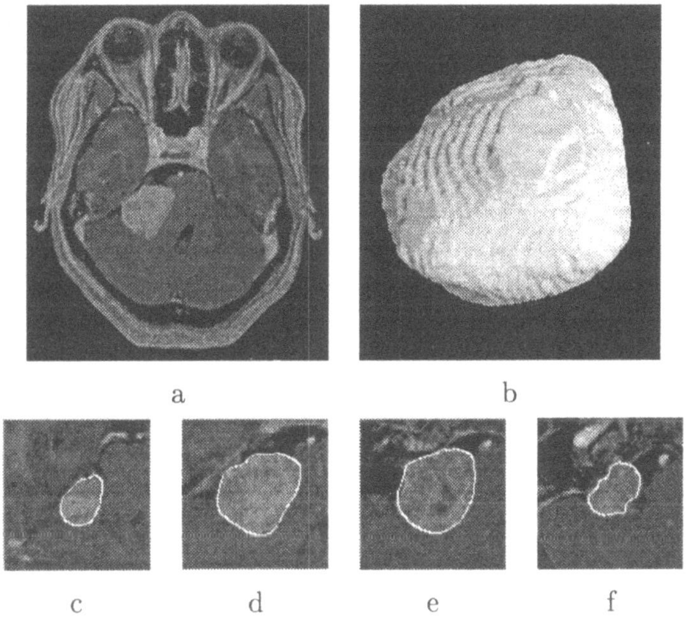

Figure 5. (a) One slice of an MR image of the brain (351 × 391). (b) Tumor segmented automatically from the 3D MR image by the image partitioning procedure (33 slices) (c,d,e,f) Overlay of the contours of the segmented tumor on some selected slices.

In order to reduce the oversegmentation of the watershed partitioning, we use the Minimum Description Length criterion to selectively merge watershed regions based on their image intensity distribution similarity. The outcome of the merging process depends on the scheme that is used to code the image data and the region boundary information. For instance, region merging is favored if a larger bin size is used when constructing the region histograms, because all histograms will be more compact and more similar.

The merging procedure implicitly constructs a hierarchy of image partitionings of different granularity, such that the boundaries of each partitioning at any level form a subset of the boundaries of the partitionings at all lower levels. Instead of only using the final partitioning obtained by merging for the segmentation of the objects of interest in the image, one or more of the intermediate partitionings in the hierarchy may be used too, for instance one with large granularity to quickly but roughly segment the object of interest, and another with smaller granularity for more detailed editing.

As illustrated by the examples shown in this paper, facilitating the segmentation of generic objects by constructing a partitioning of segmentation primitives is a powerful technique. The clinical usefulness of the method is currently being evaluated on some applications from clinical practice. Further work will focus on strategies to automate the extraction of the object of interest from the image partitioning. The segmentation of different object types will most likely require different strategies, resulting in different specialised object-finders for each type [9].

Acknowledgements

This research is supported by the Belgian National Fund for Scientific Research (NFWO). The images were provided by the Department of Radiology, UZ Gasthuisberg, Leuven, Belgium. The algorithms were developed using the Application Visualization System (AVS) from Advanced Visual Systems Inc. on an IBM RS/6000 workstation. D. Delaere and E. Bellon contributed to the implementation of the paintbrush tool.

References

1. E. Artzy, G. Frieder, G. T. Herman, "The theory, design, implementation, and evaluation of a three-dimensional surface detection algorithm", *Computer Graphics and Image Processing*, Vol. 15, pp. 1-24, 1981.
2. D. V. Beard, D. Eberly, B. Hemminger, S. Pizer, R. Faith, C. Kurak, M. Livingston, "Interacting with image hierarchies for fast and accurate object segmentation", *Proc. SPIE Medical Imaging VIII, Newport Beach, CA*, Vol. 2167, pp. 10-17, 1994.
3. J. Canny, "A computational approach to edge detection", *IEEE Trans. on PAMI*, Vol. 11, no. 2, pp. 679-698, 1986.
4. A. C. F. Colchester, "Network representation of 2D and 3D images", in *3D Imaging in Medicine*, K. H. Höhne, H. Fuchs and S. Pizer, pp. 45-62, Springer-Verlag, 1990.
5. T. M. Cover, J. A. Thomas, "Elements of information theory", J. Wiley, 1991.
6. R. O. Duda, P. E. Hart, "Pattern classification and scene analysis", J. Wiley, 1973.
7. D. Eberly, "A differential geometric approach to anisotropic diffusion", in *Geometry driven diffusion in computer vision*, B. M. ter Haar Romeny, Kluwer Academic Publishers, 1994.
8. D. Eberly, S. M. Pizer, "Ridge flow models for image segmentation", *Proc. SPIE Medical Imaging VIII, Newport Beach, CA*, Vol. 2167, pp. 54-64, 1994.
9. F. A. Gerritsen et al., "Some requirements for and experience with Covira algorithms for registration and segmentation", *Technical Report*, XLB 072-94054.1, CDS Advanced Development, Philips Medical Systems Nederland, 1994.
10. D. Gordon, J. K. Udupa, "Fast surface tracking in three-dimensional binary images", *Technical Report*, No. MIPG117, Medical Image Processing Group, University of Pennsylvania, 1987.
11. L. D. Griffin, A. C. F. Colchester, G. P. Robinson, "Scale and segmentation of grey-level images using maximum gradient paths", *Image and Vision Computing*, Vol. 10, no. 6, pp. 389-402, 1992.
12. Y. G. Leclerc, "Constructing simple stable descriptions for image partitioning", *Intern. J. Computer Vision*, 3(1), pp. 73-102, 1989.
13. F. Maes, D. Vandermeulen, P. Suetens, G. Marchal, "Computer-aided interactive object delineation using an intelligent paintbrush technique", presented at *CVRMed'95*, Nice, April 3-5, 1995.
14. J. Rissanen, "Minimum-description-length principle", In *Encyclopedia of Statistical Sciences*, Vol.5, pp. 425-527, Wiley: New York, 1987.
15. L. Vincent, P. Soille, "Watersheds in digital spaces: an efficient algorithm based on immersion simulations", *IEEE Trans. on PAMI*, Vol. 13, no. 6, pp. 583-598, 1991.

DECISION SUPPORT FOR IMAGE INTERPRETATION: A MAMMOGRAPHY WORKSTATION

Paul TAYLOR
Imperial Cancer Research Fund, 61 Lincoln's Inn Fields, London, England

This paper presents an approach to the design of decision aids for medical image interpretation. The approach is based on a model decision making. The model is applied to three decisions in mammography: the classification of abnormalities, diagnosis and the selection of a course of action. A system based on the model draws on both image processing and knowledge based techniques to help improve clinical decision making. A prototype has been developed as a decision aid for radiologists reading mammograms taken as a result of referrals from the UK breast screening programme.

1. INTRODUCTION

Technical advances are continually improving and extending our ability to obtain information about the body in the form of images. More and more of these are captured digitally and presented to clinicians via computer, making computer aids for image interpretation a natural goal of medical informatics research. Much of this research has considered the application of image processing techniques to the detection and classification of abnormalities on medical images. Many successful techniques have been developed but few have been tested to a clinically acceptable standard and none are known to have entered routine use. Some researchers are now arguing for a greater concentration on the problems of applying the developed techniques to clinical practice [1] and for an examination of how they can be made to work in concert with the radiologist to improve his or her decision making. This paper is intended as a contribution to that examination.

A different approach to providing computer aids for radiologists has been to develop knowledge based decision aids which provide advice on diagnosis and treatment decisions. Papers have been published reporting optimistic results [2,3,12,14] but, again as far as is known, none has entered widespread clinical use. This absence of take-up is not restricted to decision aids for image interpretation but affects clinical decision aids and medical expert systems in

Y. Bizais et al. (eds.), Information Processing in Medical Imaging, 227–238.

general. Recent papers [6] have discussed the failure of this research to change clinical practice and a number of explanations have been proposed. Some researchers suggest that both effective performance and user acceptability require a co-operative model of decision making in which the abilities of the user and the machine are matched [10]. The work described here embodies such a model.

A model of medical decision making has been developed at the Imperial Cancer Research Fund and used as the basis for a number of knowledge based decision support systems. The next sections contain a brief summary of this model and an analysis of the different decisions involved in medical image interpretation, these provide the basis for the architecture described in the rest of the paper, which allows image processing and knowledge based systems to be combined in decision support systems.

2. A MODEL OF MEDICAL DECISION MAKING

There are a variety of computer systems which could provide information for a radiologist interpreting medical images: databases of comparison images, image processing, knowledge based systems and numerical decision aids. In each of these kinds of system the mechanism by which support is provided depends very much on the intellectual tradition within which the system was developed. Thus it is the nature of the information that the system designer deals in, rather than the model of the decision maker's task, which has determined the design of the system.

The aim in this project is to explore how a system, which may ultimately be able to call on various knowledge sources, could be based on a model of the decision making process. Fox et al. [4], set out a framework in which decision making is viewed as comprising four activities:

proposing candidate options,
generating arguments for and against the candidates,
establishing relationships between candidates,
evaluating the relative merits of the alternative candidates.

Consider, as an example of a decision , the diagnosis for a woman presenting with a breast lump. A candidate option would be breast cancer and arguments for and against breast cancer could be constructed once other evidence was obtained. A family history of breast cancer is an argument for, relative youth is an argument against. These activities can be represented as a set of rules forming a generic model of decision making.

This approach to decision support requires that knowledge about the kind of decision, about the medical domain and about the particular case be expressed as declarative statements or propositions. Thus if the kind of decision is diagnosis, the domain breast cancer and the particular case a woman who has received a mammogram, the following knowledge would be represented:

knowledge about the kind of decision:
(1) proposal criteria *of* diagnoses *include* causes *of* diagnostic problems
(2) proposal criteria *of* diagnoses *include* kinds *of* supported diagnoses
(3) supporting arguments *of* diagnoses *include* signs
knowledge about the domain:
(4) kinds *of* breast cancer *include* intraductal carcinoma
(5) causes *of* microcalcifications *include* breast cancer
(6) signs *of* breast cancer *include* microcalcifications
(7) signs *of* intraductal carcinoma *include* elongate microcalcifications
knowledge about the case:
(8) mammogram *shows* microcalcifications

One rule from the decision procedure states that:
(9)*IF* Decision class *of* Patient *includes* Problems
and proposal criteria *of* Decision class *includes* Aspect
and Aspects *of* Problem *include* Candidate
THEN candidate options *include* Candidate.

This is a rule used to generate candidate options in a decision, it consults the knowledge about the case (8) to find out what the problem is (microcalcifications), and the knowledge about the decision (1) to find out how solutions can be suggested and the knowledge about the domain (5) to find the candidates, in this case the causes of microcalcifications, which include breast cancer.So, if a patient's microcalcifications are indicated as being a diagnostic problem, this rule can be used with the knowledge given above to generate the information that for this patient:
(10) considered options include breast cancer (from 8,1 & 5)
(11) considered options include intraductal carcinoma (from 10, 2 & 4)
Other rules of the decision procedure are used to construct the arguments for and against the candidates. One such rule will identify that breast cancer is a

supported diagnosis, and also that it is important to establish whether or not the microcalcifications are elongate.

The model is a departure from classical decision theory, firstly because it allows the options to be generated dynamically rather than requiring that they be enumerated before the decision process starts, and secondly because the support for the various options is represented in arguments rather than in straight probabilities. This approach has a number of advantages. If the view of decision making as comprising activities of proposal, argument, relation and evaluation is an appropriate one, and it seems that human decision making must somehow involve processes that perform these activities, then, using the this model to organise information guarantees that the presentation will be relevant to the user. In addition, presenting information in the form of arguments for or against proposed candidates offers a way of handling uncertain information which doesn't rely on numerical estimates of probability which, in domains where data is scarce, may be unreliable. Probabilities may be used in the evaluation of arguments, where the data is available [9]. Finally, since the model is represented in logic, a mathematically rigorous treatment of the method is possible.

A further important advantage of the model is its generality. Qualitatively different kinds of information can be presented via the same mechanism; a decision support system can be fully integrated into a clinician's work routine since the same system and the same interface can provide decision support to all the clinician's computer based activities. The model has already been used as the basis for a number of prototype decision support systems aimed at General Practitioners and oncologists [7]. In the next section we analyse how it model can be applied to provide decision support in mammography.

3. DECISION SUPPORT FOR MAMMOGRAPHY

In Britain women between 50 and 65 are invited to attend for a set of mammograms every three years. This screening programme generates an enormous demand for image interpretation expertise, with 1.5 million mammograms taken every year and each being inspected by two radiologists. There are a limited number of signs for which the radiologist must look when interpreting an image. Certain of these are not readily distinguished from chance arrangements of normal tissue while others occur in benign and malignant forms which are easily confused by the non-expert.

The screening programme has been the focus for the majority of work on computer aids for mammography. However, the screening mammogram is only the first step in a process of investigation and treatment, throughout which the investigation of preference will normally be mammography. The role of the investigation and the nature of the interpretation task is rather different outside the screening situation. The screening programme has to handle a very large number of images, the vast majority of which are normal; a single decision is made on the basis of a rapid inspection. With recalled women the expected incidence is different, the cost-benefit ratio of incorrect decisions changes and a greater variety of future courses of action is available.

The treatment of breast cancer is generally performed following fixed guidelines or protocols. These determine the course in which a sequence of interventions, treatments, investigations and monitoring activities must be performed. Care of patients requiring a complicated sequence of actions is often the responsibility of a team of people making decisions about scheduling, dose control, diagnosis of side effects and evaluation of disease progression.

Three different decisions are made in the interpretation of mammograms: the classification of a potential abnormality, a provisional diagnosis and the selection of a course of action. To apply the model of decision making described above, the relevant knowledge about each decision, about the medical domain and about the patient must be represented .

For each type of decision, the classes of entities which serve as the focus for decision making, the means by which candidate solutions are proposed and the means by which arguments are constructed must be elaborated.

Decision	Focus	Proposal Criteria	Arguments
image feature classification	image feature	interpretation, kinds	properties
diagnosis	sign	causes, kinds	signs, symptoms
selection of the next action	current action	next action	indications

Table 1: knowledge of kinds of decision

In the implemented prototype we have concentrated on diagnosis and the selection of the next action, which may be a further investigation or a plan of more or less frequent screening.

Elements of the work described here are drawn from earlier work on image processing in mammography [16] and decision support for treatment planning [15]. The author visited clinics at the Fondation Bergonie, Bordeaux, King's College Hospital and Guy's Hospital, London. Information obtained in the course of these visits was used to draw up scenarios for a mammography workstation. These were distributed to a small number of clinicians to help establish the requirements for a prototype system.

4. MAMMOGRAPHY WORKSTATION

The prototype allows users to inspect images and to use a system of pull down menus to construct a radiologist's report which is added to the patient record. The patient record consists of a set of assertions describing both the patient's medical history and the planned treatments and investigations. The status of the elements of the plan is checked and updated whenever new information is entered on the patient record. At any point the user may select an item from the patient record as an appropriate subject for decision support. The decision involved might be the classification of an abnormality, diagnosis or the choice between actions in the plan. The knowledge bases containing the representation of the treatment plan and of the possible diagnoses are both available for inspection via browsers. Image processing routines can be requested via the image display to provide additional information.

The system was developed on Sun Workstations using Prolog and C. Prolog is a logic programming language and supports a declarative style of programming appropriate for the knowledge based approach to decision support. A forward-chaining production rule language implemented in Prolog [5] was used to provide a form of object oriented programming. Prolog provides an interface to C, used for the image processing functions.

The system consists of four components: the user interface, the storage of patient data, the decision support system and the image processing routines. The architecture is shown in Figure 1. Rounded boxes represent the displays which make up the User Interface. Information stores are shown as cylinders while square boxes represent the program modules. The arrows show the flow of data, for example information is entered via the image display onto the patient record. The task and patient manager use the information in the patient record and the protocols knowledge base to keep the patient record up to date. The decision procedure consults the decision specifications and, via the knowledge source manager, the other information sources to create a volatile database which is used to generate the decision support screen.

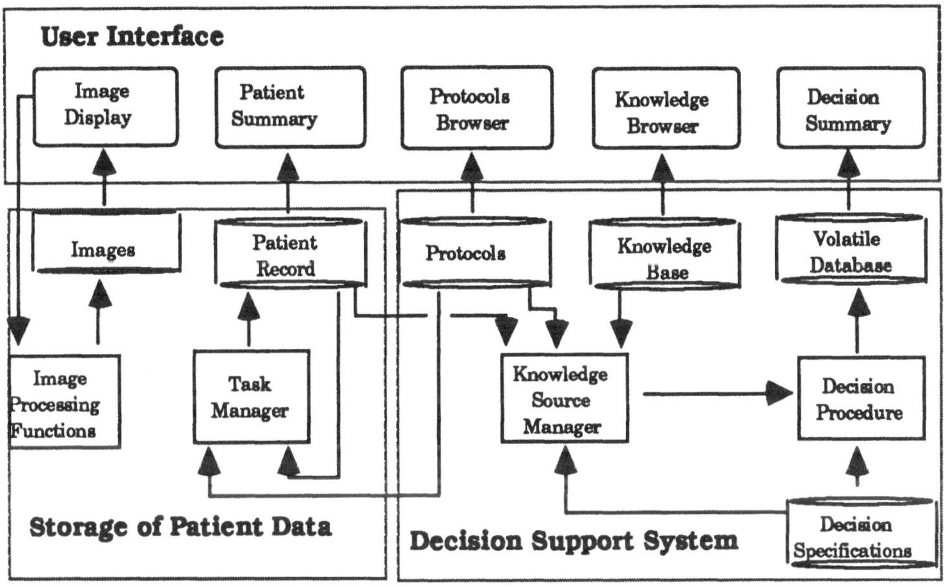

Figure 1: a diagram of the mammography system

The design of the system reflects the need to develop a generic architecture: the information stores are distinguished to allow knowledge to be added about different protocols and other medical domains. The design also embodies assumptions about how a workstation could be used in practice: the storage and management of information is handled by a self-contained sub-system which could be an interface to a hospital information system. The interface is also a separate system, since different imaging modalities have different display requirements. The four components are now discussed in turn.

4.1. The User Interface

At the top-level the user has a set of buttons through which he or she can access the screening mammograms which currently require interpretation, the list of patients awaiting clinical examination and the mammograms of patients referred from the screening programme for further investigation. Principal elements of the interface are: the patient record window, which is displayed whenever a patient's image is interpreted or the results of a clinical examination are entered; the image display windows in which images are displayed and which give access to the system's image processing facilities and to menus used to construct the radiologist's report; the browsers which allow the user to inspect the contents of the knowledge bases and the decision summary window, used to display the reasoning generated by the decision procedure.

The menus used in the construction of the radiologist's report are generated from the knowledge base. The knowledge base contains information about what sort of thing may be found on the radiologist's report: mammographic signs, mammographic flaws and image features; about which kinds of abnormalities are found on mammograms and what properties they can have.

4.2. Storage of Patient Data

Three sets of facts are recorded for each patient. Firstly a concise summary including the patient's name, sex, age and the protocol or treatment plan being followed. The second set of facts represents the protocol. This includes both the actions which are to be performed and the history of actions performed within the current protocol. The third set of facts describe the clinical data. In addition to these three sets of facts, image data is stored.

The protocol is represented as a hierarchy of plans and actions. Some plans include decisions, following a decision the sequence may split and one of two or more plans or actions may be performed. The protocol is represented on the patient record as a series of assertions, one for each plan or action. The status of a plan may be either *relevant*, *pending*, *started*, *performed*, *abandoned* or *discarded*. The rules which define these statusÕs, and which govern the transitions between them form the basis of the task manager, which is invoked whenever new information is entered into the patient record.

The facts which record the clinical data observed of a patient include simple assertions, complex assertions and references. A simple assertion would be a statement that some entity exists (e.g. a breast tumour). A complex assertion states a value, or range of values for a property of that entity (e.g. the breast tumour is 25mm across). A reference is a pointer to an entity, it asserts that some entity exists in a particular relation to the patient (e.g. screening mammograms are stored in file m001).

4.3. The Decision Support System

This part of the system includes the decision procedure, a module called the knowledge source manager and the knowledge sources. The decision procedure contains a declarative definition of the rules making up the generic model of decision making described in Section 2. These rules consult the knowledge sources. The knowledge source manager determines which knowledge source should be queried. In this prototype, knowledge about the kinds of decision is stored in decision specifications, knowledge about the domain is stored in the knowledge base and also in the protocols and knowledge about the patient is stored in the patient record. Ultimately the decision procedure should be able to use the image processing to consult some of the knowledge about the patient contained in the image.

The knowledge base consists of a set of facts, for example:

causes of breast tumour include breast cancer

which can be considered an *object-attribute-value* triple in which 'breast tumour' is the *object* and 'causes' the *attribute* for which 'breast cancer' is the *value*. Since 'breast cancer' also appears as an object, the triple can be seen a link, or relation, connecting two objects.

The system could have a range of knowledge bases for different applications. Currently one knowledge base has been compiled with sufficient facts to implement the protocol for breast screening. This knowledge base currently contains 533 facts. The system contains a single protocol: 'screening protocol', which is stored as a file of facts using the same format as the knowledge base.

The decision specifications contain the facts about the different tasks. These represent the information required by the generic decision procedure. Each decision specification indicates the nature of the focus, the links which can be used to propose candidate solutions and the links which are used in the construction of arguments. It also indicates which knowledge base is to be used by the decision procedure in conjunction with the decision specification.

The information generated by the decision procedure is stored in an object which is specially created for the purpose every time the decision procedure is invoked to consider a fresh problem. This object is termed the volatile database, since information stored in it is not retained permanently. A new volatile database is created every time the decision procedure is invoked with a new problem. The decision procedure works by proposing candidates, then constructing arguments for and against, and then looking for the evidence required by the various arguments. Supported candidates are used to propose further candidates, normally sub-types or refinements of the initial candidate and the process is repeated. The volatile database is used to record the initial candidate, and the refinements. Where the evidence required for an argument is found, a fact recording the argument and noting the evidence is stored in the volatile database. Where the required evidence is not found, the argument is not recorded but the system notes the existence of an 'other sign', a relevant piece of information which is not currently known.

For each candidate the system records the type of decision, the starting point of the decision and the link which connected the candidate to the starting point. For each argument the system records the kind of decision, the kind of argument (confirming, supporting, diminishing or excluding) and the candidate and evidence which the argument links.

4.4. The Image Processing Routines

The facilities offered provide for simple manipulation of the display and some analysis. The most sophisticated analysis provides an indication of the amount of tissue in the image [16]. This uses a technique which places a rectangle in the breast area, on the basis of an automated segmentation of the image. The texture of the image in the rectangle is then measured using a technique called `local skew', developed at Imperial Cancer Research Fund, in which the asymmetry of the histogram is measured in small tiles.

Local skew is indicative of the amount of tissue in the area, since the histograms are made up of pixels with low intensity values (representing fat) and pixels with higher intensity values (representing tissue). The distributions overlap and the shape of the resulting histogram reflects the proportion of dense tissue in the breast. The skew of the histogram, was found to be more accurate than the variance, the mean or fractal measures in predicting dense tissue. Measuring the amount of dense tissue in the breast is useful partly because dense tissue makes mammograms difficult to interpret and so knowing that a mammogram is dense may allow a more careful scrutiny to be planned and partly because there is a relationship between the amount of dense tissue and risk, which may be clarified by having a quantitative measure of density [11].

5. CONCLUSION

An architecture within which knowledge based decision support could be combined with image processing has been presented. The implemented prototype, although it provides some image processing, draws exclusively on the knowledge base when constructing decision support displays. A truly useful system would take much more information from image processing. Swett et al. [14] outline a model of human vision consisting of a pre-attentive phase in which recognition may occur automatically, followed by an attentive phase in which conscious consideration is given to identified features and finally a decision phase in which additional information may be sought. Work on the application of image processing to radiology concentrates either on the placing of prompts on suspicious regions in images, to direct gaze in the pre-attentive phase [8] or on the derivation of information to be used in the conscious decision making phase. The architecture presented, although it accommodates both types of process, is clearly of greater interest for the decision phase.

Two extensions are proposed. The first is to include another class of decision, mentioned in Section 3, the classification of an image abnormality, such as a microcalcification. Microcalcifications appear as small white specks which may be either benign or malignant. A radiologist could indicate an area of the image where microcalcifications appear and request decision support. This could then be provided in the form of arguments for and against the benign and malignant hypotheses. The second extension would be to allow the knowledge source manager to direct certain queries to the images, thus arguments for some hypotheses could be based on photometric measures, such as the elongation of microcalcifications. In order to demonstrate the generic nature of the technology, a system to support the interpretation of CT scans in the management of neuroblastoma is also being developed.

ACKNOWLEDGEMENTS

This research was supported by the Imperial Cancer Research Fund and the Fondation Bergonie. I would like to thank John Fox, Andrew Todd-Pokropek and Saki Hajnal for their guidance and to acknowledge the assistance received from Ian Fentiman, Mike Michell, Marie-Helene Dilhuydy and Beatrice Barreau who provided clinical expertise.

REFERENCES

[1] S.Astley et al. 'Automation in mammography: Computer vision and human perception', in *Proceedings of Biomedical Image Processing* pp 716- 730,SPIE, 1993

[2] F.Bonadonna et al. 'Visual knowledge processing in computer-assisted radiology: aconsultation system', *Medical Informatics*, 17 (1), pp 11-19, 1992

[3] H.M. Cook and M.C. Fox. 'Application of expert systems to mammographic image analysis', *American Journal of Physiologic Imaging*, 4, pp 16-22, 1989

[4] J. Fox et al. 'Logic engineering for knowledge engineering', *Artificial Intelligence in Medicine*, 2, pp 323-339, 1990

[5] S. Hajnal, J. Fox, and P. Krause. 'The Sceptic user manual', *Technical Report 95*, ACL, Imperial Cancer Research Fund, London, 1989

[6] H.Heathfield and J.Wyatt. 'Philosophies for the design and development of clinical decision-support systems', *Methods of Information in Medicine*, 32, pp 1-9, 1993

[7] J.Huang et al., 'Symbolic decision support in medical care', *Artificial Intelligence in Medicine*, 5, pp 415-430, 1993

[8] I.W. Hutt, S.M. Astley, and C.R.M. Boggis. 'Prompting as an aid to diagnosis in mammography', in *Digital Mammography*, Gale et al. eds, pp 389-388, Elsevier, Amsterdam, 1994

[9] P.Krause and D.Clark. *Representing Uncertain Knowledge*, Intellect Books, Oxford, 1993

[10] R.A. Miller and F.E. Masarie. 'The demise of the Greek oracle model for medical diagnositic systems'*Methods of Information in Medicine*, 29, pp 1-3, 1990

[11] A.M. Oza and N.F. Boyd. 'Mammographic parenchymal patterns: A marker of breast cancer risk', *Epidemiologic Reviews*, 15 (1), pp 196-207, 1993

[12] B.D. Stewart et al. 'The analysis of breast abnormalities by computer', in *Digital Mammography*, Gale et al. eds, pp 241-251. Elsevier, Amsterdam, 1994

[13] H. A. Swett, M. L. Giger, and K.Doi. 'Computer vision and decision support', in *Perception of Visual Information*, Hendee, Temple eds, pp 274-315, Springer-Verlag, 1993

[14] H.A. Swett and P.L. Miller. 'Icon: A computer-based approach to differential diagnosis in radiology', *Radiology*, 163, pp 555-558, 1987

[15] P. Taylor. 'The Bordeaux Oncology Support System: Technical description',*Technical Report 161*, ACL, Imperial Cancer Research Fund, London, 1991

[16] P. Taylor et al.,'Measuring image texture to separate 'difficult' from 'easy' mammograms', *British Journal of Radiology*, 67, pp 456-463, 1994

MEDICAL IMAGE UNDERSTANDING USING ANATOMICAL MODELS : APPLICATION TO CHEST X-RAYS

Laurence WILSON[1], Matthew BROWN[1,2], Habib TALHAMI[1], Robert GILL[1], Changming SUN[3] and Bruce DOUST[4]

Ultrasonics Laboratory, Division of Radiophysics, CSIRO, 126 Greville Street, Chatswood, NSW 2067, Australia[1]

Dept. of Computer Science & Engineering, University of NSW, Australia[2]

Division of Mathematics and Statistics, CSIRO, Australia[3]

St Vincent's Hospital, Sydney, Australia[4]

ABSTRACT: A comprehensive "expert assistant" system is being developed with initial application to chest X-rays. This system is characterised by the use of explicit anatomical models for reasoning about the anatomy, and for visualising the anatomical structures identified in the image segmentation. The overall strategy is to compare a modality-independent model with the image(s) by way of an intermediate feature space. The system is implemented to identify major structures on chest X-rays, mainly using long edges. An explicit three-dimensional anatomical model forms a major system component, and the object-oriented structure of this model permits adaptation and anatomical reasoning. The system is implemented in a frame structure using LISP, and features automatic and interactive segmentation of major landmarks in chest X-rays.

1. OVERVIEW OF THE SYSTEM

A role is emerging for "expert assistant" software for medical imaging workstations [13]. Image processing, high-level concepts and prior knowledge are essential components of such systems. Medical image processing systems are diverging from methodologies drawn from computer vision since many of the underlying assumptions of computer vision are not applicable in medical imaging. For example, the segmentation of a scene as objects v. ground is rarely appropriate, especially in the case of projection modalities. Multiple views of the anatomy are frequently obtained by disparate imaging modalities where data fusion cannot be achieved by image registration. But at an expert human level they can be perceived as representations of the same underlying anatomy. It is the underlying thesis of this paper that anatomical reasoning is the basis of modality-independent knowledge, and that explicit anatomical

Y. Bizais et al. (eds.), Information Processing in Medical Imaging, 239–250.

models are an essential component of a medical image understanding system, and therefore of an expert assistant system.

Anatomical knowledge is commonly introduced into medical image segmentation by means of elastic matching to anatomical atlases [1]. When performed in two dimensions such systems commonly encounter difficulty with anatomical variations, particularly those in which three-, rather than two-dimensional deformations are required for a complete description. For these reasons, this work has addressed the use of three-dimensional anatomical models in medical image interpretation. Such modelling as a strategy in image interpretation sets out to capture aspects of the human expert's reasoning, particularly those aspects which occur at the "expert" level, rather than at the level of simple object segmentation. It is believed that the reasoning process of clinicians takes place in "anatomical space", where concepts such as spatial relationships, connectivities, ranges of normality define the metric. The hierarchical, systematised nature of anatomical knowledge is compatible with computer knowledge databases [12], while links to visualisations and interactivity are readily incorporated. Systematic concepts which may be incorporated into an anatomical model include tissue and organ types described through inheritance. For spatial reasoning, the model must be deformable in an anatomically consistent fashion, maintaining, for example, invariant connections between organs. Thus it is not essential that the model represents a detailed normal atlas; however, the model must be deformable to actual anatomies. In practice, an initial state of the model close to average anatomy minimises the "distance" of the excursion in anatomical space from the initial point to an instantiated anatomy.

For compactness, the model needs to be hierarchical and parameterised. One way of achieving this is through Lindenmayer systems ("L-systems") [7,8], an iterative description of complex structures previously applied to plants and described in detail in Section 3. This allows complex structures to be described using a small number of anatomically significant parameters. Shape descriptions need to incorporate a trade-off between the compactness of parametric descriptions and the explicitness of vertex descriptions.

The ability to describe anatomical relationships in natural language [10] is achieved by an object-oriented structure. The principal invariance property is connectedness, that is, the connections between organs must be maintained. A second property, defined as adjacency, refers to contact between organs which may allow them to slide over one another. The description needs to be embedded in a knowledge system where relationships more complex than spatial relationships may be explored, including functional relationships invoked in testing disease hypotheses.

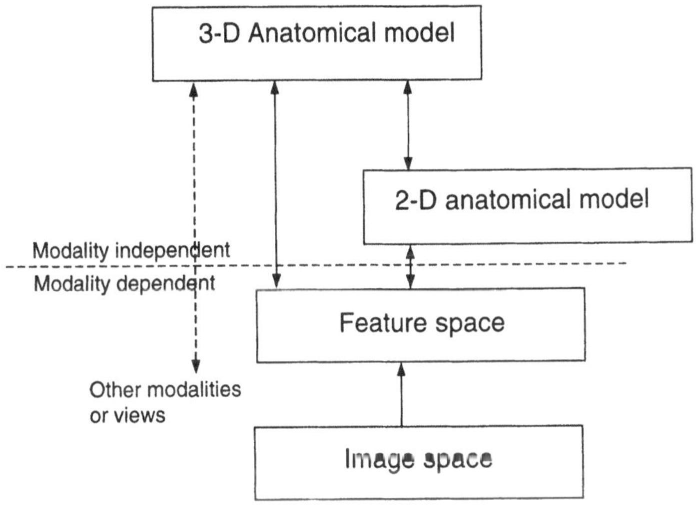

Fig. 1. The relationships between the pixel-based image space, "anatomical space" as defined by two and three dimensional models, and the intermediate feature space.

While several systems exist for describing anatomy using object-oriented models (e.g. Schubert et al., [12]), linking such a system to medical images is a distinguishing feature of the present system (see also [10]). A prime requirement of the system is that the model itself is independent of imaging modality. This requires the intermediate stage of a modality-dependent feature space description which contains symbolic information about image-space features such as edge character and location, textures and multiple relationships among them (Fig. 1). Image segmentation proceeds on an iterative basis, with refinements to the segmentation being based on crude models, which progressively refine the 3-D model.

The system has initially been implemented on posterior-anterior chest X-rays [2]. This clinical problem was chosen as an ideal application for an "expert assistant" because of the large throughput of chest X-rays in most radiology practices [5]. The X-ray projection process superimposes contributions from the whole of the anatomy, requiring 3-D anatomical knowledge. While many lung diseases are diagnosed on the basis of texture changes within the lung fields, the approach in this project has been to restrict the scope to abnormalities more readily described as anatomical or structural variations. This has effectively restricted the diagnoses to identifying abnormalities of long, smooth edges.

2. IMPLEMENTATION

The software has been implemented on a Silicon Graphics Iris Indigo workstation, using LISP for high-level control, and Iris Inventor for model visualisation. A user interface has been designed using the GL Library and resembles a standard PACS user interface, so that interactivity can be investigated by clinicians unfamiliar with image processing interfaces. "Features" not available in conventional workstations, but made possible by the image-understanding capabilities, include anatomical visualisation, automatic and interactive image segmentation, a report on abnormalities of major structures and automatic measurement of lung field sizes and the cardio-thoracic ratio.

Extensive use is made of frame structures for knowledge management [9]. Features of the frame structure include the following:
(1) Each frame is associated with a specific entity such as an organ, organ spatial relationship, image feature etc.
(2) Slots, which contain specific pieces of information related to an instantiation of the frame such as spatial coordinates, names etc.
(3) Procedures which modify slots, create frames according to defined algorithms or translate model knowledge into constraints in blackboard frames.

The frame structure is ideal for organising information in several domains in the current system, and specific applications include the following:
(1) "Blackboard" architecture [9] which is implemented for the feature space description. This is a network of frames with variable slots, in which, for example, edges may be identified from their binary relationships (e.g. "Edge A is to the left of edge B") with these relationships themselves occupying slots. The relationships are stored as fuzzy membership values of sets.

(2) The anatomical model itself stored in a frame-based structure. Slots include name of organ, pointers to parameters defining the organ shape, information about connectivities and adjacencies, X-ray attenuation and collision resolution procedure. Procedures include:
(1) Modifying anatomy of a particular organ to match an image feature;
(2) Modifying nearby organs to maintain adjacency and connectivity;
(3) Generating feature-space information from the model, such as edge positions.
Images are derived from conventional X-ray films digitised using a Lumisys 100 digitiser which records images at up to 50 µm spatial resolution, 12 bit dynamic range.

3. ANATOMICAL MODEL

Organ shape descriptions have received much attention in the literature [3,4,11]. The shape descriptions used for image matching have some strict requirements:
(1) They should be "natural" in order to express variations in anatomy, even at the level of basic topology.
(2) They must be expressible in terms of a small number of parameters which can be varied to match the image to organ shape. Editing organ shape should be intuitive: for example, an extended organ (such as a blood vessel or a stomach) should be movable without affecting its diameter.
(3) Deriving image parameters from the model must be computationally efficient, so that an iterative matching process can be invoked. For projection imaging, this requires rapid re-calculation of the edge locations; for cross-sectional imaging, rapid calculation of arbitrary sections is essential.
(4) A vertex representation of the anatomy must be derivable from the parametric description. This assists both in visualisation, and in invoking the adjacency, connectedness and collision procedures which maintain anatomical consistency.

Most organs are defined in terms of a natural medial axis, usually the embryological axis from which organ growth actually took place. The axis shape can be defined parametrically as a spline curve, or iteratively through L-systems [7,8]. At several points along the medial axis, cross-sections are defined. These are parameterised using Fourier descriptors in polar coordinate space. The axial variation of the Fourier descriptors themselves may be parameterised as, for example, polynomial curves, although this removes the local nature of individual parameters.

Positions, relationships and shapes of organs are influenced by gravity, internal pressures, elasticity and growth patterns. Organ shapes may be influenced by adjacent organs. For example, the shape of the lung is determined almost exclusively by the size and positions of surrounding organs. The concept of organ "indentation", where organs adopt shapes determined by those of the surrounding organs is a well-established anatomical concept, as is "displacement", where organs are moved from their original position, retaining their shape. The space-filling property of anatomical structures implies that each part of an organ surface is shared between two organs, and after collision and space-filling constraints have been satisfied, the vertex representations of organ shapes is highly redundant (in theory by a factor of 2). This used when defining a rank order of organs such that organ i usually indents or displaces organ $i+1$. In the chest, this is achieved by the sequence:

$i=$	1	2	3	4	5
Organ type	Bones	Diaphragm	Heart	Vessels	Lungs

Lung shape descriptors are therefore not included, the lung surfaces being defined by the mediastinum, diaphragm and costal pleurae.

Organ shapes and connectivities may be described in a compact and parametric form using an L-systems description, as used by Lindenmayer for describing plant growth [7,8]. This is an iterative symbolic description in which elements such as generalised cylinders are progressively added to form a complex system, the accretion process mimicking physical growth. This approach has been found to be particularly useful in describing skeletal structures, as shown in Fig. 2.

Organs with more complex shapes can be described by a combination of the L-system representation, with cross-sectional boundaries represented by Fourier contour descriptors. Parametrically the x and y coordinates of each contour in the set can be represented as follows:

$$x = r(c,s)\cos(\theta(c,s))$$
$$y = r(c,s)\sin(\theta(c,s)) \qquad \text{(Eq. 1)}$$
$$z = s$$

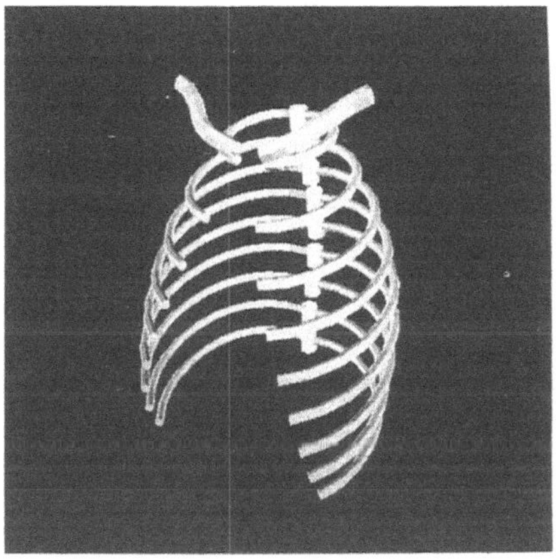

Fig 2. Model of the skeletal structures in the chest, generated using an iterative 9-parameter L-systems approach.

where r is the distance from the centroid and is a function of two parameters: c for points on the contour plane, and s for points on the surface resulting from the ensemble of contours or slices. To achieve a more parametric representation of an individual contour $r(c)$, the discrete Fourier descriptors are obtained as:

$$a_k = \frac{1}{M}\sum_{m=0}^{M-1} r(c)\exp(-2j\pi km / M) \qquad k = 0,\cdots,M-1 \qquad \text{(Eq. 2)}$$

where M is the number of vertices in the contour, and a_k are the Fourier descriptors of the contour. M can be small, typically 4, for relatively simple objects such as the heart. The resulting volume can be represented by the matrix:

$$O = \begin{bmatrix} a_{0,1} & \cdots & a_{M-1,1} \\ \vdots & \cdots & \vdots \\ a_{0,N} & \cdots & a_{M-1,N} \end{bmatrix} \qquad \text{(Eq. 3)}$$

where N is the number of slices. Further economy of parameters can be achieved if each individual column in O is modelled by a polynomial. A model of the heart, derived from this model using a total of 12 parameters, is shown in Fig. 3.

All parameters associated with the prototype anatomy (based on normal CT data) and the instantiated anatomy are stored in a frame structure for incorporation into the feature-space reasoning described below.

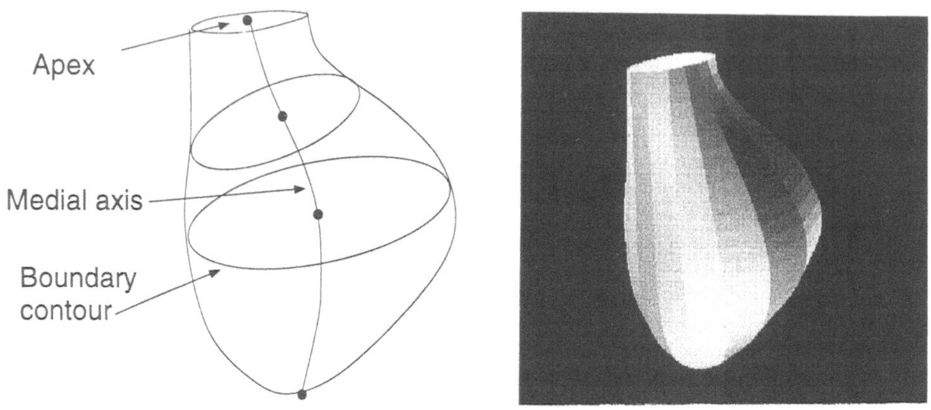

Fig 3 (Left) Fourier-based shape description. The heart surface is described by a series of curves, with a specific orientation and scale relative to its predecessor, starting at the apex. (Right) Rendered 12-parameter model of the heart.

4. FEATURE EXTRACTION AND FEATURE SPACE REASONING

The architecture for reasoning about image data is shown in Fig. 1, i.e. a 2D model used for large-scale landmarking, and a 3D model used for finding more subtle structures such as hila and pulmonary vessels. The features identified during the major landmarking are the edges forming the boundaries of the lung fields, i.e. mediastinum, diaphragm, apices and costal margins. In the 2D model these are represented by line segments to describe shape, and pairwise spatial relationships. Lengths and orientations of line segments and their relationships are described using fuzzy membership values into the "normal" set, which translate into confidence values of normality. The feature space currently contains only edges since complete or partial edges are a significant feature used by radiologists in interpreting projection-based images.

Translating image-based information to feature space begins with midline determination based on a minimum-cost path across the image, locating the mediastinum. Pixel-based processing then produces a large number of candidate edges. The first stage is a Laplacian-of-Gaussian (LoG) operation, followed by a tracking operation on the zero-crossing LoG pixels which establishes the presence of "long" edges. Labelling these edges involves checking their locations and spatial relationships against their positions predicted by the 2D model.

Reasoning in feature space is achieved by a blackboard architecture implemented as a network of frames where each edge (feature) predicted by the model has its own frame in the blackboard (Fig. 4). The blackboard frame stores knowledge about position, length and orientation of edges derived from the model, as well as spatial relationships to other edges. These relationships form the links between frames in the network and, as an edge is identified, the links indicate which other frames' edges are to be updated based on the new information.

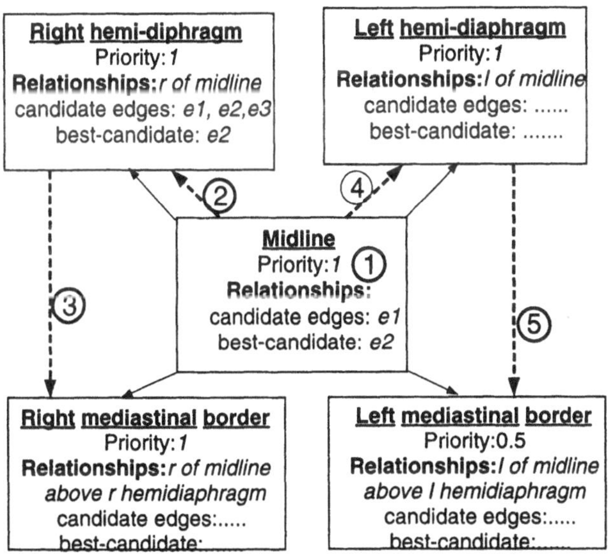

Fig. 4 A network of frames comprising a "blackboard" for image segmentation. The solid arrows represent influence relationships. Dotted arrows represent the order of segmentation. The network is shown immediately after identification of the right hemi-diaphragm. "Priority" refers to the number of established relationships necessary for segmentation of a feature.

Outlines in X-rays are often partial or indistinct due to projection and superposition, so it is difficult to identify edges with certainty. The inference engine is able to entertain multiple edge-labelling hypotheses and that which seems most consistent can be selected using confidences based on fuzzy membership values, and then changed if necessary in the light of further information. Since each constraint imposed by the model generates a confidence value, we can also be specific as to the nature of abnormalities and if the inference engine is not confident about a decision it has made, then it can report the exact reason for this lack of certainty.

The final stage consists of active contour ("snake") refinement of the edges [6]. Edges resulting from the LoG operation are tortuous and affected by overlying structures such as ribs. The active contours refine the edge positions based on the energy function :

$$E_{total} = E_{internal} + E_{image} + E_{model} \qquad \text{(Eq. 4)}$$

Internal forces are heavily constrained for low curvature, and the edge features are selected so that "cusps" (e.g. the costo-phrenic junctions) lie at the junctions of the snakes. External forces are (1) image-based, which attract the snake to areas characterised by large gradients, and (2) model-based, influencing the contour position directly from the currently instantiated two- or three-dimensional model. This is the most direct way of allowing a three-dimensional anatomical model to influence image segmentation. Alternatively, snake contours may be entered manually at the conclusion of the automatic segmentation.

5. CURRENT SYSTEM AND FUTURE DEVELOPMENTS

The current system relies more heavily on two-dimensional than three-dimensional models. However, full 3-D anatomical descriptions have been incorporated into the frame structure, with generation of two-dimensional image space (synthetic X-rays) and feature space (symbolic descriptions) of the three-dimensional anatomical instantiations, and realistic visualisations. A high-level two-dimensional feature space description has been incorporated, which currently provides:

(a) Segmentation of right and left hemi-diaphragms, heart, costal margins, apices.

(b) Report on nature (sharpness etc), location and size of the above features.

(c) Interactive refinement of selected edges using active contours.
(d) Automatic measurement of dimensions such as cardio-thoracic ratio.
In a preliminary sample of 14 patients with abnormal lung anatomy, no feature was misidentified, even with major abnormalities (such as in Fig. 5). However, the system occasionally reported a failure to identify a structure, usually due to an edge being very indistinct.
The result of a segmentation is shown in Fig. 5.

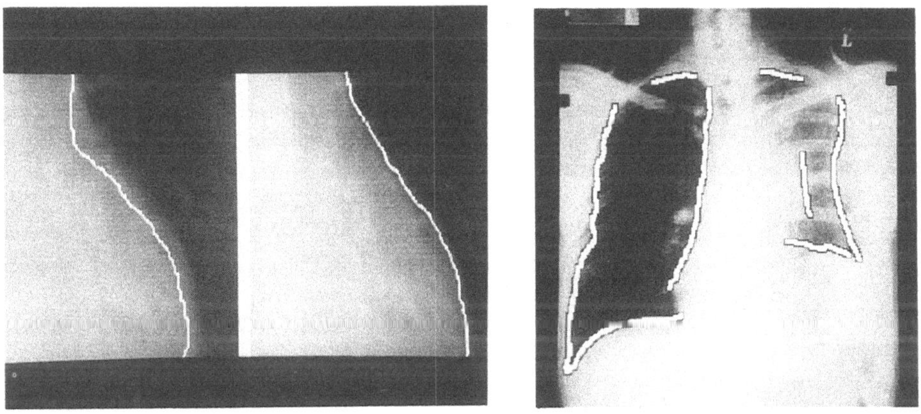

Fig 5. (Left) Refinement of the heart outline using an active contour.
(Right) Result of an image segmentation, showing a normal right lung and left lung affected by consolidation and displaced boundaries. Apices, hemidiaphragms, mediastinal and lateral boundaries are correctly identified.

The three-dimensional model is being introduced progressively to the system; at present it guides the identification of heart boundaries, by optimising the model shown in Fig. 3, and subsequently contributing to the model-based energy terms in Eq. 4.

Thus the system has been successful in coarse-scale image segmentation in chest X-rays in the presence of abnormalities and has already been received well by radiologists. More complete implementation, with full incorporation of a 3-D model, will allow the system to realise its full capabilities. Further enhancement will include other views (notably lateral X-rays) and other modalities, which will take full advantage of high-level, three-dimensional anatomical knowledge incorporated into an expert assistant workstation.

ACKNOWLEDGMENTS

The authors thank George Kossoff, David Robinson, Mark Berman, Leanne Bischof, Yuchong Jiang, Rosemary Irrgang, John Hiller, Claude Sammut, Tatjana Zrimec and an anonymous reviewer for their contributions to this work.

REFERENCES

[1] F.L. Bookstein and W.D.K. Green. "A feature space for derivatives of deformations" in *Information Processing in Medical Imaging*, Barrett and Gmitro eds, pp 1-16, Springer-Verlag, Berlin, 1993.
[2] M.S. Brown, R.W. Gill, T. Loupas, H.E. Talhami, L.S. Wilson, B.D. Doust, L.M. Bischof, E.J. Breen, Y.Jiang and C. Sun. "Model-based interpretation of chest X-rays", in *Computer Applications to Assist Radiology*, Boehme, Rowberg and Wolfman, eds, pp 344-349, Symposia Foundation, Carlsbad, 1994.
[3] T.F. Cootes, A. Hill, C.J. Taylor, and J. Haslam. "The use of active shape models in locating structures in medical images", in *Information Processing in Medical Imaging*, Barrett and Gmitro eds, pp 33-47, Springer-Verlag, Berlin, 1993.
[4] S.A. Cover, N.F. EzQuerra and J.F. O'Brien, "Interactively deformable models for surgery simulation", *IEEE Computer Graphics & Applications*, pp 68-75, November 1993.
[5] K. Doi, M. L. Giger, H. MacMahon, K.R. Hoffmann, R.M. Nishikawa, R.A. Schmidt, K.-G. Chua, S. Katsuragawa, S. Sanada, H. Yoshimura, C.E. Metz, S.A. Montner, T. Matsumoto, X. Chen and C.J. Vyborny. "Computer-aided diagnosis: development of automated schemes for quantitative analysis of radiographic images", *Seminars in Ultrasound, CT and MRI*, 13, pp 140-152, 1992.
[6] M. Kass, A. Witkin and D. Terzopoulos, "Snakes: active contour models", *Int J Comput. Vis*, 1, pp. 321-331, 1987.
[7] A. Lindenmayer, "Mathematical models for cellular interaction in development", Parts I and II., *Journal of Theoretical Biology*, 18, 280-514, 1968.
[8] P. Prusinkiewicz and J. Hauau, *Lindenmayer systems, fractals and plants*. Springer-Verlag, 1989.
[9] E. Rich, *Artificial Intelligence*, McGraw Hill, Singapore, 1983.
[10] G. P. Robinson, A. C. Colchester and L. D. Griffin, "Model-based recognition of anatomical objects from medical images" in *Information Processing in Medical Imaging*, Barrett and Gmitro eds, pp 197-211, Springer-Verlag, Berlin, 1993
[11] C. Roux, V. Burdin and C. Lefevre, "Geometrical models for the analysis of 3D anatomical shapes, application to bone structures" *Proceedings of SPIE*, 1905, pp 660-671, 1993.
[12] R. Schubert, K.H. Höhne, A. Pommert, M. Riemer, Th. Schiemann, and U. Tiede, "Spatial knowledge representation for visualisation of human anatomy and function." in *Information Processing in Medical Imaging*, Barrett and Gmitro eds, pp 168-181, Springer-Verlag, Berlin, 1993.
[13] B.K. Stewart, "Next-generation PACS focus in intelligence", *Diagnostic Imaging International*, pp 81-84, June 1994 .

A QUANTITATIVE COMPARISON OF RESIDUAL ERROR FOR THREE DIFFERENT MULTIMODALITY REGISTRATION TECHNIQUES

Paul F. HEMLER [1,2], Petra A. VAN DEN ELSEN [2,3], Thilaka S. SUMANAWEERA[2], Sandy NAPEL[2], John DRACE[2], John R. ADLER[2]

[2]*Stanford University Medical Center, Stanford University, Stanford, CA*
[3]*Computer Vision Research Group, Dept. of Radiology and Nuclear Medicine, Utrecht University/University Hospital Utrecht, Heidelberglaan 100, 3584 CX Utrecht, the Netherlands*

ABSTRACT

This paper presents a novel technique to quantify and compare residual error throughout the brain, resulting from multimodality image registration. We introduce and use a reference data set and quantification methodology to measure the accuracy of three different registration techniques: stereotactic frame-based, correlation-based, and surface-based. We also use our quantification technique to determine the sensitivity of each registration method to distortion in magnetic resonance images, decreased resolution, and partially overlapping volumes.

1. INTRODUCTION

A variety of techniques are available to register image sets of different modalities, (see [16, 7] for extensive literature reviews). A common requirement of all methods is the identification of similar objects in both sets, which are used to determine a transformation relating the image sets. Objects which have been used successfully for registration include external markers [17, 13], internal landmarks [5, 2, 1], surfaces [8, 6, 4], and geometric features [14]. There are no standard techniques to quantitatively measure and, therefore, compare residual or registration error among these different techniques. Proposed error measures include visual assessment [14], stating a quantity minimized during the registration process [8], or deriving a known transformation using a single image set [6]. Visual assessment is subjective and non-quantitative, making it impossible to accurately compare different registration techniques. Minimized quantities do not necessarily reflect registration error inside the volume of interest [3]. Using a system to derive a known transformation is limited to single modality registration. Moreover, differences between transformations are not intuitively related to registration error. In addition to the lack of a unified registration error measure, there are no standard data sets available where an unbiased, quantitative comparison can be performed.

[1]Correspondence should be addressed to: Paul F. Hemler, Department of Neurosurgery, Stanford Medical Center, 300 Pasteur Drive, Stanford, CA 94305-5327, USA, email: hemler@flamingo.stanford.edu

Y. Bizais et al. (eds.), *Information Processing in Medical Imaging*, 251–262.

We describe a CT and MR reference data set and methodology for quantifying registration error throughout the brain. We used this error measure to quantitatively compare particular implementations of three different registration techniques: stereotactic frame-based [13], voxel intensity correlation [15], and surface-based [4]. We also used the reference data set to quantify the sensitivity of each registration technique to MR distortion, reduced through-plane resolution, and partially overlapping volumes.

2. METHODS

2.1 Reference data set

Four rigid glass tubes containing a dilute copper sulfate solution visible in both CT and MR were inserted into the calvarium of a cadaver. Three of the four tubes were inserted at oblique angles to the transaxial plane; one was nearly perpendicular. Figure 1 shows a projection of the tubes on an axial and sagittal plane. The purpose

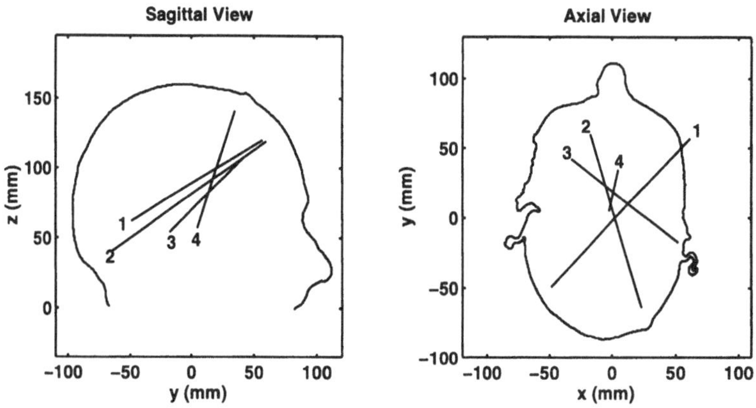

Figure 1: Two views showing the positions of the tubes used in quantifying registration error superimposed on sagittal and axial contours respectively. Each tube extends further in both directions and intersects the skull at two locations.

of the internal tubes was for error assessment of each registration technique and experiment. A BRW (Radionics Inc., Burlington, MA) stereotactic frame and fiducial system was also externally mounted.

High resolution CT [2], (approximate voxel size of $0.7 \times 0.7 \times 1.0$ mm^3), and MR [3],

[2]CT scans were made with a General Electric HiSpeed Advantage CT Scanner (GE Medical Systems, Milwaukee, WI) in helical mode, with 1mm collimation, a table speed of 1mm/s, a reconstruction interval of 1mm, an in-plane field-of-view (FOV) of 345mm with a 0.67mm in-plane resolution, and a through-plane FOV of 180mm.

[3]MR scans were made with a General Electric Signa (GE Medical Systems, Milwaukee, WI) 1.5T scanner using a 3D GRASS sequence with a 30 degree flip angle, a TR of 33ms, a TE of 12ms, 1 NEX, a 280mm in-plane FOV, an in-plane FOV of 280mm with a 1.09mm in-plane resolution, and a through-plane FOV of 124mm.

(approximate voxel size of $1.1 \times 1.1 \times 1.0\,mm^3$) scans were performed sequentially, immediately following tube and frame placement. The CT scan was started just inferior to the bottom of the nose and continued to the most superior portion of the head. The MR scan started just inferior to the nasion and continued superiorly, covering a subset of the CT volume. A more detailed description of the cadaver preparation and reference data set formation can be found in [3].

MR images contain geometric distortions arising from two major sources [11]: gradient field non-linearities and magnetic field inhomogeneities. The former depends only on the scanner while the latter depends both on scanner and imaged object. Some scanner manufacturers provide a 2D in-plane correction for gradient field non-linearities, while patient-dependent magnetic field inhomogeneities are generally ignored. We have improved the geometric accuracy of MR images by applying two correction methods: a 3D correction based on gradient field equations [11] and a magnetic field inhomogeneity correction based on magnetic field maps [12]. The reference data set contains both *uncorrected* and *corrected* MR image volumes.

The internal tubes in CT were semi-automatically removed (*edited*) by first interactively identifying a point near the center of each tube. A threshold intensity level 30% less than the brightness value of the identified point was then used to mask tube pixels in a 30×30 region. This threshold was sufficient to identify tube pixels without detecting spurious pixels elsewhere. Partial volume effects caused some pixels near the tube periphery to be below the threshold value. We dilated the mask by two pixels and found that the tube was completely masked. Intensity values in the 30×30 region outside the mask were randomly chosen as replacement values for those pixels under the mask. This filling method not only replaced tube pixels with nearby brain values, but it also preserved the intensity distribution of the region. We found that the replacement pixels contained a higher spatial frequency then surrounding tissue and tube locations could still be distinguished when viewed with sufficiently high contrast. We filtered the replaced pixels with a gaussian filter, $\sigma = 0.5$, and found it difficult to determine the presence of any tubes. A pixel replacement feature was incorporated for editing pixels near bone or skin surfaces. In this case, a 3×3 neighborhood was replaced with another interactively chosen neighborhood. We *edited* the external stereotactic frame from the MR images using a similar semi-automated technique (Figure 2).

2.2 Error quantification

The tubes in the reference data set generated easily identifiable landmarks on each slice. We determined the centers of all landmarks using a procedure described in [3]. In all three volumes we calculated the best fitting straight line to the tube center location. Figure 3 shows a plot of the distance of each tube-center location to the

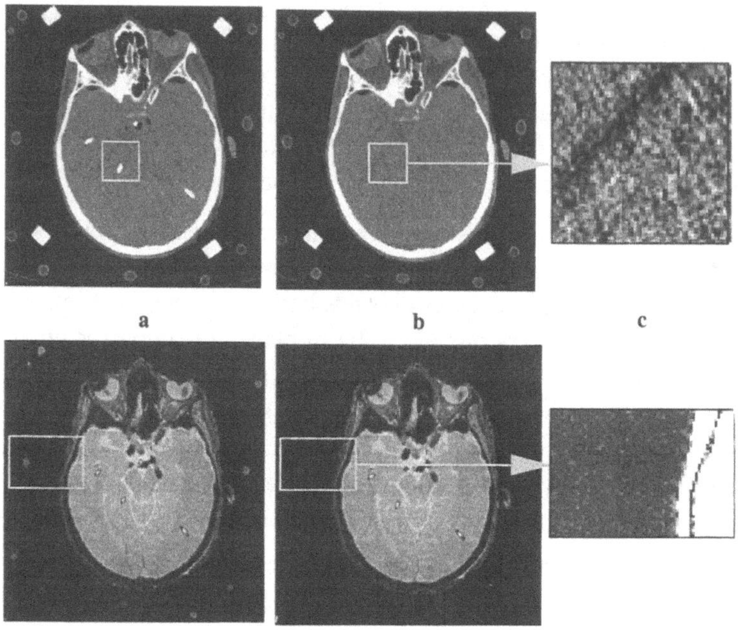

Figure 2: Original and edited CT cadaver images. (a) Original, (b) edited, (c) high contrast region around some edited pixels.

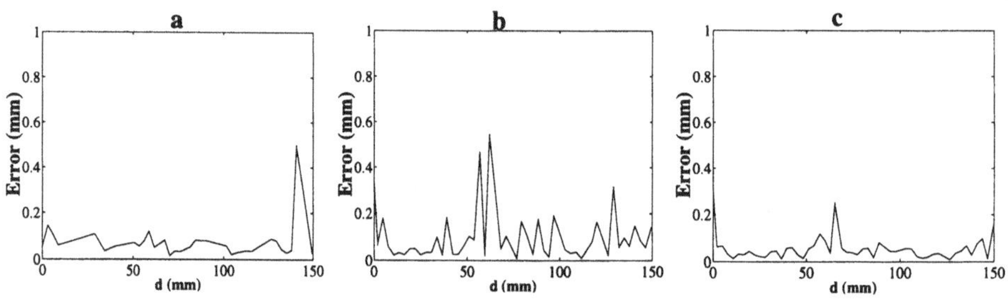

Figure 3: Error plots of detected samples to the best fitting-line for tube 1. (a) CT samples to CT lines, (b) *Uncorrected* MR samples to MR lines, (c) *Corrected* MR samples to MR lines. Note, our MR distortion correction technique reduces the distance between samples and best fitting-line.

best fitting line as a function of distance along the line for tube 1.

Any transformation mapping MR to CT can be used to transform the MR tube-centers into CT space. Plotting the perpendicular distance between each transformed MR tube-center and corresponding CT line as a function of distance along the line provides a measure of registration error at many locations throughout the brain. We refer to these plots as *distance plots*.

Distance plots show a composite error arising from misregistration, incorrect tube-center locations, and MR distortions. The error arising from incorrect tube-center locations in CT is small because distance plots are constructed using the best-fitting line, which averages this error. We use tube-centers determined in MR in the distance plots, (instead of samples along the best-fitting line in MR space), to preserve local distortions that may be present. To estimate the magnitude of the error caused by incorrect tube-center locations in MR, we constructed a distance plot using samples of the best-fitting MR line for the distortion corrected MR.

2.3 Registration techniques

Frame-based registration: We used the externally mounted stereotactic fiducial system to determine the transformation between the *unedited* CT and MR image sets by fitting a geometric model of the external fiducial system to the image data. We determined frame rod centers in both modalities by fitting ellipses [10] to the boundaries [18] of the corresponding rods. The geometric model, consisting of the rod axes, was then fit to the centers of the ellipses using the optimization procedure of Levenberg and Marquardt [9], resulting in the frame-based transformation.

Voxel intensity correlation registration: Bone was approximately segmented in the *edited* CT image set by applying a delayed ramp mapping. Image intensity below 500 HU was set to zero, and above this value a linearly increasing intensity scale was used. The matching transformation between this preprocessed CT volume and the *edited* MR volume was then found using a hierarchical grey value correlation scheme that minimizes the cross-correlation between the preprocessed CT volume and the unprocessed MR volume [15]. Thus, a strong signal obtained from bone in CT was matched with weak signals obtained from bone in MR, resulting in the correlation-based transformation.

Surface-based registration: We registered the skin surface of the *edited* CT volume to that of the *edited* MR volume [4]. A threshold value to isolate the skin surface in each modality was first identified using the full-width at half-maximum technique. The skin surface was then automatically identified as the largest boundary intersecting a vertical line that traverses the image plane half way across each axial brain section. A transformation minimizing the perpendicular distance between a piecewise-linear surface approximation in CT and surface points in MR was then determined using an iterative, non-linear optimization algorithm developed by Levenberg and Marquardt [9].

2.4 Registration experiments

Using the reference data set and quantification methodology, we performed a variety of experiments for each registration technique. In all experiments, residual error was measured with the highest accuracy by determining tube sample locations for computing distance plots in the highest resolution images.

MR distortion: For each registration technique, we performed two experiments to reveal sensitivity to gradient field non-linearities and inhomogeneity distortion in MR images. In the first experiment we registered the CT data with the *uncorrected* MR data. The MR images in this data set were extracted directly from the scanner, and correspond to typically-generated images. In the second experiment, all CT data and all *corrected* MR data were used.

Lower through-plane resolution: We evaluated the sensitivity of each registration method to lower through-plane resolution data in two experiments. In the first, we simulated 3mm data in both CT and *corrected* MR, by averaging pixel values on every three consecutive slices. Similarly, for the second experiment we simulated 5mm data by averaging every five consecutive slices.

Partially overlapping volumes: Because CT uses ionizing radiation, scanning is often confined to a specific region. MR does not suffer from this drawback, and protocols frequently image larger volumes. As a result, clinically acquired multimodality scans typically do not image exactly the same volume. To investigate sensitivities of the registration methods to this imaging scenario, we performed two experiments with data having a smaller percentage of overlap between CT and MR volumes then the previous experiments. In the first experiment, the most superior 76, 1mm CT slices were used and in the second the most inferior 87, 1mm CT slices were used. In both cases all *corrected* MR images were used.

3. RESULTS AND DISCUSSION

The goal of this work is only to quantify and compare the residual error of each registration method. However, there are many other important criteria for evaluating different registration methods including: speed, user-friendliness, amount of interaction, special scanning requirements, and retrospective utility. The stereotactic frame-based method is precise, requires minimal user interaction, and is computationally efficient. The large degree of patient discomfort reduces clinical applicability, and retrospective registration between images without a frame is impossible. The grey value correlation method requires very little interaction, and is devoid of user subjectivity. Volume correlation and interpolation are repeatedly performed by this method, contributing to a high computational expense. The surface-based technique requires user interaction and has a high computational cost for large data sets. The final transformation is dependent on an initial estimate and this sensitivity must be further evaluated.

MR distortion: We found the frame-based registration technique resulted in significantly larger residual error when using *uncorrected* MR data compared to the error found using *corrected* MR data. Neither the correlation or surface-based method showed a similar sensitivity to MR distortion. For all experiments using high-resolution CT and *corrected* MR images, we found a similar amount of residual error for the frame and correlation techniques. The error for the surface-based technique was slightly higher. With both lower-resolution data sets, the correlation and surface-based techniques resulted in approximately the same amount of residual error, while the error for the frame-based technique was significantly less. In both lower-resolution cases, the residual error for all registration techniques was less than the voxel dimension.

The mean and maximum distance of all distance plots for each tube, registration method, and experiment are shown in Table 1. The following sections describe the results of the different registration techniques by showing distance plots for one (tube 1) representative tube.

Figure 4 (a) and (b) and Table 1, (run 1 and 2), show a dramatic decrease in error after distortion correction for the frame-based registration technique. This plot suggests that the distorting effect of the material in the fiducial system of the stereotactic frame in MR imaging can adversely affect registration accuracy. The similarity between corresponding distance plots suggest that correlation and surface-based registration are less sensitive to the MR distortions corrected by our system. This may be because both methods use a global cost function which average local MR distortion. Additionally, there is minimal MR distortion inside the brain, which is where we measured the residual error. Grey value correlation used more data than the surface-based approach, which may account for the lower residual error of the former approach compared to the latter, as shown in Figure 4 (a) and (b).

The sensitivity of the distance plots to inaccurate MR tube-center locations can be estimated by constructing an additional distance plot using samples of the best fitting MR line, instead of the original MR samples, for the corrected MR set. Table 1, (run 2*), shows the average and maximum distance for a distance plot constructed in this fashion. The average error in this plot is about the same as the average error found using the original MR tube samples. The maximum error decreases only slightly.

Lower through-plane resolution: Figure 4 (c) and (d) and Table 1, (run 3 and 4), show that registration error using the stereotactic frame hardly changes as through-

		Frame		Correlation		Surface	
Run	Tube	μ	ρ	μ	ρ	μ	ρ
1	1	1.97	2.51	0.50	0.80	1.23	1.75
	2	1.47	2.00	0.64	1.42	1.21	1.80
	3	2.14	2.49	0.53	0.91	1.14	1.37
	4	2.55	3.30	0.62	1.10	0.44	1.10
2	1	0.41	0.96	0.30	0.82	1.18	1.89
	2	0.53	1.06	0.61	1.11	1.13	1.57
	3	0.44	1.10	0.37	0.61	1.05	1.37
	4	0.53	0.99	0.59	0.90	0.36	0.64
2*	1	0.42	0.79	0.30	0.53	1.18	1.58
	2	0.50	0.88	0.64	1.00	1.15	1.42
	3	0.45	0.99	0.36	0.38	1.05	1.24
	4	0.43	0.97	0.57	0.81	0.30	0.41
3	1	0.30	0.70	1.03	1.99	1.37	2.11
	2	0.40	0.83	1.30	2.67	1.28	1.91
	3	0.29	0.80	1.26	2.43	1.16	1.39
	4	0.49	0.93	0.81	1.53	0.36	0.58
4	1	0.30	0.89	1.07	2.70	1.49	2.73
	2	0.51	0.97	0.84	1.30	1.41	1.95
	3	0.44	1.04	0.68	1.68	1.29	1.81
	4	0.47	0.84	0.82	1.50	0.47	0.69
5	1	0.48	0.92	0.42	1.06	0.53	0.89
	2	0.43	1.08	0.57	1.00	0.77	1.03
	3	0.26	0.70	0.59	0.90	0.72	0.95
	4	0.58	1.03	0.73	0.98	0.50	0.91
6	1	0.73	1.30	0.81	1.32	1.37	1.99
	2	0.51	1.06	0.77	1.42	1.24	1.80
	3	0.56	1.06	0.69	0.95	1.14	1.37
	4	0.60	1.11	0.62	0.92	0.31	0.50

Table 1: Shows the mean, μ, and the maximum, ρ, distance between all transformed MR tube samples and CT lines for three registration methods. In run 1, the entire CT data set and *uncorrected* MR set was used. In run 2, the same CT set and *corrected* MR data set was used. Run 2* used the same data as run 2, but the distance plots were computed by resampling a best fitting-line to the MR samples, instead of using the original MR samples. In run 3 and 4, simulated 3mm and 5mm CT and *corrected* 3mm and 5mm MR images were used. In run 5, the most superior 76, 1mm CT slices and entire *corrected* high resolution MR set was used. In run 6, the most inferior 87 CT slices and entire *corrected* MR set were used.

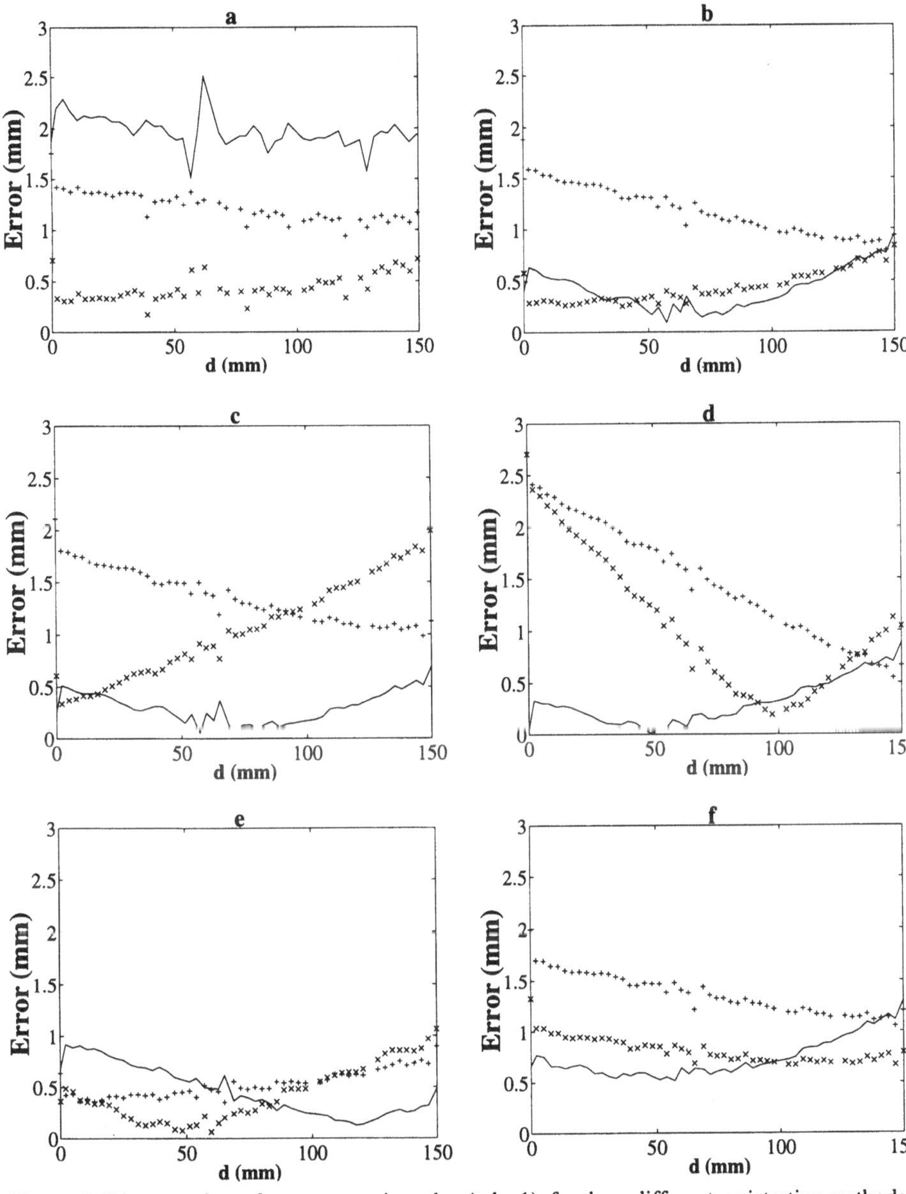

Figure 4: Distance plots of a representative tube, (tube 1), for three different registration methods. In the plots, the solid line corresponds to distance plots for frame-based transformations, the x is for the correlation approach, and the + is for the surface-based method. Plot (a) MR images before distortion correction, and (b) is after correction. In plot (c) 3mm and (d) 5mm data was simulated in both CT and *corrected* MR. Plot (e) shows the results of registering the most superior 76 CT slices and (f) the most inferior 87 CT slices with the entire *corrected* MR data set.

plane resolution decreases. This is because blurring due to lower resolution occurs uniformly in all directions. The center of mass of the stereotactic fiducial system is not sensitive to this type of blurring. Since target location accuracy is limited by the largest voxel dimension, it is important to note that Figure 4 (c) and (d) do not imply submillimeter target localization using a stereotactic frame with 5mm image data; it only indicates submillimeter CT to MR registration. The error in both correlation and surface-based methods increased with larger slice spacing but remained below the voxel dimension. This suggests that either technique performs well with lower resolution image data.

Partially overlapping volumes: Figure 4 (e) and (f) and Table 1, (run 5 and 6), suggest that a large portion of the residual error in previous plots for the surface-based technique is due to the most inferior portion of the head. This may be because the complex surfaces in the lower portion, e.g., the nose, are not adequately represented by the piecewise-linear approximation employed by this system. The smaller error in tube 4 (the tube nearly perpendicular to the transaxial slices), indicates that most of the error is in the through plane direction. The correlation method shows a slight increase in error when registering only the most inferior portion of the head.

In both subvolume experiments, registration error using the stereotactic frame remained constant. This result suggests that because of the known geometry of the frame, an adequate transformation can be determined even from a subsection.

4.CONCLUSIONS

The reference data set and quantification methodology presented in this paper provide an excellent basis for quantifying and comparing residual error throughout the brain for different multimodality registration techniques. When measuring the registration error along the internal tubes, we conclude the following.

- Frame-based registration was highly impacted by MR spatial distortion. Our MR distortion correction technique significantly improved the residual error for this registration method.
- Both grey value correlation and the surface-based approach were insensitive to the types of MR distortion corrected by our technique; the error changed only marginally.
- Both the frame-based and grey value correlation methods generated similar error for the high resolution, distortion-corrected MR image set. The maximum error was equivalent to the voxel dimension. The error of the surface-based approach was slightly higher and most of the error was in the through-plane

direction.

- The accuracy of the frame-based registration technique was relatively insensitive to through-plane resolution and partially overlapping volumes when using the corrected MR data set.
- Both grey value correlation and the surface-based approach were similarly accurate with low resolution data and the maximum error was always within the voxel dimension.
- A large portion of the error for the surface-based approach seems to be caused by difficulties in modeling the inferior portion of the volume.

It is our intention to obtain similar reference data sets using a variety of acquisition protocols. These additional sets should prove useful in further evaluating the sensitivities of different registration techniques.

5. ACKNOWLEDGEMENTS

This research was funded in part by the Lucas, Whitaker, and Murray Foundations, the Department of Veterans Affairs and the Netherlands Organization for Scientific Research (NWO). A special thanks for the extraordinary effort of Professor John F. Dolph of the Department of Anatomy, Anne Sawyer of the Radiological Sciences Laboratory, Department of Radiology, and Linda McGraw of the Department of Radiology at Stanford University. This work could not have been accomplished without their cooperation and expert assistance.

6. REFERENCES

[1] R. Boescke, T.Bruckner, and G. Ende. Landmark Based Correlation of Medical Images. *Phys. Med. Biol*, 35(1):121–126, 1990.

[2] A. Evans, S. Marret, L.Collins, and T. Peters. Anatomical-Functional Correlative Analysis of the Human Brain using Three Dimensional Imaging System. In *Proceedings SPIE Vol 1092 Medical Imaging III: Image Processing*, pages 264–274. SPIE Press, Bellingham, WA, 1989.

[3] P.F. Hemler, S. Napel, T. Sumanaweera, R. Pichumani, P.A. van den Elsen, D. Martin, J. Drace, and J. Adler. Registration Error Quantification of a Surface-Based Multimodality Image Fusion System. *Medical Physics, accepted pending revision*, 1995.

[4] P.F. Hemler, T. Sumanaweera, P.A. van den Elsen, S. Napel, and J. Adler. A System for Multimodality Image Fusion. In *Proceedings of the Seventh Annual IEEE Symposium on Computer-Based Medical Systems*, 1994.

[5] D.L.G. Hill, D.J. Hawkes, J.E. Crossman, M.J. Gleeson, T.C.S. Cox, E.E.C.M.L. Braccy, A.J. Strong, and P. Graves. Registration of MR and CT Images for Skull Base Surgery using Point Like Anatomical Features. *British Journal of Radiology*, 64:1030–1035, 1991.

[6] H. Jiang, R. Robb, and K.S. Holton. A New Approach to 3-D Registration of Multimodality Medical Images by Surface Matching. In R.A. Robb, editor, *Proceedings SPIE Vol 1808 Visualization in Biomedical Computing*, pages 196–213. SPIE Press, Bellingham, WA, 1992.

[7] C. R. Maurer and J. M. Fitzpatrick. A Review of Medical Image Registration. In R.J. Maciunas, editor, *Interactive Image-Guided Neurosurgery*, chapter A Review of Medical Image Registration, pages 17–44. American Association of Neurological Surgeons, 1993.

[8] C.A. Pelizzari, G.T.Y. Chen, D.R. Spelbring, R.R. Weichselbaum, and C.T. Chen. Accurate Three-Dimensional Registration of CT, PET, and/or MR Images of the Brain. *Journal of Computer Assisted Tomography*, 13(1):20–26, 1989.

[9] W.H. Press, S.A. Teukolsky, W.T. Vetterling, and B.P. Flannery. *Numerical Recipes in C The Art of Scientific Computing*. Cambridge University Press, 1992.

[10] T. Sumanaweera, G. Glover, S. Song, J. Adler, and S. Napel. Quantifying MRI Geometric Distortion in Tissue. *Magnetic Resonance in Medicine*, pages 40–47, 1994.

[11] T.S. Sumanaweera. *Segmentation and Distortion Correction in Medical Imaging*. PhD thesis, Stanford University, Dept. of Electrical Engineering, Stanford, CA 94305, 1992.

[12] T.S. Sumanaweera, G. Glover, T. O. Binford, and J. R. Adler. MR Susceptibility Misregistration Correction. *IEEE Transactions on Medical Imaging*, 12(2):251–259, 1993.

[13] T.S. Sumanaweera, G.H. Glover, P.F. Hemler, P.A. van den Elsen, D.Martin, J. R. Adler, and S. Napel. MR Geometric Distortion Correction for Improved Frame-Based Stereotaxic Target Localization Accuracy. *Magnetic Resonance in Medicine, Accepted pending revision*, 1994.

[14] P.A. van den Elsen, J.B.A. Maintz, E.J.D. Pol, and M.A. Viergever. Automatic Registration of CT and MR Brain Images using Correlation of Geometrical Features. *IEEE Transactions on Medical Imaging, in press*, June 1995.

[15] P.A. van den Elsen, E.J.D. Pol, T. Sumanaweera, P.F. Hemler, S. Napel, and J. Adler. Grey Value Correlation Techniques used for Automatic Matching of CT and MR Brain and Spine Images. In R. Robb, editor, *Proceedings SPIE Vol 2359 Visualization in Biomedical Computing*, pages 227–237, Bellingham, WA, 1994. SPIE Press, Bellingham, WA.

[16] P.A. van den Elsen, E.J.D. Pol, and M.A. Viergever. Medical Image Matching - A Review with Classification. *IEEE Engineering in Medicine and Biology*, 12(1):26–39, March 1993.

[17] P.A. van den Elsen and M.A. Viergever. Marker-guided Multimodality Matching of the Brain. *European Radiology*, 4(1):45–51, 1994.

[18] S. J. Wang and T. O. Binford. Local Step Edge Estimation: A New Algorithm, Statistical Model and Performance Evaluation. In *"Proceedings of the Image Understanding Workshop"*. DARPA, 1993.

AUTOMATED MULTI-MODALITY IMAGE REGISTRATION BASED ON INFORMATION THEORY

A. COLLIGNON, F. MAES, D. DELAERE, D. VANDERMEULEN, P. SUETENS AND G. MARCHAL
Laboratory for Medical Imaging Research, ESAT-Radiology K. U. Leuven, Kardinaal Mercierlaan 94, 3001 Heverlee, and U. Z. Gasthuisberg, Radiology, Herestraat 49, 3000 Leuven, Belgium.

Abstract. We propose an information theoretic approach to the rigid body registration of 3D multi-modality medical image data. The "mutual information" of grey-value pairs is proposed as a new matching criterion. The new voxel similarity based (VB) registration algorithm is applied to real world images of the human head (CT, MR, and PET volumes), and compared to stereotactic and 2 other voxel based registration solutions. Our results show that subvoxel accuracy can be obtained completely automatically and without any pre-segmentation.

1. Introduction

Image registration refers to the geometric alignment of two or more images that were generated at a different time or by different imaging modalities. Multi-modality image registration is a fundamental task in three-dimensional (3-D) medical image processing and has numerous applications. Medical diagnosis for instance benefits from the complementarity of the information in images of different modalities. In radiotherapy planning, dose calculation is based on the CT data while tumor outlining is often better performed in the corresponding MR scan. The analysis of brain function requires the correlation of functional information from PET data with anatomical information from the corresponding MR data, etc.

A general overview of image registration techniques has been compiled by Brown [1]. Two other articles survey image registration techniques and their

Y. Bizais et al. (eds.), Information Processing in Medical Imaging, 263–274.

applications in medical imaging [9], [13]. From these surveys we learn that the bulk of registration algorithms are either frame based, point landmark based (using external and/or internal landmarks), surface based, or voxel based. Point landmark and surface based registration algorithms are typically used for the alignment of anatomical scans, while voxel similarity based registration algorithms seem to be more robust for alignment problems involving functional scans ([7] and [15]), although many people are using surface based registration algorithms in the latter case also.

Voxel based (VB) registration algorithms optimize a function measuring the similarity of all possible pairs of voxel grey-values (or derived features, e.g. gradients). The main advantage of VB methods is that feature calculation is straightforward or even absent when only grey-values are used and thus, in theory, accuracy is not limited by segmentation errors as in surface based algorithms. Simple correlation of grey-values works only for different modalities if there exists a *linear* mapping between grey-values [1], which is unfortunately not always the case. Nonetheless, interest in VB registration algorithms has recently revived by the success of Woods' AIR package in clinical practice [16], which has been made more flexible by Hill et al. [7]. Their basic assumption is that the grey-value mapping should be a function (e.g. to each MR grey-value correspondends at most one PET-value). They accomplish this heuristically by demanding that a global measure of the variance of the grey-value ratios, with the function's domain variable as the ratio's denominator, be minimal. We recently compared these and some other VB matching criteria [2]. All of them are heuristic in nature. In the same work we first suggested an information theoretic approach to the design of a non-heuristic multi-modality matching criterion and we proposed to use the entropy of the joint probability distribution $p(g_1, g_2)$ of the combined grey-values of all common voxel pairs in the two images to be matched:

$$E(g_1; g_2) = - \sum_{g_1, g_2} p(g_1, g_2) \log \left(p(g_1, g_2)\right) \tag{1}$$

In this article we use a basic theorem of statistical inference theory to derive a new version of our entropy based VB matching criterion: the mutual information or relative entropy of the joint probability distribution of the grey-values of two (or more) scans. This criterion is more robust with respect to the partial overlap problem and has some interesting mathematical properties as well. Furthermore it is based on the original grey-values only and requires no segmentation at all! Therefore, our new VB algorithm is more robust than any of the surface based registration algorithms and than any of the heuristic VB registration algorithms. We show results obtained with real world data in section 3. Those results are compared with that of stereotactic and 2 other VB registration methods. First, however, we will present the theory and discuss the implementation issues in detail. In the discussion of the results we discuss robustness and accuracy of our approach, and compare it to other VB algorithms.

2. Method

2.1. THEORY

In the sequel we restrict the geometric transformations to *rigid body transformations* T_α (α consists of 3 translations and 3 rotations). The rigid body assumption is well satisfied inside the skull in 3-D scans of the head, if abstraction is made of scanner calibration problems (that can lead to inaccurately known voxel sizes) and problems of geometrical distortions (e.g. in MRI), both of which can be dealt with by careful calibration and scan parameter selection respectively. The applicability of the mutual information matching criterion is not restricted to rigid body registration, however.

The selection of the new VB matching criterion (like in Woods [15], Hill [8], and Collignon [2]) is based on the observation that the two-dimensional grey-value histogram or scatter plot of the common voxel pairs of a pair of registered images will be dispersed by misregistration as a function of $\mid \alpha^* - \alpha \mid$, where α^* is the registration solution.

We will now present a new measure of scatter-plot dispersion that also has an interesting mathematical interpretation: relative entropy or mutual information of the scatter-plot. Intuitively, we expect the dispersion to be minimal when the interdependence of the grey-values of both images to be registered is maximal. Two signals are maximally interdependent when they are related by a one-to-one mapping, i.e. when g_1 is a function of g_2 and vice versa. This is seldom so for multi-modality data. As it happens, the mathematical definition of mutual information [3]:

$$S(X;Y) = \sum_{x,y} p(x,y) \log \left(\frac{p(x,y)}{p(x)p(y)} \right) \tag{2}$$

also known as Shannon information of two random variables X and Y with marginal and joint probability distributions $p(x)$, $p(y)$ and $p(x,y)$ resp., satisfies the following properties (for a proof see [11]):

1. Symmetry: $S(X;Y) = S(Y;X)$
2. $S(X;Y) = 0 \Leftrightarrow (X$ and Y are independent$)$
3. Non-negative and bounded: $0 \leq S(X;Y) \leq \min(S(X;X), S(Y;Y))$
4. Invariance: (T is a one-to-one mapping) $\Rightarrow S(X;T(X)) = S(X;X)$

Interpretation of these properties confirms our intuition. Mutual information measures the degree of interdependence, which is bounded below by complete independence and bounded above by one-to-one mappings. We conclude that a mathematically sound matching criterion can be obtained by substitution of X and Y for the grey-values g_1 and g_2 of the images to be registered in (2). Moreover, it allows for much more general grey-value relationships than do the matching criteria proposed by Woods and Hill et al. or than any of the other currently available heuristic ones (see [2] for an overview).

2.2. IMPLEMENTATION

The algorithm can be summarized by the following equation:

$$\alpha^* = \arg \min_{\alpha} \left(-\sum_{s} p(f_1(T_\alpha s), f_2(s)) \log_2 \left[\frac{p(f_1(T_\alpha s), f_2(s))}{p(f_1(T_\alpha s))\, p(f_2(s))} \right] \right) \quad (3)$$

where s is a sample of coordinates inside the volume of feature image f_2, $f_1(T_\alpha s)$ is the set of grey-values from feature image f_1 sampled at the rigidly transformed coordinates $T_\alpha s$, and $p(f_1, f_2)$ is obtained by normalizing the 2-D scatter-plot, $H(f_1, f_2)$, of the grey-values of that part of sample s that falls in the common volume of f_1 and f_2, into a joint probability distribution as follows:

$$p(f_1, f_2) = \frac{H(f_1, f_2)}{N} \quad (4)$$

where N is the number of coordinates in the overlapping part of sample s, and where the features $f_i = \frac{(g_i - g_i^{min})}{n_i}$ are rescaled versions of the original grey-values, n_i being the rescaling factors. Due to the rigid body assumption we have restricted T_α to a superposition of a 3-D rotation and a 3-D translation: $T_\alpha s = R_\alpha s + t_\alpha$.

2.2.1. *Sample selection*

Feature values for image f_2 are known for a finite number of discrete voxel positions only. We need to decide how many and which voxel grey-values to include in our calculation of the mutual information matching criterion. Either the coordinates of all voxels of f_2, or a subset, or a superset thereof are used in the calculation of the mutual information for every α considered by the optimisation algorithm. In case a superset of the voxel coordinates is used, we need to select an interpolation method to find the grey-values at coordinates that do not coincide with the voxel coordinates of f_2. In order to keep the speed performance as high as possible we have simply used nearest neighbour interpolation to obtain samples from f_2. The aim of sub-sampling is to increase speed performance. It should be noted that some continuity of grey-values is assumed if we expect the mutual information to be well-behaved in case of sub-sampling. The grey-value continuity assumption implies that $p(f_1(\alpha))$ will vary smoothly with α. This additional assumption is not a restriction as long as sub-sampling factors (defined as the fraction of samples taken per voxel size, in a certain direction) do not become too small. The aim of super-sampling is to increase accuracy. This can only work if the same continuity assumption is valid in both images, or equivalently if a meaningful interpolation method can be defined in both images.

2.2.2. *Interpolation*

Due to the continuity of the registration parameters α, regardless of whether the sample s is a sub-, or a super-set of all voxels in f_2, the transformed sample coordinates will usually fall in between voxel coordinates of f_1, and thus interpolation is needed to obtain values from f_1. An interpolation method other than nearest neighbour interpolation is required to be able to guarantee subvoxel accuracy. The next best choice in speed is trilinear interpolation. The problem of trilinear or for that matter any other classical interpolation method is that they may introduce new grey-values, i.e. rescaled grey-values not present in the set of the original grey-values of f_1. Thus, any classical interpolation method other than nearest neighbour interpolation may introduce unpredictable changes into $p(f_1)$ as a function of changes in α. To avoid this problem we propose to use *trilinear partial volume distribution* which is dual to trilinear interpolation. It uses the same interpolation weights but instead of using them to average grey-values from f_1, they are used to distribute the voxel volume of the grey-value from f_2 *in the scatter-plot* over the grey-values of all nearest neighbours in f_1.

2.2.3. *Probability Estimation*

The evaluation of (2) requires the knowledge of grey-value probabilities which can be estimated from the grey-value scatter-plot. If quantisation of the scatter plot is too fine, application of parzen-windowing [5] can be used to reduce the effect of image noise and to increase the reliability of the probability estimates. However, parzen-windowing is rather expensive from a computational point of view. Clearly some compression of the data will benefit both the memory requirements and the quality of the probability distribution estimates. It is expected that small grey-value distortions of the observations do not induce large deviations in the registration solution. Based on these observations we propose to use simple rescalings of the images, f_1 and f_2, instead of the original grey-values, g_1 and g_2. For practical purposes it should be noted that rescaling is equivalent to binning of the grey-value histogram, which can be considered to be a crude approximation of parzen-windowing.

2.2.4. *Search Strategy*

In the search for the optimal value of α we have taken a pragmatic approach: 1) we assume we are able to select a meaningful initial value inside the search engine's attraction pool of the optimal value, 2) we then use a locally converging optimisation algorithm as the search engine. The implementation details are as follows: 1) Put the origins of the physical world coordinate systems in the image centres and assign zero values to those registration parameters that represent the relative translations of these coordinate systems as initial estimates. This is meaningful under the assumption that both images represent approximately the same volume of interest. 2) Apply rotations that are multiples of 90° so that the images are approximately identically oriented. This can be automated based on knowledge about the scanner and the scanning sequence. 3) We have used Powell's locally converging optimisation algorithm [10] for maximization of the mutual information matching criterion. The first 2 to 3 iterations may be performed using nearest neighbour interpolation for its speed, while later iterations may be performed using more complex interpolation for more accuracy. 4) For the same reasons we also suggest to use a subsample of the floating image during the first iterations, and a super-sample during the final iterations.

3. Results

3.1. AN ARTIFICIAL TEST USING A REAL MR IMAGE

We transformed a T1-weighted MR image (from the Heidelberg reference data set, see [4] for a detailed description) over (10.39, -3.27, 2.03) degrees and over (7.06, 1.21,18.26) mm using trilinear resampling. We then used this transformed version MR^T as a reference image to be registered with the original MR image. Eventhough large parts of some slices (slices 21 to 29 of 30) of the reference image have been set to zero in MR^T due to the transformation, the images could be registered. Accuracy results are summarized in table 1 for three different interpolation methods.

3.2. COMPARISON WITH STEREOTACTIC REGISTRATION: CT/MRI/GD-ENHANCED MRI

The data set used in the previous test also contains corresponding CT and Gd-enhanced MR (MRE) images. All the images were acquired under stereotactic conditions. In this section we summarize our registration results relative to the stereotactic registration solution. Note that since MR and MRE were registered by acquisition we used MR^T instead of MR for matching with the MRE image. This test is similar to, but much more re-

TABLE 1. This table gives the registration error for our algorithm when applied to an MR and its transformed version MR^T. Transformation parameters were (10.39, -3.27, 2.03) degrees and (7.06, 1.21,18.26) mm. Voxels are (1.33 x 1.33 x 4.0) mm^3 in both images. Subsample "3.0/1.0" means that only 1 in 3 voxels are sampled in X and Y direction of the image planes, and that all voxels are used in the perpendicular Z direction. The "error"-column gives us an arithmetic average registration error for all voxels in the central quarter of all slices of the MR volume. Binning parameter values were: $n_i = 16$. For MR^T: $g_{max} = 2550$ and $g_{min} = 0$. For MR: $g_{max} = 3359$ and $g_{min} = 2$.

f_1	f_2	Interpolation	sub-sampling factors	error (mm)
MR^T	MR	nearest neighb.	3.0/1.0	0.17
MR^T	MR	trilinear	3.0/1.0	0.0048
MR^T	MR	partial volume	1.0/1.0	0.17

alistic than the one in the previous section, because: 1) the noise will be different now, 2) because of the Gd-enhancement, and 3) due to the visibility of pathology (a tumor) in the MRE volume. A true multi-modality test is of course the registration of CT with MR. Table 2 gives the results.

TABLE 2. Same as table 1, but for MR^T combined with MRE. For MRE: $g_{max} = 3582$ and $g_{min} = 2$. For CT: $g_{max} = 4094$ and $g_{min} = 2$.

f_1	f_2	Interpolation	sub-sampling factors	error (mm)
MR^T	MRE	nearest neighb.	3.0/1.0	0.51
MR^T	MRE	trilinear	3.0/1.0	0.54
MR^T	MRE	partial volume	1.0/1.0	0.40
CT	MR	partial volume	1.1/0.3	2.50

3.3. COMPARISON WITH OTHER VB REGISTRATION ALGORITHMS USING 3-D MRI/CT

We have also compared the outcome of our algorithm with that of two other VB registration algorithms by van den Elsen ([12], [14]) on a CT/MR pair. In both cases her registration was performed automatically, without user interaction or use of fiducial markers. In [12] hierarchical correlation techniques were applied to "ridge"-feature images, obtained from the original data using geometrical operators in scale space. In [14] grey-value correlation is proposed between a grey-value mapping of the CT and the original

MR volumes. The difference between the solutions found are summarized in table 3. By visual inspection it was impossible to see the differences between our solution and the grey-value correlation based method of van den Elsen unless extremely large zooming factors were used. Fig. 1 visualizes our result.

TABLE 3. This table gives registration differences between our mutual information based registration algorithm and both VB algorithms proposed by van den Elsen et al. Voxels are (0.98 by 0.98 by 1.0) mm^3 for the MRI and (0.94 by 0.94 by 1.55) mm^3 for the CT volume. As before the average error Δ_{avg} was calculated over the central quarter of all planes of f_2, i.e. of the CT-volume. $\Delta_t(c)$ is the distance between the transformed coordinates of the volume centroid of f_2. Δ_θ is the angle between the unit eigenvectors corresponding to the unit eigenvalue of the rotation matrix. The solutions using nearest neighbour interpolation were found in less than 7 minutes and sub-sampling (1 voxels in 4, in all directions) while the solutions using partial volume distribution and no sub-, nor super-sampling were found in approximately 4 hours on an IBM RS/6000 workstation.

reference	interpolation	Δ_{avg} (mm)	$\Delta_t(c)$ (mm)	$\Delta_\theta(deg)$
[12]	nearest neighbour	2.72	2.2398	3.6776
	partial volume	2.02	1.6245	0.8957
[14]	nearest neighbour	1.62	0.8563	3.8177
	partial volume	1.26	0.0582	0.8674

3.4. TESTS WITH MRI/PET

The algorithm has been applied to MRI and PET data of Multiple Sclerosis patients for improved diagnosis, and to MRI and PET data of volunteers. In the latter case the registration serves to enhance the accuracy of a functional analysis of the human brain that has been performed earlier [6] using PET data alone. Due to the relatively low resolution of the PET images, even full sample application of our algorithm is fast (approximately 6 minutes for the data shown in fig. 1).

4. Discussion

There never exists a perfect one-to-one grey-value mapping between images to be registered, even in the registered position, due to noise and often also due to the nature of the data involved in multi-modality registration, e.g. skull and background have overlapping intensities in MR while they are completely different in CT. The algorithm proposed by Woods is not

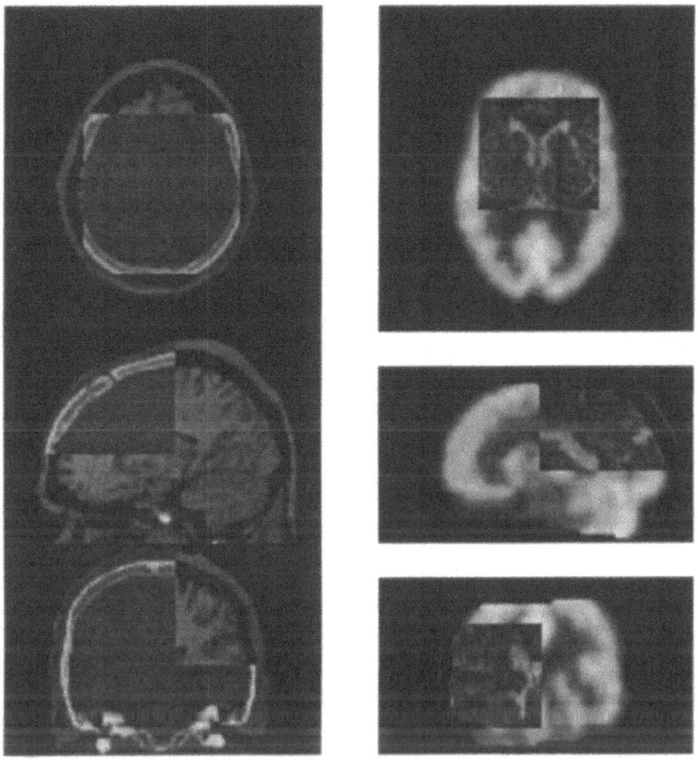

Figure 1. Left: MRI/CT combination registered using mutual information with partial volume distribution, $n_i = 16$, sampling factors equal to 1. Right: PET/MRI combination of an MS patient. The exact registration solution is unknown. This result was found to be clinically acceptable.

robust enough to handle this type of deviations from the one-to-one mapping assumption. Therefore, his algorithm requires prior segmentation of the brain from the MR image before application of his matching criterion to that part of the image. After careful inspection of the dispersion behaviour of the grey-value scatter plot Hill et al. propose to select only specific bins of the scatter plot to obtain meaningful dispersion measurements. Their approach can be automated but in a data dependent manner only. Our first attempt in [2] to increase robustness in a data independent way was based on an intuitive information theoretic reasoning only: the scatter-plot entropy as a measure of multi-modality information content or signal complexity has all the properties of a data independent measure of scatter-plot dispersion. The main problem with the entropy matching criterion, however, is its sensitivity to the problem of partial overlap, and, even worse, also to the problem that the information content of the separate overlapping parts of the images is not constant, causing it to have multiple local minima all over α-space. In this paper we have solved the latter problem also using relative entropy.

The results from sections 3.1 and 3.2 are all subvoxel accurate if the largest voxel size (mostly equals the slice distance) is taken as a reference, and if the stereotactic registration solutions are taken to be correct. The latter assumption was clearly not true for the pair of CT/MR volumes in section 3.2. In this case visual inspection showed that our mutual information based registration solution is better than the stereotactic one, which was particularly noticable in the caudal-cranial direction. This can be explained by the fact that the stereotactic localiser was fixed to the head by means of a mask and not by screws. Taking into consideration the nature of the data in these experiments we may also conclude that the mutual information is inherently robust against noise and local differences (e.g. pathological). Comparison of the results for the different interpolation methods tells us that nearest neighbour interpolation offers a very rapid solution with an acceptable, near subvoxel accuracy, even when it is used with a subsample of the floating volume, f_2. All experiments with nearest neighbour interpolation and sub-sampling were performed in less than 7 minutes. The trilinear partial volume distribution method was especially designed to overcome some edge effects, and effects of image grid interference, that are both inevitable when calculating the mutual information of a finite sample. Therefore, it does not go well with sub-sampling, which explains why no sub-sampling was used in tables 1 and 2. When the same tests were performed using the sampling factors that were used with nearest neighbour and trilinear interpolation the search algorithm got stuck in a local minimum far from the registration solution. Our implementation of the trilinear partial volume distribution has approximately the same complexity as that of trilinear interpolation which was about twice as slow as nearest neighbour interpolation. This factor two together with the differences in the in-plane sampling factor explains why the partial volume distribution based experiments in tables 1 and 2 took 90 minutes. The results from section 3.3 show us that it is possible to register images of different modalities without using any prior knowledge about the grey-value mapping between parts of the image, and yet obtain essentially the same registration results. E.g. both methods proposed by van den Elsen require the application of feature extraction algorithms to make simple correlation work. This approach makes her algorithms data dependent. The results obtained in section 3.4 have not yet been compared to that of other algorithms. However, our algorithm is more user-friendly than Woods' AIR software [16] because it does not require the prior segmentation of the brain surface from the MR images. Our algorithm does not require the selection of a part of the histogram, as does Hill's more robust version of Woods' algorithm.

In all our experiments so far, using the search procedure described in section 2.2.4 we have observed translations up to 2.5 *cm* and rotations up to 11° that were all recovered fully automatically. Thus, for our practical purposes the mutual information criterion offers a well behaved optimisation function in a reasonably large region around the initial estimate of the registration parameter values. In theory we need to prove that it is strictly quasi-convex to guarantee its unimodality. Further experiments will be needed to obtain an estimate of the range of quasi-convexity around the global optimum.

5. Conclusion

The new voxel similarity based (VB) registration algorithm has been applied successfully to real world images of the human head (CT, MR, and PET volumes) which satisfy the rigid body assumption well. Comparison of the registration outcome with that of stereotactic and 2 other VB registration solutions indicates that mutual information is a subvoxel accurate matching criterion and that it is robust in multi-modality situations.

The experimental results indicate that the mutual information matching criterion is a well-behaved multi-modality 3-D registration criterion, and that we have found satisfactory solutions for all implementation problems. Most importantly, we have found that due to the finiteness of the sample of grey-value pairs involved in the calculation of the mutual information matching criterion the use of partial volume distribution of grey-values is required to obtain the most accurate results.

6. Acknowledgments

This work is part of COVIRA (Computer Vision in Radiology), project A2003 of the AIM (Advanced Informatics in Medicine) programme of the European Commission. Special thanks go to prof. C. Schiepers, and dr. P. Dupont from the PET Center of the Department of Nuclear Medicine, at the University Hospital Gasthuisberg of Leuven for supplying us with the PET/MRI data and for visually inspecting our results. Special thanks also go to Petra van den Elsen for our informal cooperation. Her volumetric MRI was acquired in the Hospital "Medisch Spectrum" in Enschede, with special thanks to dr. G. Wilts. The high resolution CT was acquired in Utrecht University Hospital by prof. dr. F.W. Zonneveld, Drs. D.N. Velis, and P.A. van den Elsen coordinated the multimodal acquisition.

References

1. L.G. Brown, "A Survey of Image Registration Techniques", *ACM Computing Surveys*, Vol. 24, No. 4, pp. 325-376, December 1992.
2. A. Collignon, D. Vandermeulen, P. Suetens, G. Marchal, "3D Multi-Modality Medical Image Registration Using Feature Space Clustering", accepted for presentation at the *CVRMed'95 conference* on april 3-5, 1995, and publication in the *Lecture Notes in Computer Science* series, Springer-Verlag.
3. T.M. Cover, J.A. Thomas, *Elements of Information Theory*, John Wiley & Sons, Inc., New York, 1991.
4. AIM Programme, Project A2003, COVIRA Computer Vision in Radiology, Deliverable (2) D1/1.1 Set of raw images and clinical description, 26 June 1992, distributed by the Commission of the European Communities DG XIII, Brussels.
5. R.O. Duda, P.E. Hart, *Pattern Classification and Scene Analysis*, Stanford Research Institute, Menlo Park, CA, USA, A Wiley-Interscience Publication, 1973.
6. P. Dupont, G.A. Orban, R. Vogels, G. Bormans, J. Nuyts, C. Schiepers, M. De Roo, L. Mortelmans, "Different Perceptual Tasks Performed with the Same Visual Stimulus Attribute Activate Different Regions of the Human Brain: A Positron Emission Tomography Study", *Proc. Natl. Acad. Sci., USA*, Vol. 90, pp. 10927-10931, December 1993, Neurobiology.
7. D.L.G. Hill, D.J. Hawkes, N.A. Harrison, C.F. Ruff, "A Strategy for Automated Multimodality Image Registration Incorporating Anatomical Knowledge and Imager Characteristics", *13th Int. Conf., Information Processing in Medical Imaging, IPMI'93*, Flagstaff, Arizona, USA, June 1993, Lecture Notes in Computer Science, Vol. 687,pp. 182-196, Springer-Verlag, Berlin Heidelberg.
8. D.L.G. Hill, C. Studholme, D.J. Hawkes, "Voxel Similarity Measures for Automated Image Registration", *Proc. SPIE Vol. 2359, Visualization in Biomedical Computing*, SPIE Press, Bellingham, WA, Rochester, MN, October 4-7, 1994.
9. C.R. Maurer, J.M. Fitzpatrick, "A Review of Medical Image Registration", *Interactive Image-Guided Neurosurgery*, R.J. Maciunas (Ed), Park Ridge, IL, American Association of Neurological Surgeons, 1993, pp. 17-44.
10. W.H. Press, B.P. Flannery, S.A. Teukolsky, W.T. Vetterling, *Numerical Recipes in C, The Art of Scientific Computing*, Cambridge University Press, 1988.
11. I. Vajda, *Theory of Statistical Inference and Information*, Kluwer Academic Publisher, Dordrecht, 1989.
12. P.A. Van den Elsen, *Multimodality matching of brain images*, Ph.D. Thesis, Utrecht University Thesis, June 1993.
13. P.A. van den Elsen, E-J.D. Pol, M.A. Viergever, "Medical Image Matching - A Review with Classification", *IEEE Eng. in Medicine and Biol.*, pp. 26-38, March 1993.
14. P.A. Van den Elsen, E.J.D. Pol, T.S. Sumanaweera, P.F. Hemler, S. Napel, J. Adler, "Grey value correlation techniques used for automatic matching of CT and MR brain and spine images", *Proc. SPIE Vol. 2359, Visualization in Biomedical Computing*, pp. 227-237, SPIE Press, Bellingham, WA, Rochester, MN, October 4-7, 1994.
15. R.P. Woods, J.C. Mazziotta, S.R. Cherry, "MRI-PET Registration with Automated Algorithm", *JCAT*, 17(4):536-546, July/August, 1993.
16. R.P. Woods, *Automated Image Registration (AIR): User's Manual and Program Documentation*, Division of Brain Mapping Neuropsychiatric Institute, UCLA School of Medicine, 8/7/1993.

3D-2D PROJECTIVE REGISTRATION OF ANATOMICAL SURFACES WITH THEIR PROJECTIONS

Fabienne BETTING, Jacques FELDMAR
INRIA SOPHIA, Projet EPIDAURE
2004 route des Lucioles, B.P. 93
06902 Sophia Antipolis Cedex, France.
Email : Jacques.Feldmar@sophia.inria.fr

Abstract

Some medical interventions require knowing the correspondence between an MRI/CT image and the actual position of the patient. We present in this paper a new technique to perform this task without artificial markers. We find the **3D-2D projective transformation** which maps a 3D object onto a 2D (video, X-Ray) image of this object, searching for the position of the 3D object so that its projection corresponds to the *occluding contour* or *silhouette* detected in the 2D image. Our technique does not depend on the initial relative positions and deals with the occlusion and the outliers. Results are presented on real medical data.

1 INTRODUCTION

Some medical interventions require knowing the correspondence between an MRI/CT image and the actual position of the patient. Examples occur in neurosurgery, radiotherapy and interventional radiology but also in video surgery (laparoscopy). The goal can be to localize a brain tumor during the intervention, to guide the radiation sources or to visualize the position of the catheter (or endoscope) in a 3D image of the patient.

One of the most difficult tasks is to register, if possible in real time, a video or X-Ray image of the patient (intra-operative image) with the pre-operative image (MRI or CT). Basically, one must **find the 3D-2D projective transformation (composition of a rigid displacement and a perspective projection) which maps a 3D object onto a 2D image of this object.**

This problem can be solved with artificial markers visible in both images, but this produces unacceptable constraints (e.g. pre-operative images must be taken on the same day as the intervention, stereotactic frames are very painful and can prevent a free access by the surgeon, etc...).

Y. Bizais et al. (eds.), Information Processing in Medical Imaging, 275–286.

Recently, computer vision techniques have been proposed to solve this registration problem without artificial markers. A review of this research field can be found in [4]. The different approaches can be classify in *reconstruction/3D-3D rigid registration* ([5, 11, 3]), *correlation* ([10, 8]) and *silhouette-based* techniques ([7, 6]). Our approach belongs to the latter class. It resembles the technique introduced in [7]. We find the position of the 3D object so that its projection corresponds to the occluding contours detected in the 2D image. But Lavallee's approach assumes that it is possible to determine globally the interior and the exterior of the occluding contour. This is often a hard task in practice. The use of the vector normals allows us to avoid this problem. Moreover, we do not make any assumption about the initial position of the 3D object. Joshi also nicely used the occluding contour in [6] to find the pose of 3D objects, but his scheme requires second order derivatives, which can significantly reduce the robustness of the approach.

The paper is organized as follow. In section 2, we present the geometry of the registration problem. Then, in section 3, we explain how we find an estimate of the searched projective transformation. In section 4, we explain how we find an accurate transformation, dealing with the occlusion and the outliers in a rigorous way. Finally, in section 5, we present how we proceed when we do not know the calibration parameters of the 2D sensor.

2 GEOMETRY OF THE REGISTRATION TASK

The input data are the 3D (MRI,CT) pre-operative image and the 2D (video,X-Ray) intra-operative image. We can modelize the 2D sensor as a pinhole camera. Hence, the 2D image is a perspective projection of the 3D anatomical object, from an unknown point of view. A point M in the 3D image corresponding to a point m in the 2D image is so that $m = Proj(\mathbf{R}M + \mathbf{t})$, where $\mathbf{R}$ is a rotation matrix, $\mathbf{t}$ a translation vector and $Proj$ a perspective projection. In sections 3 and 4 we assume that the 2D sensor is calibrated (this means that $Proj$ is known) and we search for $(\mathbf{R}, \mathbf{t})$. In section 5, we search for both $Proj$ and $(\mathbf{R}, \mathbf{t})$.

In order to find this transformation, we extract from the MRI/CT image the surface S of the anatomical object and we extract the 2D contour c of this object in the 2D image. This contour is called the *occluding contour* or *silhouette*. The basic idea of our approach is to use a **fundamental property of occluding contour**:

If a point M on a 3D surface S is so that projection point $m = Proj(M)$ lies on the occluding contour c, then the normal vector $Normal_{3D}(M)$ to S at point M is equal to the normal vector $\vec{n}$ of the plane P defined by $(m, O, \vec{t})$, where $\vec{t}$ is the tangent vector to the occluding contour at point m. We call $Normal_{2D}$ the function which associates the normal $\vec{n}$ to the point m and we note $\vec{n} = Normal_{2D}(m)$.

We explain in the sequel of this article how we use this property to find the 3D-2D projective transformation mapping the 3D image onto the 2D image.

3 FINDING AN INITIAL ESTIMATE OF THE PROJECTIVE TRANSFORMATION

A well-known result is that knowledge of three points M_1, M_2, M_3 on the 3D object and of the three corresponding 2D points m_1, m_2, m_3 in the 2D image yields four rigid displacements $(\mathbf{R}, \mathbf{t})$ which satisfy $m_i = Proj(\mathbf{R}M_i + \mathbf{t})$ (see for example [9]). Note that there are six equations and six unknowns. Of course, it is not possible to select all the 3D and 2D triplets and to explore the possible combinations in order to find an estimate of the transformation. We have to select from the 3D surface and 2D curve two subsets of triplets which can match each other. In order to reduce the number of potential matches, we use two projective properties of surfaces:

Property 1

Two 2D points m_1 and m_2 on the occluding contour which share the same tangent line are projections of two points M_1 and M_2 on the surface which share the same tangent plane.

proof : All tangent lines to the surface at points M_1 and M_2 lie in the plane (O, m_1, m_2).

Property 2

Let m_1 and m_2 be two 2D points on the occluding contour and let M_1 and M_2 be two points on the 3D surface such that $m_1 = Proj(M_1)$ and $m_2 = Proj(M_2)$. The angle between $Normal_{2D}(m_1)$ and $Normal_{2D}(m_2)$ is equal to the angle between the normals $Normal_{3D}(M_1)$ and $Normal_{3D}(M_2)$ on the surface at points M_1 and M_2.

proof: The two planes defined respectively by $(m_i, Normal_{2D}(m_i))$ and $(M_i, Normal_{3D}(M_i))$ are identical.

Hence, to find the initial estimate $(\mathbf{R}_0, \mathbf{t}_0)$ of the rigid displacement so that

$Proj \circ (\mathbf{R}_0, \mathbf{t}_0)$ is the projective transformation which transforms S with the 2D occluding contour c, we use the following automatic algorithm:

1. Choose (m_1, m_2, m_3) on the 2D curve c such that: a) m_1 and m_2 share the same tangent line; and b) the angle α between $Normal_{2D}(m_1)$ and $Normal_{2D}(m_3)$ is as close as possible to π.
2. For each triplet (M_1, M_2, M_3) on the 3D surface S such that a) M_1 and M_2 share the same tangent plane; and b) the angle between $Normal_{3D}(M_1)$ and $Normal_{3D}(M_3)$ is α, compute the projective transformations which maps (M_1, M_2, M_3) onto (m_1, m_2, m_3). Stop if a computed projective transformation is a correct initial estimate (see [4] for details about the verification procedure).
3. If no solution was found, return to point 1.

More detail about the way we compute the pair of points sharing the same tangent planes or the same tangent lines and about the algorithm can be found in [4]. We note that the complexity is quasi-linear in the number of points describing the surface. This algorithm yields an estimate of the searched projective transformation and we use it to initialize an energy minimization algorithm and get an accurate transformation.

4 EXTENSION OF THE ICP ALGORITHM TO 3D-2D SURFACE REGISTRATION

4.1 Energy function minimization

The previous estimate of the projective transformation corresponds to, first, the application of a rigid displacement $(\mathbf{R}_0, \mathbf{t}_0)$ and, then, the application of the projection $Proj$. In order to get a more accurate projective transformation, we propose a distance between the points on the surface and the points on the occluding contour and minimize an appropriate energy function.

Let m be a point on c and let M be a point on S. We define the 5D distance:

$$d(m, M) = (\beta(m - Proj(M))^2 + (Normal_{2D}(m) - Normal_{3D}(M))^2)^{1/2},$$

where β is a normalization coefficient related to the 2D contour diameter.

The energy corresponding to this new distance is:

$$E(\mathbf{R}, \mathbf{t}, Match) = \sum_{m_i \in c} d(m_i, \mathbf{R}Match(m_i) + \mathbf{t})^2,$$

where Match is a function which associates a point on S with each point on c. This energy measures how well the "fundamental property of occluding contour" (subsection 2) is verified.

We use a two-step algorithm to minimize this energy. This algorithm can be understood as an extension of the Iterative Closest Point algorithm ([2, 12]) (the ICP algorithm was introduced to **rigidly** register surfaces represented by a set of 3D points, when a prior estimate of the displacement is available: this is the 3D-3D rigid registration problem). The first step is the minimization of E with respect to $Match$, assuming that $(\mathbf{R}, \mathbf{t})$ are fixed. The second step is the minimization of E with respect to $(\mathbf{R}, \mathbf{t})$ assuming that $Match$ is fixed. We now describe the algorithm which consists of two iterated steps, each iteration j computing a new rigid displacement $(\mathbf{R}_j, \mathbf{t}_j)$:

- Step 1: we associate with each point m_i on c a point M_i. To do so, we first compute the set

  $$CloseNormal(m_i) = \{M \in S / Angle(\mathbf{R}^t_{j-1} Normal_{2D}(m_i), Normal_{3D}(M)) < \epsilon\},$$

  where ϵ is a given threshold (note that the computation of this set is fast because we compute, as a preprocessing, a 3D kd-tree on the normals to the surface S). Next we compute the point N which minimizes $d(m_i, \mathbf{R}_{j-1}M + \mathbf{t}_{j-1})$ for all M in $CloseNormal$. In order to ensure that E decreases at this step 1 (this is the condition required to prove the convergence of the algorithm), we compare $d(m_i, \mathbf{R}_{j-1}N + \mathbf{t}_{j-1})$ and $d(m_i, \mathbf{R}_{j-1}M_i\_Prev + \mathbf{t}_{j-1})$, where M_i_Prev is the point associated with m_i in the previous iteration (the convention is that $d(m_i, \mathbf{R}_0 M_i\_Prev + \mathbf{t}_0)$ is infinity.). If $d(m_i, \mathbf{R}_{j-1}N + \mathbf{t}_{j-1})$ is smaller than $d(m_i, \mathbf{R}_{j-1}M_i\_Prev + \mathbf{t}_{j-1})$, the matched point M_i becomes N, otherwise it remains as M_i_Prev.

- Step 2: compute the rigid displacement $(\mathbf{R}_j, \mathbf{t}_j)$ which minimizes

  $$\sum_{m_i \in c} d(m_i, \mathbf{R}M_i + \mathbf{t})^2$$

The demonstration of the convergence of this algorithm is straightforward, since at each step 1 and 2, E decreases. Hence, we find an accurate transformation when each point on the occluding contour has a correspondent on the 3D surface. But, this is often not the case, especially because of the segmentation error.

4.2 Dealing with outliers

In step 1 of the two-step algorithm, we associate each point of c with a point on S. But because of segmentation error or because of occlusion, some points on c do not have any homologous point on S. Thus, given a point m_i on c, and a point M_i on S we have to decide whether (m_i, M_i) is a plausible match. This is very important because, if we accept incorrect matches, the found transformation will be biased (and therefore inaccurate), and if we reject correct matches, the algorithm may not converge towards the best solution.

As proposed in [1], we make use of the extended Kalman filter (EKF). This allows us to associate to the six parameters of $(\mathbf{R}_j, \mathbf{t}_j)$ a covariance matrix $\mathbf{S}_j$ and to compute a generalized Mahalanobis distance δ for each pair of matched points (m_i, M_i). This generalized Mahalanobis distance, under some assumptions on the noise distributions and some first-order approximations, is a random variable with a χ^2 probability distribution. By consulting a table of values of the χ^2 distribution, it is easy to determine a confidence level ϵ for δ corresponding to, for example a 95% probability of having the distance δ less than ϵ. In this case, we can consider the match (m_i, M_i) as *likely* or *plausible* when the inequality $\delta < \epsilon$ is verified and consider any others as *unlikely* or *unplausible.*

This distinction between plausible and unplausible matches implies a change in the second step of the algorithm. Instead of computing the rigid displacement $(\mathbf{R}_j, \mathbf{t}_j)$ which minimizes the least squares criterion

$$\sum_{m_i \in c} d(m_i, \mathbf{R}M_i + \mathbf{t})^2,$$

we recursively estimate the six parameters of $(\mathbf{R}_j, \mathbf{t}_j)$, and the associated covariance matrix which minimizes the criterion

$$\sum_{m_i \in c \text{ and } (m_i, M_i) \text{ is plausible}} d(m_i, \mathbf{R}M_i + \mathbf{t})^2.$$

More details about the meaning of "plausible or not" and about the EKF can be found in [4].

4.3 Results

We now present an application of the framework presented in this article for surface registration using occluding contours.

Figure 1 presents the data to register: a 3D model of a mannequin head extracted from a 3D CT-scan image (left), and a 2D video image of this mannequin head (right). Figure 2 (left) shows the initial estimate, and Figure 2 (right) shows the result of the minimization stage. The CPU time is 10 seconds on a DEC alpha workstation.

For the initial estimate, the average error is 2.1 pixels for spatial distance and 2 degrees for the difference in normal orientation. Even if some points are rejected because they are considered as outliers (they will be considered later during the iterative process), this shows that the initial estimate is already quite good. For the final transformation the average error is 0.76 pixels and 0.17 degrees which demonstrates that it is accurate. We plan to attach a stereotactic frame to the mannequin head and perform the experiment again in order to have a reference measure of the error and then try to better demonstrate the accuracy of the recovered transformation.

Note that when we dispose of several cameras, and when the relative positions of these cameras are known, it is possible to integrate into our framework several camera views. The criterion is, in this case, the sum for each image of the criterion defined in the previous subsection. We note that the use of at least two cameras is probably a necessary condition to get accurate results, especially because of the depth error.

5 THE NON-CALIBRATED CASE

5.1 Finding the eleven parameters

In the two previous sections, we assumed that we knew the calibration matrix attached to the 2D sensor. Even if this seems to be a reasonable assumption because in practice it is possible to calibrate a camera (see for example [1]), or because it is quite often possible to know the calibration parameters for a radiology system, we now explain how to proceed when we do not know the calibration matrix. It is obvious that the problem is harder. Since five calibration parameters are unknown, the problem is to find **all the eleven parameters of the projective transformation**.

We do not (yet) have a satisfactory automatic solution to find an initial

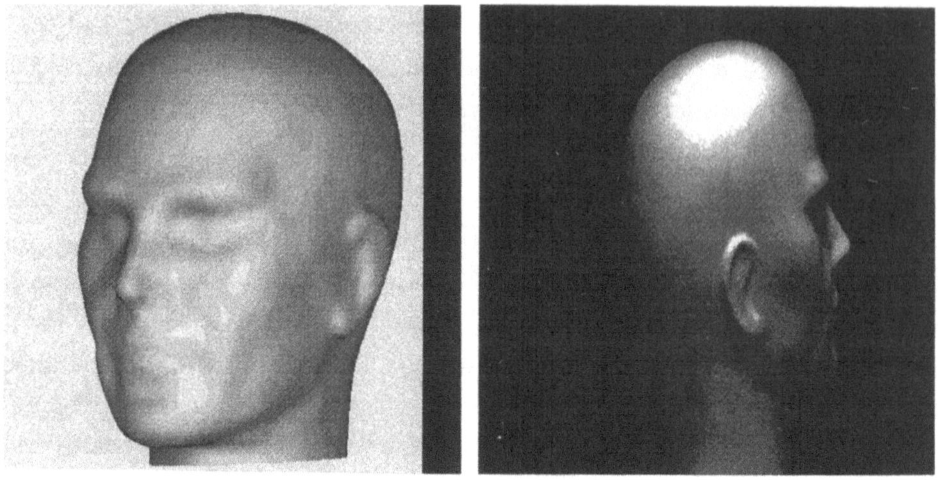

Figure 1: **Left**: The 3D model of the mannequin head. It was extracted from a 3D CT-scan image. Thanks are due to Cannes Hospital. **Right**: A video picture of the mannequin head. Our goal is to find the point-to-point correspondence between the left and the right objects.

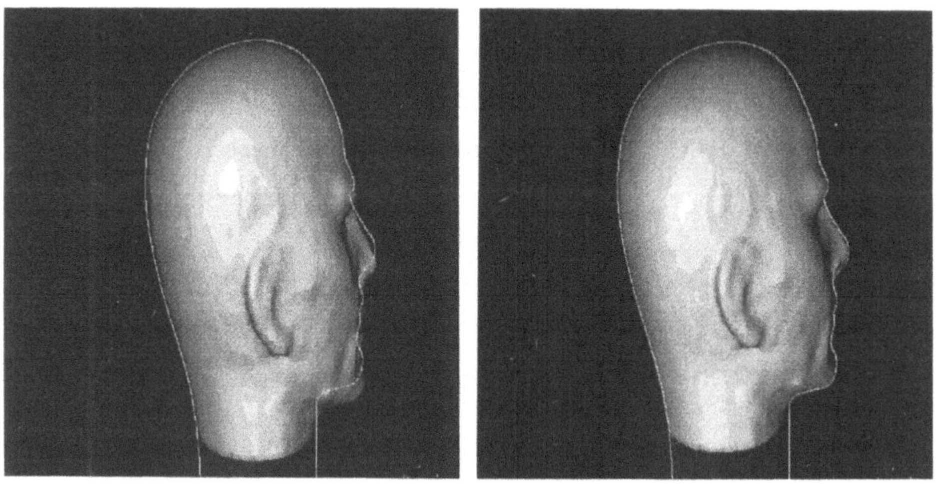

Figure 2: **Left**: the initial estimate of the transformation: we projected the 3D model using the initial transformation. The contour superimposed in this image is the contour extracted in the video image. **Right**: The final transformation: we projected the 3D model using the transformation found after minimization. You can observe that the contour which comes from the video image exactly corresponds to the contours of the 3D model. This demonstrates that we recovered the correct projective transformation.

estimate of the projective transformation. In practice, we find the estimate manually. Using an appropriate interface, it is possible to select interactively pairs of corresponding points on the 3D object and the 2D object. Selecting at least six pairs of corresponding points, it is possible, with a least squares minimization, to determine an initial estimate.

From this estimate of the projective transformation, we want to find the best one with respect to the eleven parameters. This problem also arises when the camera calibration is not very accurate, a common occurrence in practice. To do this, we minimize the same criteria as the one described in section 4. But instead of minimizing it with respect to six parameters, we minimize it with respect to the eleven parameters of the projective transformation. The minimization algorithm is almost the same. We observe that these 11D minimizations is more sensitive to the initialization conditions than the 6D case. But, because our initial estimates are not too crude, it works well in practice. This allows us to find an accurate projective transformation even if the camera calibration is not perfect, or when the initial estimate has been found interactively.

5.2 Results

We have a 3D CT-scan image of a skull and a 2D X-Ray image of it (figure 3, right). We do not know the calibration parameters of the radiography system. We first extract the skull surface in the 3D image (figure 3, left). Then, we design six pairs of corresponding points on the 3D skull surface and the X-Ray image. This yields the estimate of the projective transformation (figure 4, left). The average error is 1.5 pixels for spatial distance and 1.9 degrees for the difference in normal orientation. Then we run the minimization algorithm to recover accurately the 11 parameters of the projective transformation. The average error for the resulting transformation is 0.79 pixels and 0.7 degrees which demonstrates that it is accurate.

6 CONCLUSION

We have presented a framework to perform 3D-2D projective registration. We first showed how to use the bitangent lines and planes to find automatically an estimate of the transformation. Then we proposed a definition of a distance between the 3D object and the 2D-image and we presented an extension of the Iterative Closest Point algorithm to minimize this distance

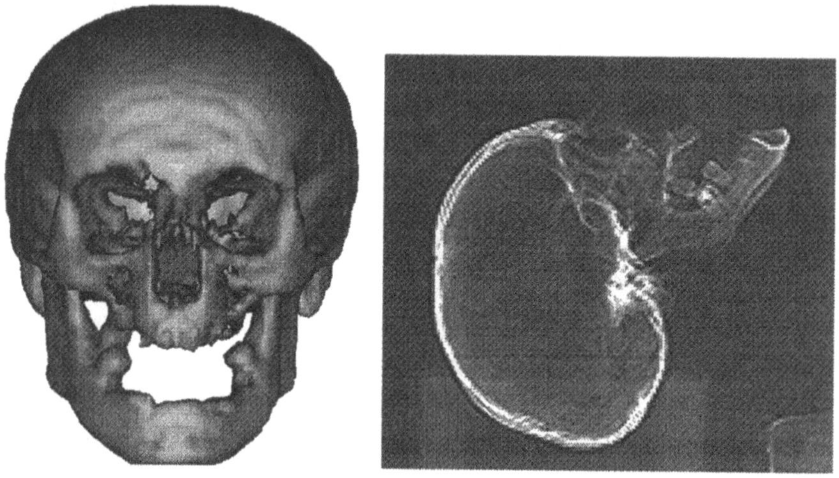

Figure 3: **Left** : The skull surface extracted from the 3D CT-scan image. **Right**: The skull X-Ray image. The goal is to recover the 3D-2D projective transformation (11 parameters) mapping the 3D surface (left) onto the 2D image (right), without any calibration information.

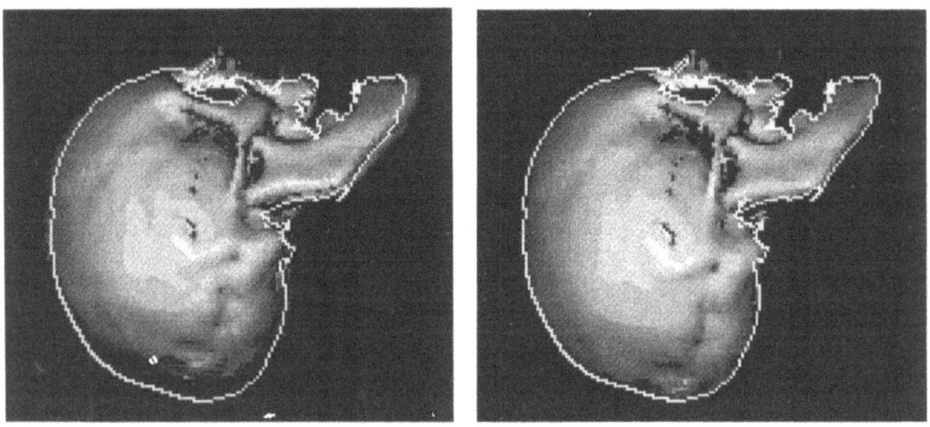

Figure 4: **Left**: the transformation found designing 6 pairs of corresponding points on the 3D surface and the 2D image. **Right**: the result of the minimization. You can observe that the contour which comes from the X-Ray image exactly corresponds to the contours of the 3D model (except small areas where the contour extraction is not perfect and which are considered as outliers). This demonstrates that we recovered the right projective transformation.

dealing with occlusion in an rigorous way.

Though we have presented results on real data which demonstrate the validity of our approach, a lot of work still has to be done. We would like to compare from a practical point of view the technique described in this paper with respect to the techniques presented by others authors ([5, 11, 3, 10, 7, 8, 6]...). For example, we plan to develop procedures to validate rigorously the accuracy of the techniques.

We believe that it should be possible to perform real-time tracking of the patient and enable the surgeon to move either the patient or the 2D sensor. Indeed, for tracking we just need to correct a projective transformation which is quite close to the right solution as explained in section 5. Hence, because the initialization would be good, it should be possible to use the extensions of the iterative closest point algorithm presented in this paper with just a few points without local minimum problem and to get fast convergence for the eleven parameters of the projective transformation.

Finally, we would like to try to better understand the 3D-2D registration problem when the 3D object is deformable. This problem arises for example in mammographic processing.

7 ACKNOWLEDGMENT

We wish to thank the members of the Epidaure and Robotvis group for their practical help and stimulating discussions. Some very interesting discussions with Mike Brady (currently invited professor in our group at INRIA) and Gerard Medioni also influenced our work.

Thanks to Dr Michel Royon (Cannes Hospital) and Dr Jean-Jacques Baudet (La Timone Hospital) for providing us data and for fruitful discussions. Finally thanks to Drs David Dean and Court Cutting for skull data.

This work was supported in part by a grant from **Digital Equipment Corporation** and by the European Basic Research Action **VIVA** (Esprit project). Future work will be also partially supported by a grant from **General Electric Medical System Europe (GEMS)**

References

[1] N. Ayache. *Artificial Vision for Mobile Robots — Stereo-Vision and Multisensory Perception.* Mit-Press, 1991.

[2] Paul Besl and Neil McKay. A method for registration of 3−D shapes. *IEEE Transactions on Pattern Analysis and Machine Intelligence*, 14(2):239–256, February 1992.

[3] A.C.F. Colchester, J. Zhao, C. Henri, R.L. Evans, P. Roberts, N. Maitland, D.J. Hawkes, D.L.G. Hill, A.J. Strong, D.G. Thomas, M.J. Gleeson, and T.C.S Cox. Craniotomy simulation and guidance using a stereo video based tracking system (vislan). In *Visualization in Biomedical Computing*, Rochester, Minnesota, October 1994.

[4] J. Feldmar, N. Ayache, and F.Betting. 3d-2d projective registration of free-form curves and surfaces. research report 2434, INRIA, 1994. Available via ftp anonymous on zenon.inria.fr, file /pub/rapports/RR-2434.ps and via WWW ftp://zenon.inria.fr/pub/rapports.

[5] W.E.L. Grimson, T. Lozano-Perez, W.M. Wells III, G.J. Ettinger, S.J White, and R. Kikinis. An automatic registration method for frameless stereotaxy, image guided surgery, and enhanced reality visualization. In *IEEE Proceedings of Computer Vision and Pattern Recognition 1994 (CVPR'94)*, Seattle, USA, June 1994.

[6] Tanuja Joshi and Jean Ponce. Hot curves for modelling and recognition of smooth curved 3d objects. In *IEEE Proceedings of Computer Vision and Pattern Recognition 1994 (CVPR'94)*, Seattle, USA, June 1994.

[7] Stéphane Lavallée, Richard Szeliski, and Lionel Brunie. Matching 3-d smooth surfaces with their 2-d projections using 3-d distance maps. In *SPIE, Geometric Methods in Computer Vision*, San Diego, Ca, July 1991.

[8] L. Lemieux and R. Jagoe. The accurate 3-d registration of stereotactic x-ray and non-stereotactic ct images. *Medical Physics*, 20, 1993.

[9] A. Linnainmaa, D. Harwood, and L.S. Davis. Pose determination of a three-dimensional object using triangle pairs. *IEEE Transactions on Pattern Analysis and Machine Intelligence*, 10(5):635, September 1988.

[10] A. Shweikard, R. Tombropoulos, J. Adler, and J.C. Latombe. Planning for image-guided radiosurgery. In *AAAI 1994 Spring Symposium Series. Application of Computer Vision in Medical Image Processing*, Stanford University, March 1994. Also in Proc. IEEE Int. Conf. Robotics and Automation, 1994.

[11] D. Simon, M. Hebert, and T. Kanade. Techniques for fast and accurate intra-surgical registration. In *First international symposium on medical robotics and computer assisted surgery*, Pittsburgh, Pennsylvania, September 1994.

[12] Z. Zhang. Iterative point matching for registration of free-form curves and surfaces. *the International Journal of Computer Vision*, 13(2):119–152, 1994. Also Research Report No.1658, INRIA Sophia-Antipolis, 1992.

MULTIRESOLUTION VOXEL SIMILARITY MEASURES FOR MR-PET REGISTRATION

Colin STUDHOLME, Derek L.G. HILL, David J. HAWKES
Radiological Sciences, UMDS, Guys Hospital
St Thomas' St. London, SE1 9RT, UK
C.Studholme@umds.ac.uk

1 INTRODUCTION

It has been shown that the accurate spatial registration and combination of different imaging modalities provides valuable additional information to the clinician [1],[3],[4]. Most robust approaches to registration are still manual, based on interactive location of corresponding points or general features. This approach is both time consuming and subject to human error. A variety of automated and semi-automated approaches have been proposed to find the transformation between MR and PET images of the head. Pelizzari [4] used a surface matching algorithm to align brain surfaces extracted from MR and PET images. Jiang [5] developed this concept further using chamfer distance minimisation. These techniques need prior identification and segmentation of corresponding surfaces requiring significant manual intervention. In concentrating on specific surfaces they are only using a small proportion of the data available which may prove a problem when using limited fields of view provided by targetted clinical scans. In addition surfaces such as the skin-air boundary can be subject to non-rigid deformation between acquisitions and, susceptibility artifacts in some MR sequences.

Apicella [2] has investigated the use of Fourier correlation to directly estimate MR-PET registration parameters without prior segmentation but this was limited to two dimensional image data. Van den Elsen [7] has investigated the use of correlation to match geometric features extracted from thin slice MR and CT images of the brain. Woods has shown [8] a voxel similarity measure derived from his PET-PET registration measure to be useful when applied to registering PET to segmented MR brain images. He reports the approach has proved unreliable when applied to unsegmented MR images.

In this paper we adapt and compare correlation, the variance measure proposed by Woods and measures based on the 2D distribution of corresponding voxel intensities proposed in [6]. We evaluate and optimise

Y. Bizais et al. (eds.), Information Processing in Medical Imaging, 287–298.

normalised measures over multiple scales in the spatial domain without prior MR segmentation. Unlike the approach in [8] we use voxel averaging of the image data as opposed to alternate pixel sampling and a different optimisation technique.

2 VOXEL SIMILARITY MEASURES

In this section we describe a number of plausible similarity measures within the framework of the two dimensional distribution of corresponding voxel values (the voxel intensity feature space). Figure 1(a) shows a 2D distribution of intensities for a manually registered MR and PET image pair. Bright areas in the images correspond to large numbers of voxels with those MR and PET intensities. Figure 1(b) and (c) shows distributions at different translational misregistrations. All the measures evaluated attempt to quantify these observed changes in some way and each makes assumptions about the nature of the distribution. Ideally we would like a measure of voxel similarity to give us a global optimum at registration and be a monotonic function of misregistration.

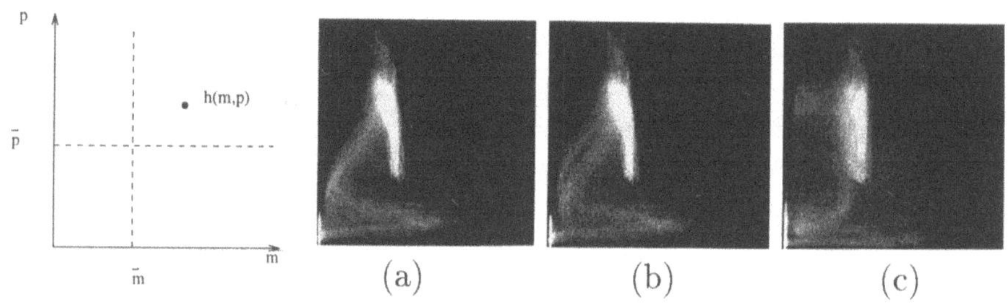

Figure 1: 2D histograms created from an MR (horizontal) and PET (vertical) image pair registered (a) and misregistered by translation along the cranio-caudal axis by 2 voxels (b) and 4 voxels (c).

2.1 Cross correlation

The basic cross correlation function is dependent both on the volume of overlap and the intensity of the images. A section of MR image may match to two structurally similar regions of PET but the match would be biased to the PET image with the larger intensity. A better measure is the Correlation Coefficient defined as:

$$\gamma = \frac{\sum_x \sum_y \sum_z [p(x,y,z) - \overline{p}][m(x,y,z) - \overline{m}]}{\{\sum_x \sum_y \sum_z [p(x,y,z) - \overline{p}]^2 \sum_x \sum_y \sum_z [m(x,y,z) - \overline{m}]^2\}^{1/2}}$$

Where MR and PET values are given by $p(x, y, z)$ and $m(x, y, z)$ respectively with means of $\overline{p}$ and $\overline{m}$ over the volume of overlap. This can be rewritten in terms of the 2D distribution of voxel values (see figure 1) as:

$$\gamma = \frac{\Sigma_{m,p}(m - \overline{m}).(p - \overline{p}).h(m,p)}{\{\Sigma_{m,p} h(m,p)(p - \overline{p})^2 . \Sigma_{m,p} h(m,p)(m - \overline{m})^2\}^{1/2}}$$

Where $h(m,p)$ is the number of corresponding voxels with values m and p in the overlap of the two images. At registration γ should be maximised. The measure is independent of the number of voxels in the overlap but assumes a predominantly linear relationship between voxel values in the two modalities.

2.2 PET intensity variance minimisation

This measure proposed by Woods [8] makes the basic assumption that for a given MR value the range of corresponding PET voxel values is a minimum at registration. If $\overline{p}(m)$ is the mean value of corresponding PET voxels for a given MR value, then the normalised standard deviation of corresponding PET values for a given MR value m is:

$$\sigma'_p(m) = \sigma_p(m)/\overline{p}(m)$$

In terms of the 2D histogram of the voxel values, the standard deviation of the distribution of PET (abscissa) intensities for each MR (ordinate) value should be minimised at registration.

$$\sigma'_p(m) = \frac{1}{\overline{p}(m)} . \sqrt{\frac{\Sigma_p h(m,p)(p - \overline{p}(m))^2}{\Sigma_p h(m,p)}}$$

A weighted sum of this normalised standard deviation for all MR values is a measure of similarity over the volume of overlap:

$$\sigma''_p = \sum_m [\sigma'_p(m) . \frac{\Sigma_p h(m,p)}{V}]$$

Where $V = \Sigma_{m,p} h(m,p)$ is the total number of voxels in the overlap. The weighting ensures that the measure is influenced most strongly by PET variance for the most common MR values.

This has been applied by Woods to segmented MR where the measure is calculated only from those voxels with MR values corresponding to brain and PET values above a certain threshold also corresponding to brain.

2.3 Moments of the distribution of values in the 2D histogram

This approach was first proposed by Hill[6]. As the images approach registration, the peaks in the 2D histogram increase in height and the regions of the 2D histogram which contain lower counts decrease in area. Thus there are more larger values in h(m,p) and fewer smaller values. This results in an increase in skewness in the distribution values of h(m,p) which can be measured by the higher order moments of this distribution. If $u(h)$ is the number of occurrences of a particular value of h for all values of m and p, then the order n moment is defined as:

$$M_n = \sum_h u.h^n$$

This can be normalised by dividing by the zero moment or mass, and by dividing $h(m,p)$ by the volume V of overlap of the two images in order to make it independent of this volume.

$$M_n = \frac{\sum_h u(h).(\frac{h}{V})^n}{\sum_h u(h)}$$

2.4 Entropy of the 2D histogram

This is a measure of the degree to which the 2D histogram deviates from a random distribution of values. This should be minimised at registration and is normalised by dividing values of $h(m,p)$ by the volume of overlap V.

$$E = -\sum_{m,p} \frac{h(m,p)}{V} \log(\frac{h(m,p)}{V})$$

3 METHOD

3.1 The data

The basis for our registration tests were a set of ten clinically acquired image pairs of the brain. All data consisted of nominally transaxial slices but with a range of orientation typical in routine clinical scanning. The PET images were all static, summed from axially smoothed dynamic ^{18}FDG acquisitions reconstructed to 31 slices of 128 by 128 voxels. The voxel size was 2.0x2.0x3.375mm and the point spread function has a full width half maximum of approximately 8.0mm. The MR data consisted of one 2D spin echo sequence (patient A) and nine 3D gradient echo sequences (patients B to J), all were T1 weighted. Large MR volumes were used to give a good range of tissue types (white matter,grey matter,skull,scalp etc) for registration.

All the image pairs were first manually registered by interactive location of between 10 and 14 corresponding point landmarks. These estimates are shown in table 1 in the form of 3 translations and 3 rotations to map each PET image to MR coordinates (where the X axis is the patient right-left, the Y axis from front to back and the Z axis feet to head). To check registration accuracy each of the PET images were transformed and resampled to MR coordinates and displayed together for inspection by colour overlay of slices and adjacent display of orthogonal cuts. All registrations were confirmed visually to be within about 3mm over the brain volume.

Manual MR-PET(^{18}FDG) Registration									
Patient	Translation (mm)			Rotation (deg.)			R.M.S.	Num.	MR Voxel
	X	Y	Z	X	Y	Z	Error	Points	Size (mm)
A	2.9	23.3	8.4	11.7	-1.9	-0.5	3.9	12	0.86x0.86x2.5
B	-2.6	1.8	7.6	24.2	1.1	8.7	3.6	12	0.90x0.90x1.5
C	-9.6	-9.8	22.8	21.6	-0.8	6.1	3.0	14	0.94x0.94x1.5
D	-3.5	7.0	15.4	16.1	-2.7	3.8	2.9	14	0.94x0.94x1.5
E	-1.8	18.3	0.6	11.3	0.2	0.6	2.7	12	0.94x0.94x1.5
F	-3.4	-11.0	0.0	18.4	-0.8	5.9	3.5	12	0.94x0.94x1.5
G	1.0	7.0	-8.7	18.6	4.3	-5.7	3.1	11	0.90x0.90x1.2
H	2.1	11.8	0.4	14.1	-0.9	-0.4	4.3	14	0.86x0.86x1.2
I	3.5	12.8	-7.3	18.3	0.5	3.7	2.9	10	0.86x0.86x1.2
J	0.7	22.6	-4.9	12.7	-3.0	2.3	2.5	11	0.86x0.86x1.2

Table 1: Transformation parameters estimated by manual point landmark identification.

3.2 Multi resolution sampling

In these tests all the data were resampled to $2mm^3$ voxels. Sampling down all MR data to this resolution ensures that we reduce the impact of variable MR sampling, leaving only differences in image content and field of view. The PET image data was first resampled to cubic voxels using linear intensity interpolation between slices. Multiple resolution versions of the images were created by voxel averaging by a factor of two to give 4mm, 8mm and 16mm versions of the image. Intermediate resolutions were created by tri-linear interpolation from the next highest resolution. This simple resampling and voxel averaging approach was favoured because of speed of operation. The MR data being of anisotropic resolution was first subsampled down by voxel averaging by the nearest integer factor and then by tri-linear interpolation to $2mm^3$ resolution.

3.3 Evaluation of measures

A given rigid body transformation defines a mapping of one discretely sampled set of voxels onto the other. We must employ some form of interpolation to decide what PET value corresponds to a particular MR voxel for a given transformation. In our implementation we used nearest neighbour interpolation. This is the most efficient in terms of processing time, which can be important when evaluating measures for large numbers of voxels, but can introduce sampling artifacts. Using this scheme we first create a 2D histogram of voxel counts consisting of 128 by 128 bins onto which is mapped the range of voxel values of MR and PET. The number of bins used was a somewhat arbitrary choice. Sixty four and 256 were tried and produced similar results. The measures described in section 2 were then evaluated from this distribution.

3.4 Optimisation of measures

In previous work we have explored the use of a genetic algorithm to optimise registration measures. This is an effective approach to avoiding local minima but can be computationally expensive. In this work we take a new approach to simplify the optimisation and reduce processing time.

The search space defined by a six dimensional registration measure is difficult to visualise but it is useful to consider the source and size of possible unwanted local optimum values. If we assume the data is band limited and in fact over sampled (particularly for the PET images), the features present in the data due to the objects being imaged can only produce changes in a registration measure over a distance larger than the sampling resolution. Minima of approximately the sampling size are voxel sampling artifacts. If we use steps equal to or larger than the sampling resolution when exploring the registration measure we should avoid getting trapped by these minima. If by using a coarse enough resolution we can avoid local minima due to the image data we should be able to use a simple gradient descent to locate the true global minima.

For the results presented here we used evaluation at image resolutions of 2mm, 3mm, 4mm, 6mm, 8mm and 12mm. At each data resolution we used a translation step size equal to the data resolution and a rotational step size equal to the data resolution divided by the field of view (in radians), until no further descent was possible. To increase accuracy and speed we continued to optimise with the step size reduced to that of the next highest resolution before increasing the data resolution. Unlike other approaches we do not depend on local evaluations to calculate gradients.

An important observation about the way in which optimisation was occurring in our scheme was that a final solution at a given resolution was not always within the expected accuracy for that resolution, but these larger errors were corrected by more steps being taken at higher resolutions. One measure of optimisation efficiency is the number of evaluations of the similarity measure required which can be expressed in terms of the equivalent number of evaluations at the highest resolution (i.e. evaluations of a resolution containing 1/8 the number of voxels to transform would count as 1/8 of an evaluation) . Over the many tests carried out on the data a mean of around 100 evaluations was recorded ranging from 50 to around 150 (for those optimisations which were successful).Typical processing times were between 3 and 6 minutes on a SPARC station 5/70, dominated by evaluations at the finest resolution.

4 RESULTS

4.1 Whole head registration

In order to test the registration approaches we presented the optimisation algorithm with a set of 100 random starting estimates for each of ten pairs of images. The starting position was based on the manual estimate of registration with a random translation with a range of $+/-30mm$, and a random rotation with a range of $+/-30^o$ (both distributed evenly) added to it.

Table 2 shows the mean length of the translation and rotation vectors between the point based estimate and the final evaluation for each patient. The mean lengths were computed from the root of the sum of squares of the 3 orthogonal components of the translation and rotation vectors respectively. Table 3 shows the standard deviation of the lengths of the automated translation and rotation vectors about their means. Only optimisations which converged within the equivalent of 200 highest resolution evaluations were considered.

Optimisation of the correlation coefficient shows distinctly separate clustering for each image pair. Pairs C,G,H,I and J showed good clustering around a point near the point based estimate, while the remaining pairs clustered around points with significant translational error. Ex-

Image Pair	Measure							
	Correlation Coefficient		3rd Order Moment		Entropy of 2D Histogram		Corresponding PET Variance	
	E_{trans}	E_{rot}	E_{trans}	E_{rot}	E_{trans}	E_{rot}	E_{trans}	E_{rot}
A	22.63	8.25	40.7	32.3	21.9	12.6	2.48	2.21
B	6.80	2.22	36.0	22.1	13.3	10.3	3.35	4.87
C	3.51	2.10	29.7	22.2	20.1	17.9	2.88	4.03
D	19.64	5.18	41.5	35.9	16.5	26.0	1.83	3.50
E	28.34	4.94	43.5	29.7	12.9	12.5	0.92	3.94
F	24.31	8.68	31.0	25.6	12.4	13.7	3.99	5.43
G	0.99	2.90	35.0	19.0	6.5	3.9	1.27	3.10
H	1.29	2.91	52.3	28.2	26.7	16.1	1.85	3.05
I	2.37	3.86	45.8	24.5	29.4	13.5	2.64	3.47
J	1.50	2.68	37.2	24.7	22.3	9.41	1.90	1.33

Table 2: Mean lengths of translation(mm) and rotation(deg) error vectors between successful optimisations and the manual estimate for 100 random starts for each of the ten patients.

amination of the MR images and histograms showed that poor results occurred with those MR images with strong grey/white differentiation or scalp fat signal, resulting in a partial inverse correlation with the PET image. It was observed that optimisation at lower resolutions provided good results which then were degraded at higher resolutions. When fine details such as scalp and grey/white boundaries are blurred their effect on the overall correlation may be less.

Maximisation of the moments of the distribution provided poor results dependent on starting estimate. The assumption of maximum skewness occurring at registration is violated at larger values of misregistration where appreciable amounts of brain and scalp tissue overlie values for air in the corresponding image. In the 2D histogram this results in a wide range of intensities in one modality corresponding to the narrow range of values for air in the other modality (vertical and horizontal lines passing through the origin). This effect can be avoided by simply not including intensities corresponding to air, but at the cost of reduced tolerance to larger misregistrations. For these results we have used a minimum MR threshold of 10% of the maximum value and a PET threshold of 15% of maximum value. In table 2 we show the results for the third order moments, the 2nd and 4th order moments produced similar results.

Maximisation of the entropy of the 2D histogram produced poor overall results which were similar to those for the moments. The same minimum thresholds were used as for the moments to produce the results shown in tables 2 and 3.

	Measure							
Image Pair	Correlation Coefficient		3rd Order Moment		Entropy of 2D Histogram		Corresponding PET Variance	
	σ_{trans}	σ_{rot}	σ_{trans}	σ_{rot}	σ_{trans}	σ_{rot}	σ_{trans}	σ_{rot}
A	0.49	4.81	28.9	16.9	34.8	14.1	0.59	1.94
B	7.39	1.17	33.7	13.7	20.7	10.5	0.51	0.50
C	0.49	0.91	24.4	13.4	16.4	12.8	0.43	0.57
D	2.72	7.06	24.2	19.5	16.3	23.0	0.51	0.50
E	5.70	4.30	28.5	21.0	18.8	18.2	0.40	0.56
F	6.03	11.0	24.9	14.3	16.3	13.2	0.41	0.51
G	0.39	0.49	33.6	13.9	24.4	3.09	0.35	0.47
H	0.54	0.65	33.6	18.7	38.8	14.7	0.51	0.56
I	0.40	0.62	34.0	14.7	34.4	13.9	0.46	0.52
J	0.62	2.78	28.2	18.5	37.2	12.2	0.47	0.55

Table 3: Standard deviations of lengths of translation(mm) and rotation(deg) vectors for the results of successful optimisations for 100 random starts for each of the ten patients.

Measure			
Correlation Coefficient	Entropy of 2D Histogram	3rd Order Moment	Corresponding PET Variance
99.3%	39.0%	55.7%	99.6%

Table 4: Percentage of Optimisations Succeeding

Optimisation of corresponding PET variance minimisation provided the most robust results of the measures tested. The mean length of the translation and rotation vectors between the point based estimate and the final evaluation were all within $4mm$ and 5° respectively which is close to the expected tolerance of the point based estimate. Only four of the 1000 random starting positions failed to converge. The standard deviation of the mean lengths were all about $0.5mm$ and 0.5^o except for the rotational vector for patient A (1.94^o). Examination of the MR data for this patient showed considerable motion artifact during this 2D 4-slice interleaved acquisition.

Table 4 shows the percentage of failed optimisations for each measure using the criteria that if an optimisation took more than the equivalent of 200 fine resolution iterations then the search had in effect become lost. The results show that both the correlation coefficient and the PET variance measure reliably converged for the vast majority of the starting estimates while both the entropy and moment measures had a high probability of failure.

Figure 2 provides scatter plots of the translation and rotation error vectors for one patient (B). These show that for both entropy and third order moment measures there is a significant cluster near to the point

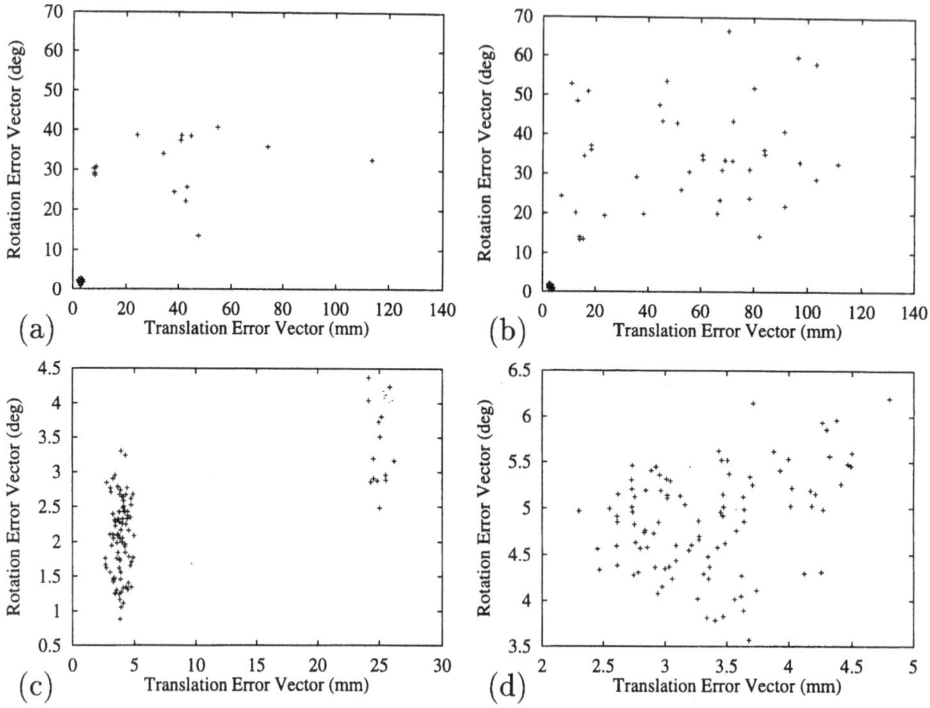

Figure 2: Mean translation(mm) and rotation(deg.) error vectors between the manual estimate and automated estimate for patient B following optimisation of entropy(a), third order moment(b), correlation(c) and PET variance(d)

based estimate. Correlation produces two clusters, one near to the point based estimate, the other over 20mm away. The PET variance measure on the other hand produces a tight cluster with a standard deviation of about $0.5mm$ and 0.5^o reasonably close to the point based estimate.

Axial	σ_{trans}	σ_{rot}
72mm(+/-30)	2.70	1.46
48mm(+/-30)	1.90	3.24
24mm(+/-30)	19.17	10.98

Table 5: The standard deviation of the translation and rotation estimates for 100 random starting positions for different MR axial extent.

4.2 Data truncation

In section 4.1 we demonstrated that the similarity measure that is most robust to changes in the starting estimate was the minimisation of the corresponding PET variance. We investigated the robustness of this algorithm to truncation of the data sets being registered. PET images are generally reconstructed to cover the entire brain (31 slices from our

CTI scanner), so truncation is not a common problem. MR images are much more likely to be targeted around a region of interest. MR image volumes with an axial extent of 72mm, 48mm and 24mm were extracted from the image data of patient J (with an original axial extent of 144mm). The registration algorithm was then run with 100 random starting estimates for each truncated image randomised with a range of $+/-30mm$ and $+/-30^o$. The standard deviations of the rotation and translation estimates are included in table 5 and show a considerable reduction in performance as the axial extent of the data is reduced from 48mm to 24mm.

5 CONCLUSION

Using a simple optimisation approach over multiple resolutions we have compared the robustness of a number of voxel similarity measures for MR and PET brain image registration when presented with a typical range of positioning differences. For whole brain MR images we have shown that the minimisation of corresponding PET variance is less effected by MR image content than correlation and is more robust to large positioning differences than simple histogram measures such as entropy and moments. Randomly generated displacements over $+/-30mm$ and $+/-30^o$ have only a small chance (9%) of lying within $+/-20mm$ and $+/-20^o$ and only a very small chance (0.14%) of lying within $+/-10mm$ and $+/-10^o$. Therefore almost all our starting estimates are near the extremes of clinically observed misregistration and our test has been purposely stringent.

The work presented here concentrates on efficiently and robustly solving the 'macro' registration problem (where misregistration is large compared to the sampling of the original data), which is useful for general image comparison in day to day clinical diagnosis. Nevertheless we have demonstrated that the PET variance measure with our multiresolution optimisation approach, as well as being very robust, produces reliable results close to those expected from point based registration. Further evaluation of the accuracy of the registration with respect to a known gold standard needs to be done. Extending the approach to handle subvoxel precision (e.g. where versions of the PET image are created at significantly higher than original sampling) may require a more rigorous approach to multiresolution construction.

Initial work in comparing the sensitivity of the approach to truncation of the MR dataset indicates that the robustness of the measure is badly effected. Other measures of registration which include more knowledge of the expected distribution may provide a better solution. More work needs to be done in looking at the registration of truncated data from a range of typical sites in the head where different tissue types and structures may be present and clinical abnormalities may dominate in the image.

Although we have applied the algorithm to a range of typical MR images being registered with PET at our site, it is necessary to evaluate these algorithms on a much wider range of MR sequences and over a larger number of images before any more definite conclusions can be drawn about their relative merits. Early work looking at the use of these multiresolution measures for the registration of functional MR to anatomical MR in the brain, and PET to CT scans in the skull base and neck show promising results.

Acknowledgements

This work was funded by UK EPSRC Grant GR/H48170. We are grateful for the support and encouragement of our clinical colleagues in this work, in particular Dr Wai-Lup Wong (Radiologist), Dr Iain Cranston, Dr M. Wong, and for the technical assistance of the Radiographic staff of Guy's and St. Thomas' Hospitals in London.

References

[1] Hill DLG,Hawkes DJ,Gleeson MJ, Cox TCS, Strong AJ, Wong W-L,Ruff CF, Kitchen ND, Thomas DGT, Sofat A, Crossman JE, Studholme C, Gandhe A, Green SEM, Robinson GP. "Accurate frameless registration of MR and CT images of the head: Applications in Planning Surgery and Radiotherapy", *Radiology* , 191, pp 447-454, 1994

[2] Apicella A,Kippenham JS, Nagel JH. "Fast multi-modality Image Matching", *IEEE Trans. on PAMI-9*,pp 698-700, 1987

[3] Evans AC, Marrett S, Torrescorzo J, Ku S, Collins L. "MRI-PET correlation in three dimensions using a volume-of-interest (VOI) atlas", *J Cereb. Blood Flow Metab.*, 11, A69-A78, 1991

[4] Pelizzari CA, Chen GTY, Spelbring DR, Weichselbaum RR, Chen C-T. "Accurate three dimensional registration of CT, PET and/or MR images of the brain", *J. Computer Assist. Tomogr.*,13,pp 20-26, 1989

[5] Jiang H, Robb RA, Holton KS. "New approach to 3-D registration of multimodality medical images by surface matching", *Visualisation in Biomedical Computing, Proc. Soc. Photo-Opt Instrum Eng*,1808,pp 196-213, 1992

[6] Hill DLH, Studholme C, Hawkes DJ. "Voxel similarity measures for automated image registration", *Visualisation in Biomedical Computing, Proc. Soc. Photo-Opt Instrum Eng*,2359,pp 205-216, 1992

[7] Van den Elsen PA. "Multimodality matching of brain images", *Utrecht University Thesis*, 1993

[8] Woods RP, Mazziotta JC, Cherry SR. "MRI-PET registration with automated algorithm", *J. Comp. Assis. Tomogr.*,17, pp 536-346, 1993

LINEAR MODELS OF ORTHOGONAL SUBSPACES & NETWORKS FROM FUNCTIONAL ACTIVATION PET STUDIES OF THE HUMAN BRAIN

Stephen C. STROTHER[123], Jon R. ANDERSON[1], Kirt A. SCHAPER[1], John S. SIDTIS[2], David A. ROTTENBERG[123]

PET Imaging and Neurology Services, VA Medical Center, Minneapolis, USA[1]

Departments of Neurology[2] and Radiology[3], University of Minnesota, USA

Abstract. We present a general linear model for obtaining quantitative measures of the "functional connectivity" of activated brain "networks" from the spatial covariance structure of [^{15}O]water PET scans. The model is based on a modified factor analysis of variance (FANOVA: a two-way ANOVA combined with singular value decomposition) procedure called the Scaled Subprofile Model (SSM). For a simple motor task variance partitions of eigenvectors from this model identified a well defined intersubject subspace that accounted for most of the nonrandom "signal" structure, contained almost none of the motor task's activation signal and had dimensions of (number of subjects - 1). Orthogonal to this intersubject subspace was an intrasubject subspace in which the largest eigenvector represented the subjects' common "network" response to the motor activation task but accounted for only a few percent of the total signal variance. Identification of a stable spatial covariance pattern representing the motor network required the use of SSM compared to unmodified FANOVA, the removal of outlying subjects' scans, and groups of subjects that were neither too big nor too small. Our results indicate that the subspaces' structure may provide cost functions for studying and optimizing experimental and methodological effects (e.g., different intersubject alignment techniques) and for building realistic data simulations of [^{15}O]water PET activation studies.

1. INTRODUCTION

A natural first-order systems/network model of brain organization [1] for [^{15}O]water PET functional activation scans is provided by linear combinations of regions with covarying responses to specific stimuli, i.e.,"functionally connected" regions. Linear combinations representing such covarying regional responses are contained in the spatial/temporal covariance structure of groups of functional activation scans.

Within a linear model framework, techniques based on two-way ANOVA combined with singular value decomposition (SVD) or principal component analysis (PCA: see Sec. 2.1 for equivalence to SVD [2]) may allow the separation of signal variation due to functional activation or disease from that

Y. Bizais et al. (eds.), Information Processing in Medical Imaging, 299–310.

due to methodological or anatomical factors, or normal within and between subject biological variation. However, few studies have attempted to model the experimental, methodological and functional sources of spatial/temporal covariance structure associated with functional activation.

During the last decade, analyses of [^{15}O]water PET functional activation studies have relied on simple parametric tests applied separately to the distributions across scans at each voxel location, e.g., voxel-based *t*-tests of activation minus baseline scan pairs [3,4]. Recent analytic approaches are based on tests of the null hypothesis that the noise structure of activation-baseline scan pairs follows a Gaussian random field model [5,6]. However, no attempt is typically made to model the signal as opposed to the noise structure or to account for the nonrandom but highly heterogeneous covariance structure observed experimentally [7,8,9].

As little is known about the general signal structure in functional activation studies we have developed an exploratory, nonparametric modelling approach that deliberately departs from the current parametric noise models. The idea of a "true" parametric population model may have limited utility for the small numbers of high dimensional (# voxels) scan vectors generally available from functional PET studies to characterize the dynamic and highly multivariate signal structure that reflects neuronal activity in the brain. Therefore, we are concerned with finding the "best fitting" model(s) supported by the available data and consistent with our experimental attempts to control the signal structure, e.g., controlled stimuli, scan registration techniques. These model(s) should be parsimonious in the sense that they do not contain more than the number of parameters necessary to "fit" the data while at the same time being "generalizable" so that they describe the structure of new image(scan) vectors, (i.e., minimize prediction error for new observations). [A similar philosophical approach to exploratory modelling is found in supervised learning using artificial neural networks (ANN); see Lautrup et al. [10] for a recent analysis that utilizes SVD with ANN's to analyze a [^{15}O]water PET study.]

2. LINEAR MODELS OF BRAIN FUNCTION

2.1. Additive global scan effects

The linear model associated with a two-way ANOVA of Q scans, each providing I regional measurements (i.e., ROI's or voxels), with one observation, v_{iq}, per cell (i,q) is given by

$$v_{iq} = p + g_q + r_i + i_{iq} + e_{iq} \qquad \textit{(Eq. 1)}$$

where p is the overall mean, g_q is the true global effect of the qth scan, r_i is the true local effect of the ith region, i_{iq} is the true interaction of the qth scan and the ith region and e_{iq} is the experimental error. *The restriction to only one observation per cell is applied because we believe it is unlikely that multiple intersubject or even intrasubject scans—given the temporal sampling rates available—may be considered to be random replicates of the same fixed brain state.* With only one observation per cell there is no way to explicitly test for the true interaction signal distinct from the error since there is no independent measure of error. In the statistical literature this is the problem of testing for nonadditivity, i.e., a significant multiplicative interaction term [2]. Interaction effects are important in functional brain imaging because they contain the linear combinations of functionally connected regions, i.e., measures of functional connectivity, that reflect underlying brain networks/systems.

Gollub [11] proposed FANOVA to solve the problem of separating true interaction terms from errors by combining Eq. 1 with a linear interaction model based on SVD. Eq. 1 becomes,

$$v_{iq} = p + g_q + r_i + \sum_{k=1}^{G} h_{ik} \sqrt{c_k}\, s_{kq} + e_{iq} \qquad \textit{(Eq. 2)}$$

where G = min(Q-1, I-1), s_{kq} is the qth element of a 1 x Q orthonormal vector ($\mathbf{s}_k^T$) of scan weights with $s_{k\bullet} = \sum_q s_{kq}/Q = 0$, h_{ik} is the ith element of an I x 1 orthonormal vector ($\mathbf{h}_k$) of regional weights with $h_{\bullet k} = \sum_i h_{ik}/I = 0$, and c_k is the eigenvalue of the kth principal component. Let v_{iq} be the (i,q) element of the data matrix X then there are three ways to obtain the vectors $\mathbf{s}_k$ and $\mathbf{h}_k$. First double-center X to obtain the elements d_{iq} of the matrix D, given by

$$d_{iq} = v_{iq} - v_{i\bullet} - v_{\bullet q} + v_{\bullet\bullet} \qquad \textit{(Eq. 3)}$$

PCA of $D^TD/(Q-1)$ provides eigenvectors $\mathbf{s}_k$ and $D\mathbf{s}_k/c_k^{1/2}$ generates $\mathbf{h}_k$; PCA of $DD^T/(I-1)$ provides eigenvectors $\mathbf{h}_k$ and $D^T\mathbf{h}_k/c_k^{1/2}$ generates $\mathbf{s}_k$. *These symmetry relationships are important for analysis of groups of scans where there are typically only a few hundred scans generating image volumes of 30k to 40k voxels per scan.* For such data sets, formation and PCA of the traditional XX^T or DD^T matrices is difficult because of their size [7] and unnecessary because they have the same rank as the much smaller X^TX and D^TD matrices. In addition, we may directly solve for the elements of the summation in Eq. 2 by using SVD of D to produce singular vectors that match the eigenvectors of DD^T and D^TD, with singular values $c_k^{1/2}$ [2]. The singular vector pairs $\mathbf{s}_k$ and $\mathbf{h}_k$ provide the optimal linear model of the interaction term (in a mean square error sense) that maximizes the variance of a possible K dimensional "signal" subspace (k = 1,...,K) and minimizes the variance of the remaining (G-K) dimensional "error" subspace (k = K+1,...,G). The acronym FANOVA will be used to refer to the additive global effects model defined above.

2.2. Multiplicative global scan effects

Moeller et al. [12] proposed the closely related Scaled Subprofile Model (SSM) for analysis of functional brain datasets (see also Moeller and Strother [13]). Allowing for both the inter- and intrasubject scans commonly obtained in PET $[^{15}O]$water brain activation studies, SSM may be written as,

$$V_{ijsr} = GSF_{jsr}\,(GMP_i + SRP_{ijsr}), \qquad (Eq.\ 4)$$

$$SRP_{ijsr} = \sum_{k=1}^{N} SSF_{kjsr}\,GIS_{ik} \qquad (Eq.\ 5)$$

where V = voxel based image activity, i = voxels, j = 1,..,(# subjects), s = 1,..,(# of scans with different stimulus states in a stimulus group), r = 1,..,(# of repeat stimulus groups per subject), GSF = global scaling factor, GMP = group mean profile, SRP = scan residual profile (containing true interactions and error) with $SRP_{i\bullet\bullet\bullet} = 0$, SSF = subprofile scaling factor, GIS = group invariant subprofile and N = (# scans - 1), assuming N << I. For a voxel-based analysis the word *profile* in GMP and SRP should ideally be replaced with the words *image volume* but we shall retain the original acronyms for consistency with earlier publications [12,13]. A logarithmic (ln) transformation is applied to the data, which–using the small signal approximation $\ln(1 + x) \approx x$ for $x \ll 1$–provides a two-way ANOVA decomposition equivalent to Eq. 1 with true interactions and errors combined in the relative residual term $(SRP_{jsr}/GMP)_i$:

$$\ln(V)_{ijsr} \approx \ln(GSF)_{jsr} + \ln(GMP)_i + \left(\frac{SRP_{jsr}}{GMP}\right)_i \qquad (Eq.\ 6)$$

Estimation procedures for the SSM parameters are described in [9,13]. Note that with the FANOVA index q replaced by the indices j•s•r, the singular vector pairs from SSM and FANOVA may both be thought of as being composed of an eigenvector of scan weights (SSF's) and principal component scores–the spatial GIS image volumes for FANOVA and $[(GIS_{jsr}/GMP)_i - (GIS_{jsr}/GMP)_.]$ for SSM. We shall use the phrase "principal component" (PC) to refer generically to this pair of vectors associated with every eigenvalue extracted by a PCA or SVD from the interaction term. The acronym SSM will be used to refer to the multiplicative global effects model defined above.

2.3. A useful signal variation partition

The following algebraic identity partitions the signal variation of a PC's eigenvector weights into useful parts defining intersubject and intrasubject variation expressions, and the intrasubject expression is further partitioned.

$$\sum_{jsr} SSF_{jsr}^2 = \sum_{jsr}\left(SSF_{jsr} - SSF_{j \bullet r} + SSF_{j \bullet r} - SSF_{j \bullet \bullet} + SSF_{j \bullet \bullet}\right)^2$$

$$\sum_{jsr} SSF_{jsr}^2 = \left[\sum_{jsr}\left(SSF_{jsr} - SSF_{j \bullet r}\right)^2 + \sum_{jsr}\left(SSF_{j \bullet r} - SSF_{j \bullet \bullet}\right)^2\right] + \sum_{jsr} SSF_{j \bullet \bullet}^2 \quad \left(\because SSF_{\bullet\bullet\bullet} = 0\right) \quad (Eq.\ 7)$$

PC variation = intrasubject variation:[about stimulus-group means + of stimulus-group means] + intersubject variation

3. METHODS

3.1. [^{15}O]water PET scanning

[^{15}O]water baseline-activation (s = 2) PET scans were acquired from 21 young normal right-handed volunteer subjects using a Siemens-ECAT 953B PET scanner operated in its 3D acquisition mode. For baseline studies, subjects had their eyes loosely patched and ears plugged with insert earphones, and were instructed to lie still and remain awake. For activation studies, the subjects performed a simple motor task consisting of sequential finger-thumb opposition between the thumb and successive digits at a rate of one opposition per second. Ten-milliliter boluses of 10-25 mCi [^{15}O]water were infused intravenously at 1 ml/sec and scanning started when the radioactivity reached the brain and continued for 90 seconds [9]. Each subject's scanning session consisted of eight ninety-second PET scans separated by ten-minute "rest" periods to allow for ^{15}O decay. The first, third, fifth and seventh scans were acquired in the "baseline state", and the second, fourth, sixth and eighth scans were acquired in the "activated state"; baseline-activation scan pairs constituted one stimulus group (r = 4). For a more detailed account of scanning and stimulus methods see [9].

3.2. Scan alignment and preprocessing

The eight scans from each subject were aligned to the first baseline-state scan using the intramodality image ratio technique described by Woods et al. [14]. For each of the 21 subjects the eight "intrasession aligned image volumes" were summed, and 21 intersubject transformations were calculated. Intersubject registration to a single simulated reference PET image volume was accomplished using our own implementation of the general 12 parameter linear transformation described by Woods et al. [15]. The simulated PET image volume was obtained by applying the sampling, resolution parameters and reconstruction algorithm from our PET scanner to projections derived from a higher resolution, segmented, reference MRI scan [16]. The simulated PET volume was transformed to Talairach space using transformation parameters derived from the reference MRI scan [17]. Aligned intrasession image volumes were normalized by the injected dose/kilogram and smoothed. These images were then transformed to Talairach-aligned volumes [16] using the 21 intersubject transformations, and voxels below 45% of each slice's maximum voxel value were discarded; only Talairach voxel locations that were nonzero in all image volumes were used in subsequent analyses.

4. RESULTS

Multiple SSM analyses were performed to study the signal variation and subspace structure of SRP (eqs. 4 & 5) as a function of number of subjects analyzed. For every SSM analysis performed the eigenvalues (i.e., an eigenspectrum), intersubject variation ($\sum_{jsr} SSF^2_{kj\bullet\bullet}$) and p-values from *t*-tests of activation-baseline group differences ($SSF_{k\bullet 2\bullet}$ - $SSF_{k\bullet 1\bullet}$) were plotted as a function of PC numbers, k = 1 to 25.

Figure 1 presents the eigenspectra from the ten SSM analyses performed for groups of 2, 4, 6, 8, 10, 12, 14, 16, 18 and 21 subjects. All spectra appear to slowly pass through (2 & 4 subjects), reach (6-16 subjects) or approach (18 & 21 subjects) a plateau of eigenvalues with values from 100 to 150 that probably represents random error effects in the SRP term. A break, or elbow, is seen in these eigenspectra plots, particularly in the spectra of the smaller groups just before or at the point where the component number equals the number of subjects being analyzed. For larger groups this break becomes less distinct and seems to move one or two components above this point.

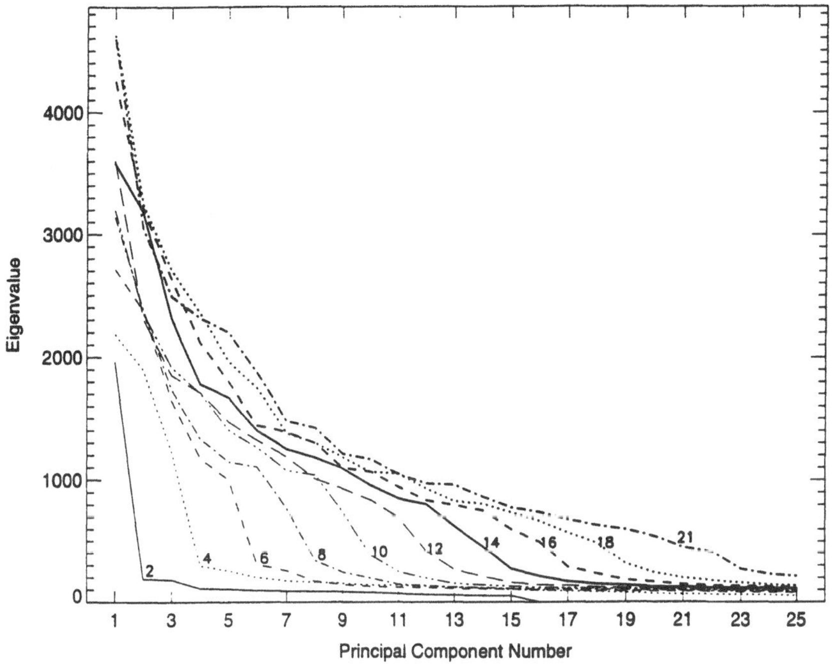

Figure 1: The eigenspectra of the 10 SSM analyses performed for groups of 2 to 21 subjects; eigenvalues are plotted against principal component numbers. The labels represent the number of subjects analyzed to produce each curve and are located at the point on the curve where the component number equals the number of subjects analyzed.

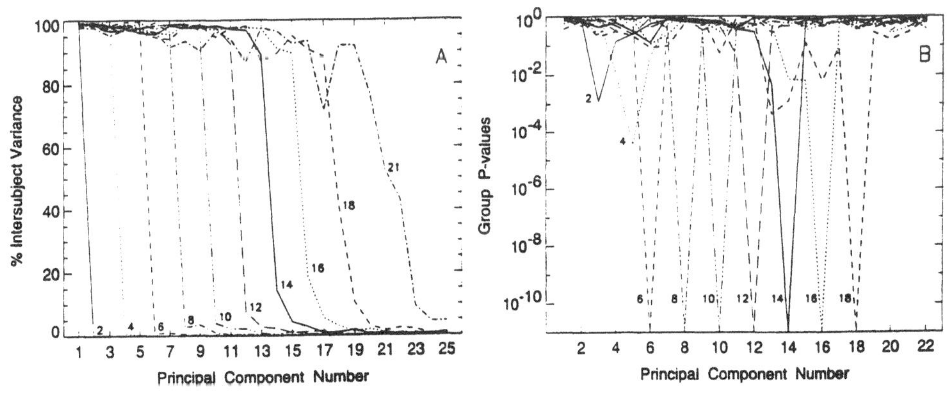

Figure 2: For SSM analyses of groups of 2 to 21 subjects with labels representing the number of subjects analyzed to produce each curve: (**A**) the % intersubject variation ($\sum_{jsr} SSF^2_{j\bullet\bullet}$) of each principal component (PC) is plotted against PC number–the labels are located at the point on the curve where the component number equals the number of subjects analyzed; (**B**) the p-values (uncorrected for multiple comparisons) from *t*-test differences between groups of activation subprofile scaling factors (SSF's) and baseline SSF's are plotted for each PC.

Figure 2 demonstrates the structure underlying the breaks in the eigenspectra in Fig. 1. They represent the abrupt transition from an intersubject to an intrasubject subspace with the intersubject subspace having a dimension of one less than the number of subjects being analyzed. As the group size increases the transition between the spaces becomes less distinct and spreads over several PC's.

For the first SSM analysis of 21 subjects "gross" intrasubject changes–perhaps due to misalignment–were identified by calculating the total stimulus-group mean variation (Eq. 7, $\sum_{kr}(SSF_{kj \bullet r} - SSF_{kj \bullet \bullet})^2$) for k = 1 to 25 for each subject. Three subjects were obvious outliers with values of 0.87, 0.66, 0.31 compared to 0.076 ± 0.044 (mean ± SD) and range [0.018,0.18] for the remaining 18 subjects. These three subjects were removed and the next SSM analysis of the remaining 18 subjects contained no obvious stimulus-group mean variation outliers: 0.113 ± 0.067 with range [0.038,0.268] for the 18 subjects. In this analysis one PC (# 18) had a very small p-value for activation-baseline group differences (Fig. 2B). Two subjects had obviously outlying variation contributions to the total variation of PC18: Outliers 0.122, 0.260 (see subjects 6 & 10, respectively, Fig. 3B) compared to 0.039 ± 0.01 with range [0.021,0.056] for the remaining 16 subjects.

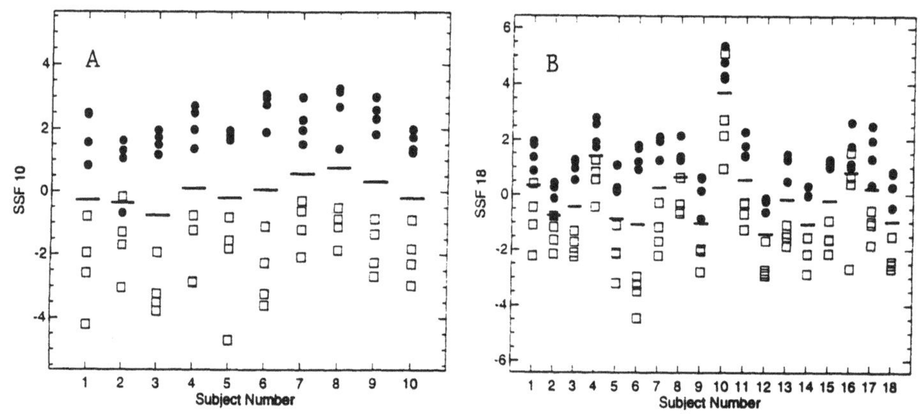

Figure 3: The subprofile scaling factor (SSF) distributions for the first component of the intrasubject subspace plotted for each subject from SSM analyses of (**A**) 10 subjects (80 scans) and (**B**) 18 subjects (144 scans). The filled circles represent scans performed with motor activation and the open squares are baseline scans without motor activation. Each subject's mean SSF value is displayed as a short black horizontal bar. In **B**, subjects 6 and 10 were outliers because of the unusually large amount of variation they contributed to the total variation of SSF18 (12% & 26%, respectively).

Subsequent SSM analyses of 16, 14, 12, 10, 8, and 6 subjects each contained a single component with a significant activation-baseline group difference occurring for the PC number equal to the number of subjects being analyzed (see

Figs. 2B & 3), at the intersection of the inter- and intrasubject subspaces (cf. Fig. 2A & B). All of these PC's with significant activation-baseline group differences had uniform variation contributions per subject without obvious outliers (see Table).

# of Subjects	2 (PC 3)	4 (PC 5)	6 (PC 6)	8 (PC 8)	10 (PC 10)	12 (PC 12)	14 (PC 14)	16 (PC 16)
Mean ± SD	0.50 ± 0.11	0.25 ± 0.16	0.17 ± 0.06	0.13 ± 0.04	0.10 ± 0.03	0.083 ± 0.023	0.071 ± 0.019	0.063 ± 0.017
Range	0.42,0.58	0.13,0.49	0.07,0.25	0.06,0.19	0.05,0.15	0.04,0.12	0.04,0.10	0.03,0.08

Table: The mean ± SD and range of the fraction of principal component (PC) variation contributed by each subject to the identified (PC #) from an SSM analysis with the given number of subjects.

SSM analyses of 2 and 4 subjects (16 and 32 scans, respectively) produced SSFs with a sharp separation between inter- and intrasubject subspaces (Fig. 2A) but the PC's with significant activation-baseline group differences occur as the second components of the intrasubject space (PC's 3 and 5 in Fig. 2B, respectively).

For SSM analysis of groups of 6 to 16 subjects Fig. 4 demonstrates that the PC's with SSF distributions that have a significant baseline-activation group difference (PC# = # of subjects analyzed, Fig. 2B) have relatively stable PC image volumes, i.e., GIS's, based on correlation with a reference image. These subgroups with 6, 8, 10, 12, 14 and 16 subjects were re-analyzed with FANOVA to test the use of the additive global effects model. The dashed line in Fig. 6 indicates that the unmodified FANOVA model produces a different and possibly less stable GIS—as a function of the number of subjects analyzed—than the GIS's from the SSM model.

In Fig. 5A compared to Fig. 2A the separation between intersubject and intrasubject subspaces is less clearly marked and breaks down more quickly with increasing group size for FANOVA compared to SSM, respectively. This mixing of the subspaces with FANOVA may account for the GIS differences between FANOVA and SSM in Fig. 6. In Fig. 5B compared to Fig. 2B there are additional PC's in the intersubject subspace that are approaching significance for activation-baseline group differences, particularly for the 12-16 subject subgroups. Fig. 2 demonstrates that compared to FANOVA in Fig. 5, SSM more clearly concentrates group mean differences into single components.

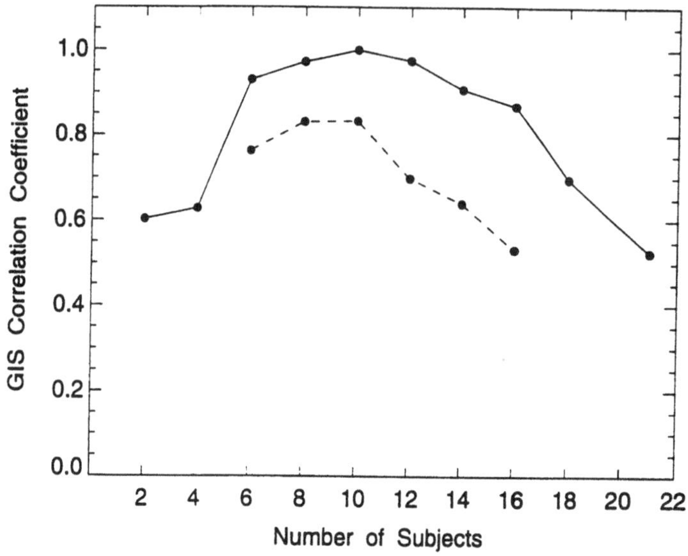

Figure 4: For SSM analysis of 2-21 subjects (solid line) and FANOVA analysis of 6-16 subjects (dashed line) the correlation coefficients of the principal component (PC) images (GIS's) for the first component of the intrasubject subspace (PC# = # subjects analyzed) correlated with GIS10 from a 10 subject SSM analysis are plotted against number of subjects analyzed.

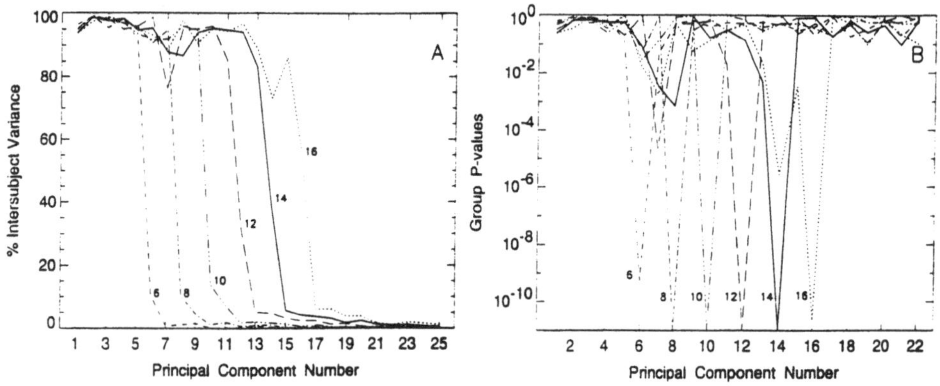

Figure 5: For FANOVA analysis of groups of 6-16 subjects with labels representing the number of subjects analyzed to produce each curve: (**A**) the % intersubject variance $(\sum_{jsr} SSF^2_{j\cdot\cdot})$ of each principal component (PC) is plotted against PC number–the labels are located at the point on the curve where the component number equals the number of subjects analyzed; (**B**) the p-values (uncorrected for multiple comparisons) from *t*-test differences between groups of activation subprofile scaling factors (SSF's) and baseline SSF's are plotted for each PC.

5. CONCLUSIONS

We have demonstrated that there is a rich spatial covariance structure in [^{15}O]water PET functional activation studies of the brain, which has little to do with the noise covariance structure introduced by reconstruction algorithms. In unpublished studies of the eigenspectra of simulated PET images that differed only in their random noise we obtained approximately flat eigenspectra with values in the 150-250 range suggesting that the plateau observed in Fig. 1 is due to reconstruction noise.

Linear models appear to be sufficient to extract stable patterns of regional interactions from activated brain networks provided a multiplicative global effects model (SSM) is used to control the boundary between the intersubject and intrasubject subspaces. This observation is important because the appendix of [9] indicates a close connection between additive global effects (FANOVA) and the most commonly used analysis procedure, ANCOVA adjusted statistical parametric mapping [4,7]. The observation that each new subject added to an analysis adds a separate dimension of intersubject signal variation, which appears to eventually change the functional activation motor component, may indicate that intersubject alignment procedures must be improved over the 12 parameter linear model used here. It will be interesting to see if nonlinear alignment can stabilize the motor activation component for larger subject groups. However, this might also be achieved by using techniques such as Fisher's linear discriminant [18] to construct an activation component from a linear combination of both inter- and intrasubspace components. Moreover, improved alignment procedures do not guarantee removal of the intersubject subspace, which may also have significant components from functional intersubject differences.

We have demonstrated that the sorting of orthogonal components according to their signal variation (i.e., eigenvalues) provides a useful tool for separating intersubject methodological (e.g., alignment) and functional signal differences from functional activation signals. The models and variation partitioning techniques introduced here appear to provide powerful tools for further exploring experimental and methodological changes in the design and analysis of functional activation brain studies.

6. REFERENCES

[1] Churchland PS, Sejnowski TJ. "Perspectives on cognitive neuroscience", *Science*, 242, pp 741-745, 1988

[2] Jackson JE. *A Users Guide to Principal Components*, Sections 10.3 & 13.7, New York, John Wiley & Sons, 1991

[3] Fox PT, Mintun MA, Reiman EM, Raichle ME. "Enhanced detection of focal brain responses using intersubject averaging and change-distribution analysis of subtracted PET images", *J Cereb Blood Flow Metab*, 8, pp 642-653, 1988

[4] Friston KJ, Frith CD, Liddle PF, Dolan RJ, Lammertsma AA, Frackowiak RSJ, "The relationship between global and local changes in PET scans", *J Cereb Blood Flow Metab*, 10, pp 458-466, 1990

[5] Worsley KJ, Evans AC, Marrett S, Neelin P. "A three-dimensional statistical analysis for CBF activation studies in human brain", *J Cereb Blood Flow Metab*, 12, pp 900-918, 1992

[6] Friston KJ, Worsley KJ, Frackowiak RSJ, Mazziotta JC, Evans AC. "Assessing the significance of focal activation using their spatial extent", *Human Brain Mapping*, (in press)

[7] Friston KJ, Firth CD, Liddle PF, Frackowiak RSJ. "Functional connectivity: The principal component analysis of large (PET) data sets", *J Cereb Blood Flow Metab*, 13, pp 5-14, 1993

[8] Strother SC, Kanno I, Rottenberg DA. "Principal component analysis, variance partitioning and 'functional connectivity'", *J Cereb Blood Flow Metab*, (in press)

[9] Strother SC, Anderson JR, Schaper KA, Sidtis JJ, Liow J-S, Woods RP, Rottenberg DA. "Principal component analysis and the scaled subprofile model compared to intersubject averaging and statistical parametric mapping: I. 'Functional connectivity' of the human motor system studied with [^{15}O] PET", *J Cereb Blood Flow Metab*, (in press)

[10] Lautrup B, Hansen LK, Law I, Morch N, Svarer C, Strother SC. "Massive weight sharing:A cure for extremely ill-posed problems", in Supercomputing in brain research:From tomography to neural networks, HLRZ, Julich, Germany, (in press), 1994

[11] Gollob HF. "A statistical model which combines features of factor analytic and analysis of variance techniques.", *Psychometrika*, 38, pp 555-569, 1968

[12] Moeller JR, Strother SC, Sidtis JJ, Rottenberg DA. "Scaled Subprofile Model: A statistical approach to the analysis of functional patterns in positron emission tomographic data", *J Cereb Blood Flow Metab*, 7, pp 649-658, 1987

[13] Moeller JR, Strother SC. "A regional covariance approach to the analysis of functional patterns in positron emission tomographic data", *J Cereb Blood Flow Metab*, 11, pp A121-A135, 1991

[14] Woods RP, Cherry SR, Mazziotta JC. "A rapid automated algorithm for accurately aligning and reslicing positron emission tomography images", *J Comput Assist Tomogr*, 16, pp 620-633, 1992

[15] Woods RP, Mazziotta JC, Cherry SR. "Automated image registration", in *Proceedings of Brain PET '93 AKITA: Quantification of Brain Function*, Uemura K, Lassen NA, Jones T, Kanno I, eds, Excerpta Medica, Amsterdam, 1993

[16] Strother SC, Anderson J, Xu X-L, Bonar DC, Rottenberg DA. "Quantitative comparisons of image registration techniques based on high-resolution MRI of the brain", *J Comput Assist Tomogr*, 18, pp 954-962, 1994

[17] Bonar DC, Schaper KA, Anderson JR, Rottenberg DA, Strother SC. "Graphical analysis of MR feature space for the measurement of CSF, gray-matter and white-matter volumes", *J Comput Assist Tomogr*, 17, pp 461-470, 1993

[18] Mardia KV, Kent JT, Bibby JM. *Multivariate analysis*, Section 11.5, San Diego, Academic Press, Inc., 1979

CONDITIONED STATISTICAL MODEL FOR FUNCTIONAL MRI STUDIES OF THE HUMAIN BRAIN

H. Benali[1], J.L. Anton[2], M. Di Paola[3], F. Frouin[1], O. Jolivet[4] R.Y. Carlier[4], J.Bittoun[4], Y. Burnod[2] and R. Di Paola[1]

U66 INSERM, CHU Pitié Salpêtrière, 75634 PARIS 1

CJF 93-03, CREARE, 9 quai saint Bernard, 75005 PARIS 2

Institut Gustave-Roussy, 94805 Villejuif Cedex, France 3

CIERM, 78 rue du Général-Leclerc, 94275 Le Kremlin Bicêtre 4

Abstract : A statistical method for detecting significant cortical activation of the humain brain as measured with functional magnetic resonance imaging (fMRI) of the brain, is presented in this paper. The method is based on the generalized fixed effect model and its asymptotic theory. The solution of this statistical model is obtained by Partial Principal Components Analysis (PPCA), which estimates the relevant part of the signal variation, free of the noise and of the unrelated artifactual signal. The use of the asymptotic distributions of the relevant part of data yields to a map of statistical significant signal variability. These statistical distributions depend on the number of pixels, and the deduced activation map is obtained without reference to any a priori physiological assumptions. The interest of our approach is illustrated with sensorimotor experiments performed at 1.5T by normal subjects, and comparison with the detected high levels signal activity map and MR angiography image is presented.

1. INTRODUCTION

Recent developments in Magnetic Resonance Imaging (MRI) enable noninvasive investigation of humain brain function [1, 2, 3]. Different MRI sequences have been used to detect MR signal enhancement in specific local brain areas [4, 5] from a series of images obtained during analysed or control tasks. The fMRI signal is related to the concentrations of paramagnetic deoxyhemoglobin in the cerebral blood. The task-induced decrease in the local concentration of deoxyhemoglobin produces an increase in the MRI signal intensity [4, 5].

Several statistical approaches have been proposed for the determination of the significant brain activity maps [4, 6, 7, 8]. All of them combine empirical observation with statistical models to detect differences in local activation patterns. One of the problems is to remove the artifactual signal changes, which exist on both control and task images, and are unrelated to the induced neuronal activation. This artifactual signal is seen in other physiological processes, gross motion, pulsatile blood, etc.[4]. As the changes in MR signal, due to activation, are small, in the range of 5-10% of signal intensity [7, 9], the artifactual signal changes make it difficult to ascribe statistical significance to the detected local activation [4].

The commonly used differencing procedure for the determination of the activity maps consists of [4] :

Y. Bizais et al. (eds.), Information Processing in Medical Imaging, 311–322.

- determining a region of interest including the suspected activity area,
- subtracting the averaged images acquired during the task from the averaged control images, of the same time series. This is the empirical observation part of the method,
- using a classical Student t-test to detect the pixel with significant brain activity. This statistical model considers every pixel as an independent variable. As the number of images required by the t-test increases, the acquisition time increases, yielding a larger number of false-positive detected pixels due to autocorrelated artifactual signal, in space and/or time [10].

A new method is introduced to overcome subtraction method drawbacks. It is based on a multidimensional geometric approach for the analysis of the relevant part of signal activation, which is first presented. The related statistical model for the MRI data set is then described. The asymptotic theory is used for the determination of the model estimators distributions. These distributions yield a significant statistical brain activity map. The practical applications of this method are illustrated in sensorimotor experiments performed at 1.5T by normal subjects. A correlation is established between "high level intensity" of the detected signal activation and the MR angiography image which is acquired in the same plane as the functional images. Finally, the underlying hypotheses of our method are compared with hypotheses of other proposed methods.

2. A GEOMETRIC MODEL

MRI activation data set can be viewed as a collection of n sample vectors x_i (i=1,n) of p control and p task interleaved components x_{it} (t=1,2p) that form the random process representing the underlying physiology. x_i represents the variation of the signal within the pixel i in the dynamic sequence (ndexed by the variable t). Assuming the time course of the activation is known, the signal x_i can be decomposed as follows:

$$x_{it} = \begin{cases} a_{it} & \text{for } t \text{ corresponding to the control} \\ b_{it} & \text{for } t \text{ corresponding to the task} \end{cases} \quad \text{(Eq. 1)}$$

Let A (resp. B) be the corresponding set of the n sample functions a_i (resp. b_i) of p components a_{ij} (resp. b_{ij}) (j=1,p).

2.1. Subtracting method

To detect the signal changes, images a_{ij} and b_{ij}, acquired during repeated trials, are averaged and the map M of the brain activity is deduced as follows:

$$m_i = \sum_{j=1}^{p} a_{ij} - \sum_{j=1}^{p} b_{ij} = \sum_{j=1}^{p} (a_{ij} - b_{ij}),$$

which yields in matrix form:

$$M = (A - B)\,\mathbf{1}^t, \quad \text{(Eq. 2)}$$

where M is the (n,1) matrix of the activation map, **1** is the (1,p) vector of ones, and ^{t} denotes the transpose.

From the geometrical point of view, the image M corresponds to the projection of the data set A without the effect of the data set B (subtracting the effect of B) onto the direction generated by the vector **1**.
This geometric approach interpretation shows two drawbacks of the method:
- the subtraction of A - B is not able to remove the effect of the control data set B from the task data set A, since the two sets are not independent.
- the projection of A - B onto the "trivial" vector **1** is arbitrary and cannot be easily related to the signal of interest.

We further define the mathematical notions of "remove the effect of" and "projection onto optimal vectors".

2.2. Conditioned data activation

When two data sets A and B are correlated, the optimum linear effect of B to A, denoted C, should be estimated. The subtraction of C from A yields the conditioned data activation D, representing the part of A which is not due to B. In other words, the correlated effects of B are removed from A. A can be decomposed into the sum of two uncorrelated data sets [11] :

$$A = C + D \text{, with } C = B\,(B^tB)^{-1}BA \text{, and}$$

$$D = A - B\,(B^tB)^{-1}BA \qquad \text{(Eq. 3)}$$

We denote by Q the orthogonal projector $Id - B(B^tB)^{-1}B$ where Id is the (n,n) identity matrix.

3. A STATISTICAL MODEL: THE GENERALIZED FIXED EFFECT MODEL

A statistical model for factor analysis of medical image sequences methods has been proposed to estimate the underlying physiological structures [12]. This model determines the optimal representation of the signal of interest, without the noise which has been considered as independent. This hypothesis is not verified for the conditioned data set. As the conditioned process induces a spatial autocorrelation. To deal with this problem, we proposed a more general statistical model. The solution of this statistical model leads to the expression of the projection onto an optimal subspace which represents the main part of the conditioned signal. To be consistent with the underlying physiological process of activation, the statistical model must state that the conditioned process D can be decomposed into its unobservable fixed part $\tilde{D}$, corresponding to the "true" part of data, and its unobservable random error σe. The generalized fixed effect model [13, 14] fulfils this condition and includes some general statistical properties of the error e.

3.1. Definition of the generalized fixed effect model

The generalized fixed effect model is defined as follows [14] :
- D is a matrix of n vectors d_i, which are n random vectors defined on a probability space, it can be written :

$$D = \tilde{D} + \sigma e \ , \qquad \text{(Eq. 4)}$$

where the vectors error e are orthogonal to $\tilde{D}$. It is the true or the fixed effect part that is of real interest. We would like to know it free of error.
- The expectation of D is :

$$E(D) = \tilde{D} \ . \qquad \text{(Eq. 5)}$$

- The variance of D can be written

$$Var\,(D) = \sigma^2 Q \otimes \Gamma, \qquad \text{(Eq. 6)}$$

where $\otimes$ is the the tensor product (or outer product) [15] and Γ is a (p,p) symmetric positive definite matrix of temporal covariance of the error e_i. It is assumed to be known.

- There exists a linear subspace $\mathbb{S}$ of $\mathbb{R}^p$, of dimension q (q<p) such that $\tilde{D}$ belongs to $\mathbb{S}$.

The subspace $\mathbb{S}$, the unobservable fixed effect part $\tilde{D}$ belonging to $\mathbb{S}$ and the parameter σ have to be estimated.

3.2. Solution to the generalized fixed effect model

The least square estimate of $\mathbb{S}$ is obtained by minimizing:

$$E[\,|D-\tilde{D}|^2_W\,] \qquad \text{(Eq. 7)}$$

where **W** is a (p,p) symmetric positive definite matrix, defining a metric of $\mathbb{R}^p$. The minimum of (Eq. 7) is reached for $\mathbb{S}$ such that $\mathbb{S}$ is spanned by the q eigenvectors u_k associated with the q largest eigenvalues λ_k of the matrix [14]:

$$\frac{1}{n} D^t D W = \frac{1}{n} A^t Q^t A\, W \qquad \text{(Eq. 8)}$$

Thus, the projection of conditioned data set D is the projection onto optimal vectors u_k.

4. ASYMPTOTIC THEORY OF THE GENERALIZED FIXED EFFECT MODEL

The estimation of the subspace $\mathbb{S}$ depends on the metric **W** which is closely related to the statistical properties of errors. The asymptotic theory enables to determine the optimal metric **W** to be used for the eigendecomposition involved in the generalized fixed effect model [13, 16, 17].

4.1. Optimal metric choice for the generalized fixed effect model

Let $P_{\mathbb{S}}$ be the orthogonal projector onto the subspace $\mathbb{S}$, the asymptotic theory (if the number of the pixels n is large) shows that for any σ, $P_{\mathbb{S}}$ is a strong estimator if and only if **W** is equal to Γ^{-1} [16].

So, the optimal subspace $\mathbb{S}$ can be computed if Γ is known or can be estimated. $\mathbb{S}$ is then the subspace in which the relevant part of the conditioned process is properly restored, without the noise. It can be estimated as follows:

$$\tilde{D} = V_q \Lambda_q U_q, \qquad \text{(Eq. 9)}$$

where,

$$V_q = D\,\Gamma^{-1}\, U_q^t\, \Lambda_q^{-1}, \qquad \text{(Eq.10)}$$

U_q is the (q,p) matrix of the eigenvectors u_k, Λ_q is the (q,q) diagonal matrix whose diagonal elements are the square roots of the eigenvalues λ_k.
The least square estimator of σ is the average of the p-q smallest eigenvalues:

$$\sigma^2 = \frac{1}{p-q} \sum_{k=q+1}^{p} \lambda_k . \qquad \text{(Eq.11)}$$

The solution of the generalized fixed effect model corresponds to the singular value decomposition involved in Partial Principal Component Analysis (PPCA) [18], where the orthogonal images V_q are the maps of optimal signal variations free of artifactual signal.

4.2. Asymptotic distributions of estimators

The covariance of the relevant part of the conditioned process $\tilde{D}$ is represented by the q first eigenvectors u_k (k=1,q). In order to identify the statistical significant covariance, we study the distribution of the eigenelements U_q, V_q and Λ_q under the following null hypothesis H_0:
- the n vectors e_i have identical normal distributions,
- the covariance matrix of a conditioned data set $\sigma^2 Q \otimes \Gamma$ is known.

General null hypotheses and demonstrations of the important asymptotic Gaussian distributions of parameters U_q, Λ_q can be found in [16]. Using these general theoretical results, the asymptotic distributions of the orthogonal images V_q are derived and a statistical significant variation of the signal can be estimated in the images.
If we denoted by u^*_k, v^*_k and λ^*_k the asymptotic random variables of the eigenelements u_k, v_k and λ_k, we have the following results :

4.2.1. *Asymptotic distribution of eigenvalues Λ_q*

Under the hypothesis H_0 the statistic $\sqrt{n}\,(\lambda_k - \sigma^2)$ converges to a Gaussian random variable $\hat{\lambda}_k$ when n converges to infinity, where $E(\hat{\lambda}_k) = 0$, and

$$Var(\hat{\lambda}_k) = \frac{2\sigma^4}{p-q} . \qquad \text{(Eq.12)}$$

Using the following normalized Gaussian random variable:

$$\frac{\lambda_k}{\sqrt{Var(\lambda_k)}} = \frac{\sqrt{n(p-k)}}{2} \frac{\lambda_k - \sigma^2}{\sigma^2} , \qquad \text{(Eq.13)}$$

we can derive the dimension q of the subspace $\mathbb{S}$ which represents the relevant part of the conditioned data set.

4.2.2. *Asymptotic distribution of eigenvectors U_q*

Under the hypothesis H_0 the statistic $\sqrt{n}\,(u_k - u^*_k)$ converges to Gaussian random vector $\hat{u}_k$ where $E(\hat{u}_k) = 0$, and

$$Var(\hat{u}_k) = \sum_{j=k}^{p} \left[\frac{\sigma^2}{\lambda_j^* - \sigma^2} + \left(\frac{\sigma^2}{\lambda_j^* - \sigma^2} \right)^2 \right] u^*_j u^{*t}_j . \qquad \text{(Eq.14)}$$

5. THE GENERALIZED FIXED EFFECT MODEL FOR MRI DATA

5.1. Optimal fMRI signal enhancement

The statistical study of the noise in MRI shows that noise e_i is uncorrelated in time (Γ=Id) and has an identical normal distribution [19]. We apply our general theoretical approach to this imaging modality. If we consider that the spatial correlation of errors is negligible, then the covariance matrix of the conditioned data set D can be modeled as

$$Var(D) = \frac{\sigma^2}{n} Q \otimes Id . \qquad \text{(Eq.15)}$$

According to the solution of the fixed effect model, the orthogonal decomposition must be performed using the metric defined by the matrix W= Γ^{-1} =Id. The subspace $\mathbb{S}$ is spanned by the q eigenvectors u_k associated with the q largest eigenvalues λ_k of the matrix $\frac{1}{n} A^t Q A$.

Consequently, the optimal decomposition of the conditioned data set D to determine the relevant part of data is that used in PPCA [18].

5.2. Statistical significant activation map

Upon estimating the q statistical significant eigenvalues using equations 13 and 14 we can derive the asymptotic Gaussian distributions of the q first eigenimages.

Using these asymptotic distributions, we assigned a statistical significant coefficient to each pixel of the conditioned images D. These pixel statistics allow to determine the statistical significant activation map.

5.2.1. *Asymptotic distribution of eigenimages V_q*

For any vector d_i which represents the conditioned signal variation within the pixel i we consider the following statistic :

$$\left[\sqrt{\lambda_k}\, v_k(i) - \sqrt{\lambda^*_k}\, v^*_k(i) \right] = d_i^t \,(u_k - u^*_k). \qquad \text{(Eq.16)}$$

Using the continuity of the scalar product and the asymptotic results (13) and (14) we derive the following results:

$\sqrt{n}\,[\,\sqrt{\lambda_k}\, v_k(i) - \sqrt{\lambda^*_k}\, v^*_k(i)\,]$ converges to a Gaussian random variable $\hat{v}_k$

where $E(\hat{v}_k) = 0$ and,

$$Var(\hat{v}_k) = \sum_{i=k}^{p} [\frac{\sigma^2}{\lambda^*_j - \sigma^2} + (\frac{\sigma^2}{\lambda^*_j - \sigma^2})^2]\ \lambda^*_j\, v^*_j(i)^2. \qquad \text{(Eq.17)}$$

6. MATERIALS AND METHODS

In this study, the investigation of the effect of sensorimotor task was reported. The subjects execute the motor sequence task, which consists of right hand fingers to thumb opposition movements [2].

An image sequence of 15 control images and 15 motor-task images was investigated.

6.1. Data acquisition

MRI images were carried out on a 1.5T Signa (GE Medical system) using a standard head coil. Anatomical images were performed to detect the hand area around the sulcus centralis. Functional scans used a gradient echo sequence (80 ms TR, 60 ms TE, 40° flip angle) and produced 5 mm thick slices. The FOV was 24 cm, and matrix 128x256. The pixel dimension was 1.88x0.94 mm. For display, the images were interpolated to 256 x 256. For each condition (control and task), the trials included 5 acquired images, each one corresponding to 20.5s. There were 20s for trial feedback between each task. Typically, 30 images were collected, 15 for each condition.

In addition, MR angiography scans using a gradient echo sequence (45 ms TR, 13 ms TE, 60° flip angle) were performed in the same plane as the functional images.

6.2. Data pre-processing

Subject motion during the experiment yields artifacts in subtraction methods, and more generally in every method which assumes the spatial stability of pixels during the time image sequence. Even a motion with an amplitude less than the pixel size can produce artifacts, as shown on simulation studies [20]. To overcome this source of error in conditioned analysis, an inter-frame motion was automatically compensated. Due to the thickness of the slice (5 mm) and the small amplitude of motion, the bidimensional approximation of an “in plane” displacement was considered to be valid. Assuming a rigid motion, three planar parameters were estimated to describe planar translation and rotation. One image in the serie (in general, the first one) was chosen as a reference image and the other images were registered according to this reference. The parameters were estimated by maximizing the correlation coefficient between the transformed image and the reference image. The maximization was performed using the Powell optimization procedure of

Harwell library [21]. To decrease the probability of getting a local maximum of the criterion, initial values for transformation parameters were chosen as the values of the parameters corresponding to the geometrical transformation which had been estimated for the previous image in the sequence. Fractional pixel displacements were taken into acount using a cubic interpolation [22].
After removing the head motion artifacts, no binary mask is used and all the pixels are taken into account.

6.3. Results

6.3.1. *Partial Principal Component Analysis*

The PPCA is conducted using the metric Γ^{-1}=Id, and the dimension q of the subspace representing the relevant part of data is deduced using equation (13). Figure 1 shows the conventional differencing image (Fig. 1a) and a typical first eigenimage V_1 (Fig. 1b).

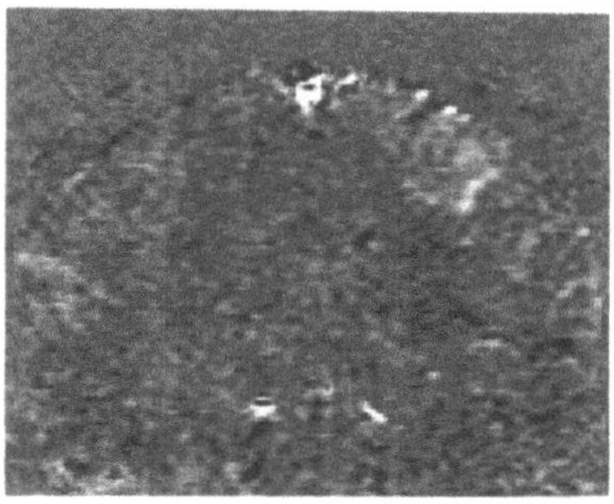

Fig. 1a: activation map obtained by the differencing procedure

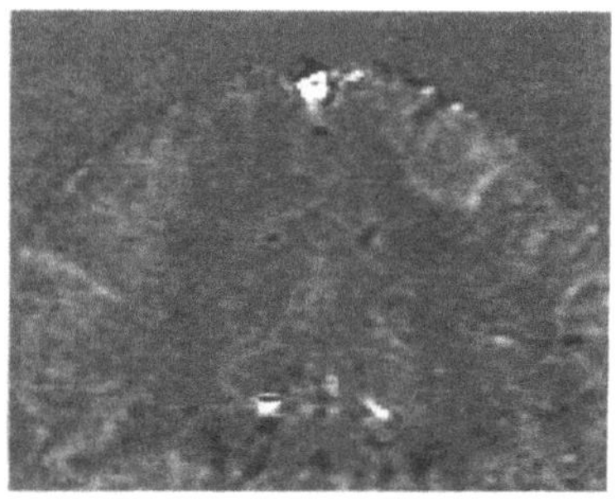

Fig. 1b: activation map corresponding to the first eigenimage V_1 obtained by PPCA

The statistical properties of the first eigenimage is used to perform a significant statistical map of signal variation for motor task.

Statistical significant map

The statistical significant motor activation map is obtained from the first eigenimage, using equation 17. Two kinds of threshold are applied to this image. Figure 2a shows the map of the higher signal intensity. The threshold value is higher than 4 which corresponds to a probability value less than 10^{-4}. The represented areas are at least of 3 connected pixels. Figure 2b displays the superimposition of the MR angiography image and of the high level signal intensity map. The correlation observed between these two images allows to assign the high level of activation to the main veinous signal. This result is equivalent to those obtained by Lai [23] and Segebarth [24].

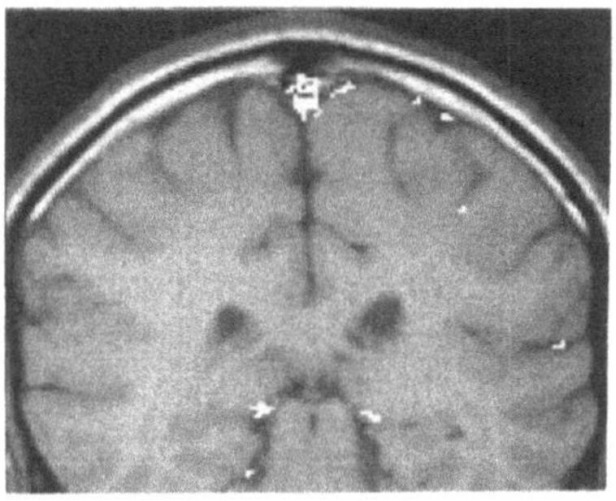

Fig. 2a: activation map for the high signal intensity superimposed on the anatomical image

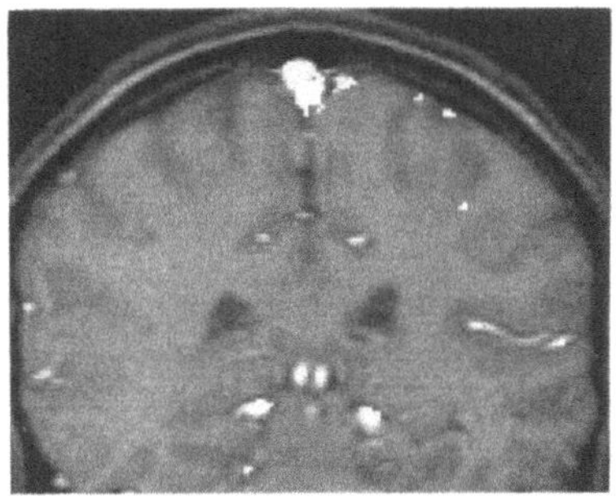

Fig. 2b: activation map of the high signal superimposed on the MR angiography image

Figure 3a shows the map of the lower signal in the range of 1.5 to 4. These thresholds correspond to the probability values of p=0.07 and p=0.0001. Only areas of at least 25 connected pixels are represented. These activated regions correspond to the contralateral primary motor and sensory hand area. This is in agreement with the classical results described in [2].

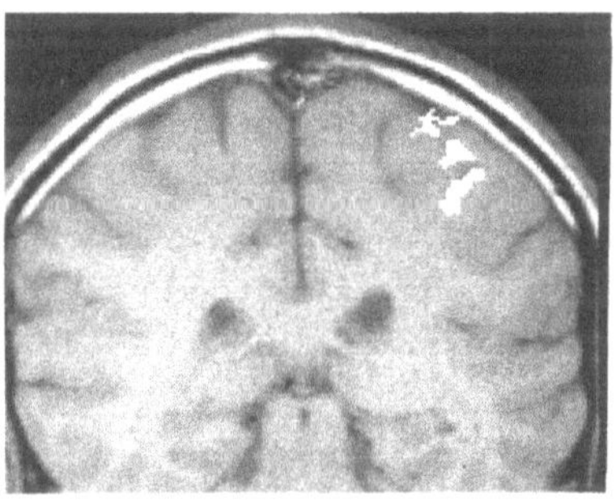

Fig. 3a: activation map of the lower signal activation intensity superimposed on the anatomical image

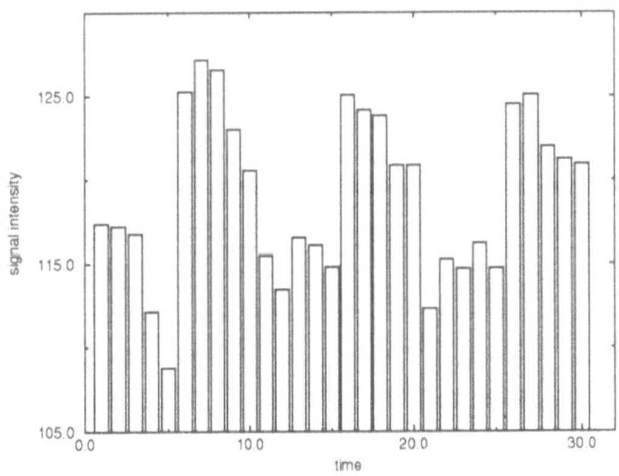

Fig. 3b: activation curve corresponding to the lower signal intensity.

Figures 3b shows the temporal signal changes within the selected pixels of the initial functional images. The on-off cycle corresponds well with the task-control paradigm.

7. DISCUSSION

The introduction of a statistical generalized fixed effect model [14] for functional magnetic resonance images, gives a theoretical basis to determine a significant cortical activation map. This model takes into account the artifactual signal [4] using the conditioned data. Then, it estimates the relevant part of the signal variability free of the noise, and finally, the statistically significant signal variability map is deduced. These steps involve several hypotheses :

- The unknown artifactual signal induces a correlation between the analysed and control images. One would like to remove this correlation part and leave only the activation signal not due to this artifactual signal. The optimal linear effect of baseline images to activation images is calculated. This is subtracted from activation images to yield the conditioned data.
- In order to determine the relevant part of the conditioned data, the statistical model requires the covariance matrix of the error to be known or to be estimated. Therefore, the statistical properties of the acquired data must be studied. In magnetic resonance imaging, it has been shown that, in time, errors have an identical independent normal distribution [19]. The matrix Γ is then identified as the identity matrix. It appears that the main part of conditioned data is obtained by Partial Principal Component Analysis (PPCA) [18]. The spatial covariance matrix of the error used in PPCA is the identity matrix. The generalized fixed effect model, provides a theoretical basis for the choice of other spatial covariance matrix to be used for PPCA.
- In order to obtain reliable statistical inferences for signal variability, we have studied the statistical properties of the eigenelements of the PPCA. Asymptotic Gaussian distributions have been derived using the central limit theorem over a very large number of pixels [17]. Therefore these distributions depend on the number of pixels and not on the amount of images. Some hypotheses must be considered to determine their associated covariance operators [13, 16] to yield specific Gaussian distributions of the eigenelements. Hence, a statistically significant signal enhancement in specific local brain area is derived using these asymptotic distributions. These adaptive Gaussian distributions give strong statistical results.

Numerous methods have already been proposed to find the cortical activity map [4, 7, 8, 10]. An automatic or a priori mask was usually performed on the image sequences in order to limit the number of examined pixels [25]. Such a mask is operator dependent.

We have discussed the efficiency of the conventional subtraction method [4]. One of the problems is to remove the artifactual signal changes with the subtraction of signals during repeated trials within a given task. The associated statistical model , t-test, treats every pixel as an independent test yielding to a large number of false-positives detected pixels [25].

As an alternative to this method, the theory of Gaussian random field has been used for PET studies [7, 8]. The methods, single filter or multifiltering detection, involve a gaussian filtering and an averaged difference image which is modeled by a 2D Gaussian distribution. Since the number of averaged images was large, the distribution of the differencing images was close to a

Gaussian distribution. As discussed above, the differencing approach is not able to remove artifactual signal from the activation images. In [10], the theory of a stationary Gaussian stochastic process has been used and adresses the problem of autocorrelation in time, due to intervening hemodynamic response function. The images were acquired every 3 seconds, yielding an autocorrelated time series. This method involves testing the autocorrelation by convolving a stimulus function by a hemodynamic function modeled as a Poisson distribution arbitrary chosen.

One of the main problems in fMRI studies is to separate the venous activation signal from the signal of the capillaries. In our results, the high activation signals are related with the main veins, and do not yield to a precise spatial area of the cortical activation. However, the low activation signal is well correlated to the spatial form of the grey matter in the suspected cortical area. Several arbitrary statistical thresholds can be chosen, each one yields to high and low signal intensities. Therefore, we can suppose that the estimated activation signal is continuously varying from main veins to capillaries. This assumption is under investigation.

Only the first eigenimage has been used to map the sensorimotor areas. More efficient activation map taking into account the spatial covariance matrix S of the errors ($S \neq Id$) will be derived from the q statistical significant eigenimages ($q>1$).

8. CONCLUSION

Several statistical methods have been proposed for the analysis of functional magnetic resonance imaging (fMRI) experiments. The most widely known subtraction method is revisited and a new statistical method is defined. The model estimates the relevant part of the signal variation free of noise and of the unrelated artifactual signal. This unknown artifactual signal is supposed to be represented on both control images and task images. The solution of this statistical model is obtained by Partial Principal Component Analysis (PPCA) using an optimal metric. This metric depends on the expression of the data variance. It is shown by means of a sensorimotor fMRI studies that the use of the asymptotic normal distribution of the eigenelements of PPCA allows the determinimation of a map of the statistically significant signal variability. It has been shown that the detected high level signal intensity is related to vein, and the detected low signal intensity is related to cortical activation. More work is needed to validate this assumption.

9. REFERENCES

[1] J.W. Belliveau, D.N. Kennedy, R.C. McKinstry, B.R. Buchbinder: "Functional mapping of the human visual cortex by magnetic resonance imaging", Sci. 254, 621-768

[2] P.E. Roland: Brain activation. New York: Wiley-Liss 1992.

[3] R.G. Shulman, A.M. Blamire, D.L. Rothman and G. McCarthy: "NMRI and spectroscopy of humain brain function", Proc Natl Acad Sci USA 89, 1837-1841, 1993

[4] P.A. Bandettini, A. Jesmanowicz, E.C. Wong and J.S. Hyde: "Processing Strategies for Time-Course Data Sets in Functional MRI of the Human Brain", MRM 30, 161-173, 1993

[5] K.K. Kwong, J.W. Belliveau, D.A. Chesler, I.E. Goldberg, R.M. Weisskoff, B.P. Poncelet, D.N. Kennedy, B.E. Hoppel, M.S. Cohen, R. Turner, H.M. Cheng, T.J. Brady and B.R. Rosen: "Dynamic magnetic resonance imaging of humain brain activity during primary sensory stimulation", Proc Natl Acad Sci USA 89, 5675-5679, 1992

[6] K.J. Friston, C.D. Frith, P.F. Liddle and R.S.J. Frackowiak: "Comparing Functional (PET) Images: The assessment of Significant Change", J Cereb Blood Flow Metab 11, 690-699, 1991

[7] K.J. Worsley, A.C. Evans, S. Marrett and P. Neelin: "A Three-Dimensional Statistical Analysis for CBF Activation Studies in Human Brain", J Cereb Blood Flow Metab 12, 900-918, 1992

[8] J.B. Poline and B. Mazoyer: "Enhanced Detection in Brain Activation Maps Using a Multifiltering Approach", J Cereb Blood Flow Metab 14, 639-642, 1994

[9] W. Schneider, B.J. Casey and D. Noll: "Functional MRI Mapping of Stimulus Rate Effects Across Visual Processing Stages", Human Brain Mapping 1, 117-133, 1994

[10] K.J. Friston, P. Jezzard and R. Turner: "Analysis of Functional MRI Time-Series", Humain Brain Mapping 1, 153-171, 1994

[11] J.S. Bendat and A.G. Piersol: Random Data. Analysis and measurement procedures. New York: John Wiley & Sons 1986.

[12] H. Benali, I. Buvat, F. Frouin, J.P. Bazin and R. Di Paola: "A statistical model for the determination of the optimal metric in Factor Analysis of Medical Image Sequences (FAMIS)", Phys Med Biol 38, 1065-1080, 1993

[13] T.W. Anderson: The 1982 Wald Memorial Lectures. "The Estimating Linear Statistical Relationships",The Annals of Statistics 12(1), 1-45, 1984

[14] H. Caussinus. "Models and uses of principal component analysis. Multidimensional data analysis", J. de Leeuv ed., DSWO Press, Leiden, 149-178, 1986

[15] M.L. Eaton: Multivariate Statistics. A vector space approach. New York: Wiley 1983

[16] J. Fine and A. Pousse: "Asymptotic study of the multivariate functional model. Application to the metric choice in principal component analysis", Statistics 23, 63-83, 1992

[17] J. Fine: "Asymptotic study of the multivariate functional model in the case of a random number of observations for each mean", Statistics 25, 285-306, 1994

[18] C.R. Rao: "The use and interpretation of principal component analysis in applied research", Sankya A(26), 329-358, 1964

[19] W.A. Edelstein, G.H. Glover, C.J. Hardy and R.W. Redington: "The Intrinsic Signal-to-Noise Ratio in NMR Imaging", Mag Res Med 3, 604-618, 1986

[20] J.V. Hajnal, R. Myers, A. Oatridge, J.E. Schwieso and I.R. Young: "Artifacts Due to Stimulus Correlation Motion in Functional Imaging of the Brain", MRM 31, 283-291, 1994

[21] M.J.D. Powell: "An Efficient Method for Finding the Minimum of a Function of Several Variables Without Calculating Derivatives", Comp J 7, 155-162, 1964

[22] P.E. Danielsson and M. Hammerin: "High accuracy rotation of images", CVGIP 54, 340-344, 1992

[23] S. Lai, A.L. Hopkins, E.M. Haacke, D. Li, B.A. Wasserman, P. Buckley and L. Friedman: "Identification of Vascular Structures as a Major Source of Signal Contrast in High Resolution 2D and 3D Functional Activation Imaging of the Motor Cortex at 1.5T: Preliminary Results", MRM 30, 387-392, 1993

[24] C. Segebarth, V. Belle, C. Delon, R. Massarelli, J. Decety, J.F. Le Bas, M. Decorps and A.L. Benabid: "Functional MRI of the human brain: predominance of signals from extracerebral veins", NeuroReport 5, 813-816, 1994

[25] N. Lange: "Variability and covariability in magnetic resonance functional neuroimaging", In: F.L. Bookstein, J.S. Duncan, N. Lange and D.C. Wilson (eds): SPIE Mathematical Methods in Medical Imaging. San Diego: SPIE , pp. 150-160, 1994

ON VARIABLE BRIGHTNESS OPTICAL FLOW FOR TAGGED MRI

Sandeep N. GUPTA , Jerry L. PRINCE
IACL, Department of Electrical and Computer Engineering, The Johns Hopkins University, Baltimore, MD 21218, USA

Abstract: The problem of computing velocity fields from tagged MR images is considered, with particular attention paid to the image brightness changes caused by tag pattern fading with time. Shortcomings of existing optical flow methods for this application are pointed out and Gennert and Negahdaripour's optical flow method is discussed. This method involves estimation of a local brightness transformation field in addition to the velocity field. Approximations for these transformations are developed, leading to a faster algorithm which does not require careful selection of the regularization coefficients. This method is validated by tests on both simulated and actual MR data, and is demonstrated to be computationally robust to inaccurate knowledge of MR parameters.

1. INTRODUCTION

Techniques for myocardial motion analysis have been widely studied and used to diagnose abnormal heart function [1], [2]. MR tagging methods [3] have shown promise for detailed non-invasive motion imaging, and several tagging methods and associated image processing algorithms have been proposed for tagged MR images [4], [5]. Our research focuses on optical flow (OF), which is a method to estimate a dense flow field between image pairs. The key assumption in most standard optical flow (SOF) algorithms is that the brightness of a material point does not change with time. As has been pointed out by Prince and McVeigh in [6], this condition does not hold in cardiac MR tagged images. In fact, the tags become brighter at a rate determined by the T_1 value of the myocardium. To use OF for this application, the authors of [6] proposed a modification to OF, which they called *variable brightness optical flow* (VBOF). In this method, the brightness change of a material point is estimated by tracking the point as it moves. Since the brightness constancy assumption is not made in VBOF, it gives better estimates of the velocity fields than SOF.

VBOF suffers from a major limitation however in that it requires prior knowledge of the parameters D_0 (proton-density), T_1, and T_2 everywhere in the region of interest. The sensitivity of VBOF to the MR parameters makes it unreasonable for practical

This research was supported by NIH grant R29-HL47405, and NSF grant MIP-9350336.

Y. Bizais et al. (eds.), Information Processing in Medical Imaging, 323–334.

application. Another problem with VBOF is that its performance degrades over time making estimates of motion over long time sequences impractical.

This work aims both at improving OF performance for long time sequences and at reducing the dependency on prior knowledge of MR parameters. We first implement the method of Gennert and Negahdaripour (GNOF) [7] and investigate its performance under tag fading conditions. This algorithm relaxes the brightness constancy assumption of SOF and allows for a brightness variation modeled by a local linear transformation. This algorithm achieves significant improvement of performance over SOF and VBOF for very long time sequences. To make the algorithm more suitable for practical use, we next use the physics of the MR imaging process to derive approximations for the brightness variations. These approximations are then used in a new algorithm which requires only approximate knowledge of T_1 in the image.

The following is a brief outline of this paper. In Section 2.1, we discuss GNOF. In Section 2.2, we derive approximations for brightness transformation fields. Experimental results are presented in Section 3. A summary of this work and a discussion is presented in Section 4.

2. METHODS

2.1. Gennert and Negahdaripour's Optical Flow

Optical flow algorithms estimate a velocity vector for each point in a scene given images of the scene at two or more different times [8]. Let $\psi(\mathbf{r}, t)$ denote the brightness of the image at the spatial point $\mathbf{r}$ at time t. Gradient-based methods typically calculate both $\psi_t(\mathbf{r}, t)$, the spatial time derivative of brightness (which is the partial derivative of brightness holding $\mathbf{r}$ constant), and $\nabla\psi(\mathbf{r}, t)$, the spatial brightness gradient. These are related to the material time derivative $\dot{\psi}(\mathbf{r}, t)$ (which is the partial derivative of brightness holding the material coordinate fixed) as follows

$$\dot{\psi}(\mathbf{r}, t) = \nabla\psi(\mathbf{r}, t) \cdot V(\mathbf{r}, t) + \psi_t(\mathbf{r}, t) . \qquad \text{(Eq.1)}$$

Here $V(\mathbf{r}, t)$ is the spatial velocity field with components (u, v), which we are trying to determine.

The usual *brightness constraint equation* is found by setting $\dot{\psi}(\mathbf{r}, t)$ to zero. This assumes that the brightness of material points does not change over time. This assumption has been shown to be valid in a typical volume-imaging medical imaging modality [9]. It does not hold in tagged MR imaging, however. To address this problem, Prince and McVeigh [6] estimate the non-zero material time derivative of brightness using tracking. Their approach has two drawbacks. First, it leads to performance degradation from accumulating reference map estimation errors. Second, it requires extensive prior knowledge of T_1, T_2, and D_0 which is typically not available.

Gennert and Negahdaripour [7] relaxed the brightness constancy assumption as follows. They modeled the brightness variation of a point in motion by a linear transformation of the form

$$\psi(\mathbf{r}+\Delta\mathbf{r}, t+\Delta t) = m(\mathbf{r}, \Delta\mathbf{r}, t, \Delta t)\psi(\mathbf{r}, t) + c(\mathbf{r}, \Delta\mathbf{r}, t, \Delta t)\,. \qquad \text{(Eq.2)}$$

Here m is called the multiplier field and c is called the offset field of the transformation. For small Δt, m is expected to be close to 1 and c close to 0. Under this assumption, a revised brightness constraint equation can be derived,

$$\psi_t(\mathbf{r}, t) + \nabla\psi(\mathbf{r}, t)\cdot V(\mathbf{r}, t) - \psi(\mathbf{r}, t)m_t(\mathbf{r}, t) - c_t(\mathbf{r}, t) = 0\,, \qquad \text{(Eq.3)}$$

where m_t and c_t are the time derivatives of the multiplier and the offset respectively.

Assuming $\nabla\psi$ and ψ_t are calculated from the images (as in SOF) and since ψ is known, this new brightness constraint equation gives a single equation at each pixel with four unknowns: u, v, m_t and c_t. As in SOF, the illposedness of this problem is solved by regularization using a variational framework. The Euler-Lagrange equations for this variational problem are [7]

$$\begin{aligned}
\nabla^2 u &= \frac{1}{\alpha^2}\psi_x\left(\psi_t + \psi_x u + \psi_y v - \psi m_t - c_t\right), && \text{(Eq.4a)}\\
\nabla^2 v &= \frac{1}{\alpha^2}\psi_y\left(\psi_t + \psi_x u + \psi_y v - \psi m_t - c_t\right), && \text{(Eq.4b)}\\
\nabla^2 m_t &= \frac{-1}{{\alpha_m}^2}\psi\left(\psi_t + \psi_x u + \psi_y v - \psi m_t - c_t\right), && \text{(Eq.4c)}\\
\nabla^2 c_t &= \frac{-1}{{\alpha_c}^2}\left(\psi_t + \psi_x u + \psi_y v - \psi m_t - c_t\right), && \text{(Eq.4d)}
\end{aligned}$$

where ∇^2 is the Laplacian operator and α^2, α_m^2, and α_c^2 are the regularization coefficients.

To solve (Eq.4), we first estimate $\nabla\psi = (\psi_x, \psi_y)$ and ψ_t using local derivative approximations (as in [6]). We then solve (Eq.4a)-(Eq.4d) numerically using the local relaxation algorithm described by Kuo et al. [10]. We use Neumann boundary conditions where the normal component of the derivative of the four unknowns is set to zero on the boundary. This method does not require any prior knowledge about the MR parameters.

It is possible to model the noise, the velocity components, and the transformation fields as Brownian processes and derive expressions for the regularization coefficients in terms of the variances of the underlying stochastic processes (cf, [11]). This would provide a theoretical framework for choosing the α's. However, in this work, we choose these parameters empirically.

Also, it should be noted that OF methods produce erroneous results at the boundaries of occluding surfaces. In our simulations, we assume that a segmentation of the image into regions of interest is available and we use that information to prevent smoothing across boundaries (see [6]). If such a segmentation is not available, our tests have shown that the performance degrades marginally.

2.2. Approximating the Fields

It is possible to apply GNOF directly to tagged MR image sequences. As we show in Section 3.1, substantial improvements over VBOF are possible. To get good velocity estimates over the entire cycle, however, it is necessary to vary α_m and α_c with time (see Section 3.1). A heuristic choice of α_m and α_c for each time is not practical. Therefore, in this section, we derive approximate expressions for m_t and c_t using the physics of the MR imaging process. We also propose setting m_t to a fixed value and estimating only u, v, and c_t. The resulting algorithm is more robust and computationally more stable and requires only knowledge of T_1 over the image domain. We note that although the following analysis is carried out for the spin-echo imaging method, we have also developed analogous expressions for SPGR imaging.

To model the brightness change effects due to tag pattern fading alone, we assume that motion is not present. From the spin-echo tagging equation, the signal at time t can be written as

$$\psi(t) = D_0 e^{-T_E/T_2}\left(1 - e^{-T_R/T_1}\right) + D_0 e^{-T_E/T_2}(\xi - 1)\left(e^{-t/T_1} - e^{-T_R/T_1}\right), \tag{Eq.5}$$

where D_0 is the proton density, T_1 and T_2 are the relaxation time constants, T_R is the repetition time, T_E is the echo time, and ξ is the tag modulation coefficient (see [6]). At time $t + \Delta t$ we have

$$\psi(t+\Delta t) = D_0 e^{-T_E/T_2}\left(1 - e^{-T_R/T_1}\right) + D_0 e^{-T_E/T_2}(\xi - 1)\left(e^{-(t+\Delta t)/T_1} - e^{-T_R/T_1}\right). \tag{Eq.6}$$

We now write $\psi(t + \Delta t)$ in terms of $\psi(t)$ using (Eq.5) and (Eq.6), yielding

$$\psi(t + \Delta t) = e^{-\Delta t/T_1}\psi(t) + D_0 e^{-T_E/T_2}\left(1 - e^{-\Delta t/T_1}\right)\left(1 - \xi e^{-T_R/T_1}\right). \tag{Eq.7}$$

Comparing (Eq.2) and (Eq.7) under the assumption of no motion, we get the following approximate expressions for $m(t)$ and $c(t)$.

$$m(t, \Delta t) = e^{-\Delta t/T_1}, \quad \text{(Eq.8a)}$$
$$c(t, \Delta t) = D_0 e^{-T_E/T_2} \left(1 - e^{-\Delta t/T_1}\right) \left(1 - \xi e^{-T_R/T_1}\right). \quad \text{(Eq.8b)}$$

If Δt is small, we can assume $m = 1 + \Delta m$ and $c = \Delta c$ where Δm and Δc are also small. Then it can be shown that to first order

$$\Delta m = -\Delta t/T_1, \quad \text{(Eq.9a)}$$
$$\Delta c = D_0 e^{-T_E/T_2} \left(1 - \xi e^{-T_R/T_1}\right) \left(\Delta t/T_1\right). \quad \text{(Eq.9b)}$$

Hence, the derivatives are obtained as

$$m_t = \frac{-1}{T_1}, \quad \text{(Eq.10a)}$$
$$c_t = \frac{D_0 e^{-T_E/T_2} \left(1 - \xi e^{-T_R/T_1}\right)}{T_1}. \quad \text{(Eq.10b)}$$

In principle, we could use these expressions to fix the values of m_t and c_t, and then solve only for the two velocity components in (Eq.4a) and (Eq.4b). This reduces the computational work involved in solving the coupled partial differential equations. In order to use (Eq.10a) and (Eq.10b), however, we must know the parameters D_0, T_1, and T_2 everywhere in the image. Also, since there *is* motion, we must track these parameters through space and time. This situation is identical to that in VBOF and is to be avoided.

A useful compromise is to fix only m_t using (Eq.10a) and solve the PDE's of (Eq.4) for u, v, and c_t. This requires the knowledge of T_1 throughout the image, but our simulations have shown that the results are robust to inaccurate values of T_1 (see Section 3.3). We therefore expect that an approximate value for T_1 in the heart muscle will yield accurate motion estimates in the myocardium.

(Eq.10) reveals that (to first order), m_t and c_t do not depend on t. Therefore, their regularization coefficients should not depend on t. This is contrary to what we have observed in our simulations (see Section 3.1), which we believe can be explained by the fact that m_t and c_t are not uniquely determined by the problem statement. This allows these two fields to change dramatically in time despite what the physics predicts. When m_t is fixed, however, the situation changes: in this case a single regularization coefficient α_c is sufficient to yield good performance at all time t. This is a very strong additional benefit gained by fixing m_t.

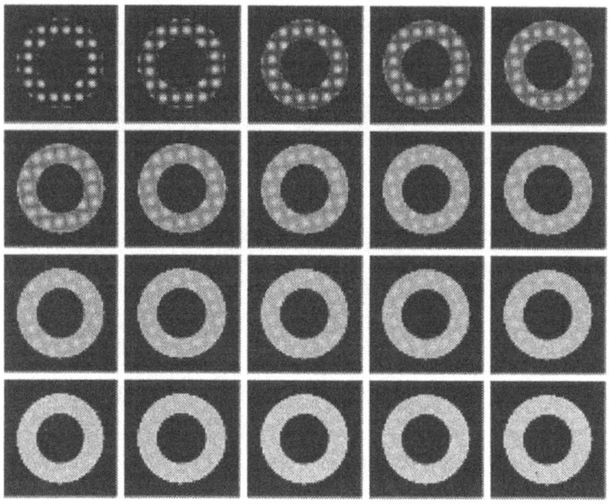

Figure 1: A simulated tagged MR image sequence for a rotating ring phantom.

3. EXPERIMENTAL RESULTS

In this section we present results from four experiments. In the first study, we show the results achievable by GNOF through empirical selection of the regularization coefficients. In the second and third studies we demonstrate the performance of GNOF using the approximate multiplier model and investigate the effect of inaccurate knowledge of T_1. These first three studies use simulated data with the same parameters as in [6]. A summary of this simulated data is given in Table 1, and a sequence of twenty images is shown in Figure 1. The first ten images are the same as those shown in Figure 4 of [6]. The last ten are included because our algorithm performs well in this range also while VBOF fails even with adequate prior knowledge. The fourth study shows the performance of our method on actual tagged MR phantom data.

3.1. Study 1: GNOF Performance

In this study, we obtain velocity estimates for all adjacent image pairs shown in Figure 1 by solving the (Eq.4a)–(Eq.4d). Since the true velocity field is known for this simulation, we empirically determined the optimal α_m and α_c for each time index where for simplicity, we set $\alpha_m = \alpha_c$. Figure 2 shows a plot of the optimally chosen regularization coefficients (α_m and α_c) as a function of time index. Using these constants and $\alpha = 10.0$, we use GNOF, SOF, and VBOF to calculate motion estimates for the image sequence. Figure 3 shows the average

Table 1: Simulated Data set Summary

Parameter	Value
Image size (nx,ny)	150 pixels
Horizontal and Vertical Pixel Spacing (dx,dy)	0.078125 cm.
Inner and Outer radius of ring (p1,p2)	2.5 cm, 4.5 cm.
Period of rotation (T_0)	10 sec.
Time interval between images (Δt)	0.05 sec.
Tag spatial frequencies (f_x, f_y)	0.75 cm^{-1}
Proton Density of phantom (D_0)	300
T_1 value of phantom	0.3 sec.
T_2 value of phantom	0.1 sec.
Repetition time (TR)	10.0 sec.
Echo Time (TE)	0.03 sec.

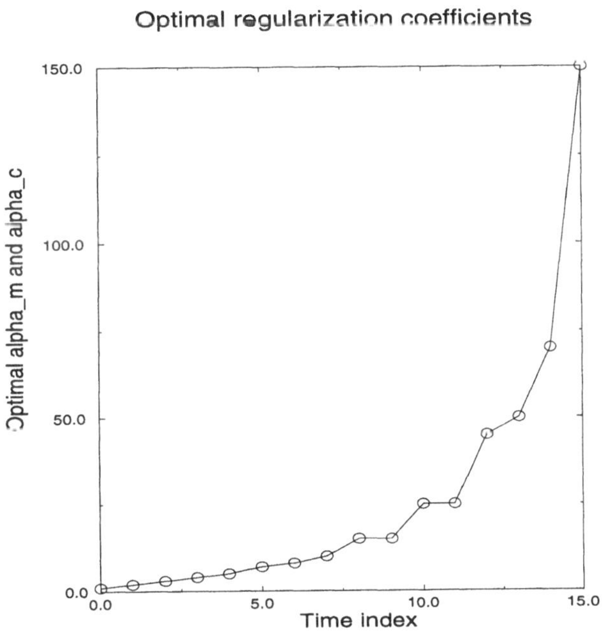

Figure 2: Optimal regularization coefficients (α_m and α_c) for use in GNOF as a function of time index.

percent velocity errors in the three methods as a function of time index. This plot shows that if the regularization coefficients in GNOF are chosen correctly, we can get good results over the full time sequence plotted. Since we use the constraint $\alpha_m = \alpha_c$, the performance shown in Figure 3 is not optimal, and we believe this explains the high error at the second time index. Towards the end of the cycle, the gradient calculations become increasingly inaccurate because the tags have faded

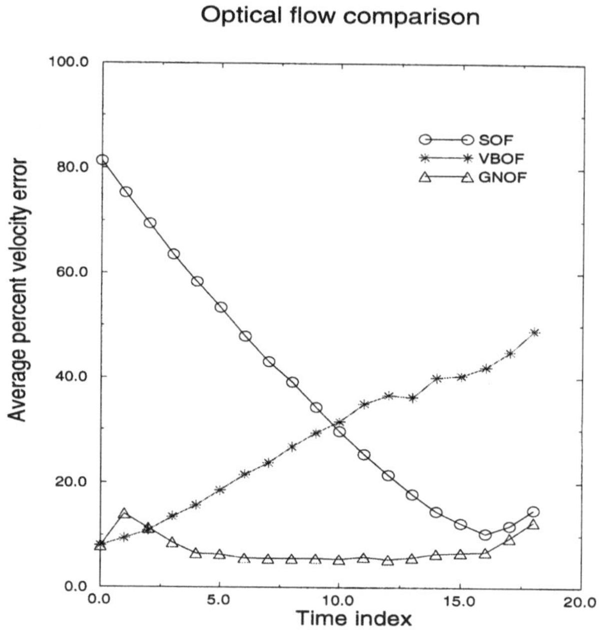

Figure 3: Comparison of the performance of SOF, VBOF and GNOF on simulated phantom data.

and the performance degrades towards the performance of SOF.

3.2. Study 2: Using the Fixed Multiplier Model

This study also uses the data shown in Figure 1. Using the known T_1 value of the simulated phantom, we use (Eq.10a) to fix m_t and then solve (Eq.4a, b, and d). Figure 4 shows the performance of this method as a function of time index. (The error bars appearing in this figure are discussed in the following section.) As was suggested in Section 2, we hold α_c constant (at 1.0) for the entire sequence. It can be observed from the plot that the performance is still significantly better than both VBOF and SOF. A comparison of these results with the results of GNOF in Section 3.1 shows that on an average there is a three percent increase in estimation error. However, this method performs approximately 1.25 times faster than GNOF and does not require careful selection of the regularization coefficients at each time index.

3.3. Study 3: Effect of T_1 errors

For the above method, we require knowledge of T_1 everywhere in the image. In this study, we investigate the effect on performance of our method if exact values

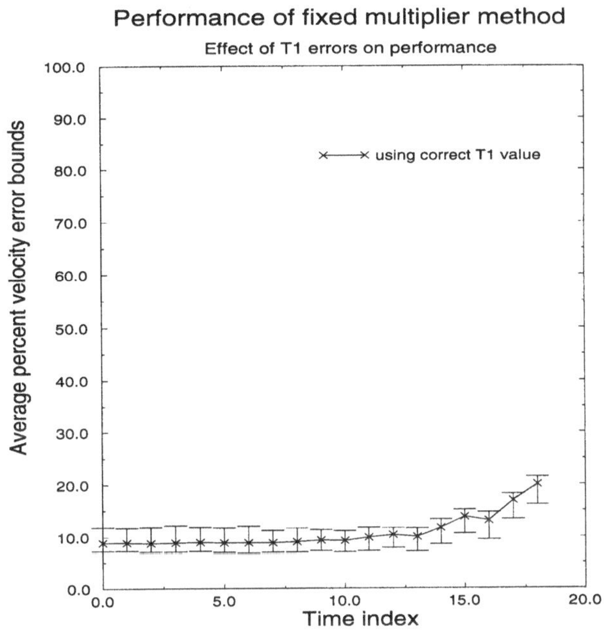

Figure 4: Curve shows the performance of the fixed multiplier method by using the known T_1 value of the phantom. Error bars are plotted for 75% variation around the correct T_1.

of T_1 are not available. We carry out repeated simulations as in Section 3.2 by using incorrect values of T_1 in (Eq.10a). Figure 4 shows the error bounds in the performance when T_1 is varied from its correct value. We have plotted the results for a 75% variation of T_1 around its correct value. The tests showed that varying T_1 by twenty percent results in a performance variation of approximately two percent. This insensitivity towards precise T_1 knowledge shows that we can expect good performance even if we have an approximate estimate of T_1 in the myocardium.

The error bars in Figure 4 also show that the performance gets *better* for certain incorrect T_1's. This is not an anomaly, as it should be recalled that our relationship between m_t and T_1 is only an approximation. It turns out that in some cases the approximation $m_t = -1/T_1$ for incorrect T_1's is closer to the optimal m_t than that predicted using the correct T_1.

3.4. Study 4: Rotating Ring Phantom

Finally, our method was validated on actual MR images of a gelatin ring phantom, tagged and imaged in an MR scanner. The data set is described in Section IV-C in [6]. We summarize it in Table 2.

Table 2: Actual Phantom Data Summary

Parameter	Value
Image size (nx,ny)	150 pixels
Horizontal and Vertical Pixel Spacing (dx,dy)	0.078125 cm.
Inner and Outer radius of ring (p1,p2)	2.55 cm, 4.44 cm.
Period of rotation (T_0)	20.0 sec.
Time interval between images (Δt)	0.175 sec.
Tag Spatial Frequencies (f_x, f_y)	1.117 cm^{-1}, 1.145 cm^{-1}
Proton Density of phantom (D_0)	188.841
T_1 value of phantom	1.65 sec.
T_2 value of phantom	0.9 sec.
Repetition time (TR)	2.40 sec.
Echo Time (TE)	0.03 sec.

We ran GNOF holding $m_t = -1/T_1$ on this data set and compared its performance to SOF and VBOF. Error results were computed using the known motion of this phantom. Figure 5 shows the plot of average percent velocity error as a function of time index for these three methods. It is observed that GNOF outperforms SOF and VBOF at every time point. All the curves in Figure 5 show erratic behavior and this is attributable to motion artifacts appearing in the image sequence. For the same reason, the performance on actual data is significantly worse than the simulation studies.

4. CONCLUSION

We have developed the use of Gennert and Negahdaripour's OF algorithm for use in tagged MRI. We have shown that it performs remarkably well and largely eliminates the problem of tag-pattern fading. We developed approximations for the brightness transformation fields which led to a faster algorithm that does not require careful selection of regularization coefficients. We also demonstrated the insensitivity of this method to the MR parameters and showed that good performance is achievable by nominal prior knowledge of T_1. We validated the method by tests on actual MR data.

In the future, it may be possible to arrive at better approximations for the brightness transformation fields. In particular, we can relax the assumption of absence of motion made in Section II-B. But this ties m_t to some other parameters like nominal velocity magnitude and tag pattern frequency. This leads to a tradeoff between performance improvement at the cost of requiring more and more prior

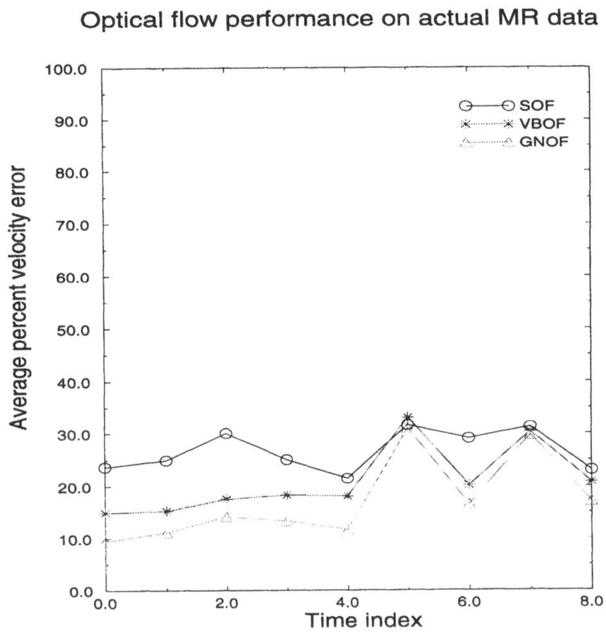

Figure 5: Comparison of the performance of SOF, VBOF and GNOF on actual MR data.

knowledge.

Another future goal is to explore a stochastic framework for this algorithm. We could incorporate the approximations developed in this work as nominal values in the stochastic process model, leading to a potentially powerful methodology.

ACKNOWLEDGEMENTS

The authors wish to thank Dr. Sundar Amartur for suggesting the use of Gennert and Negahdaripour's optical flow and Dr. Thomas S. Denney for his helpful discussions.

REFERENCES

[1] A. Lieberman, J. Weiss, and et al, "Two-dimensional echocardiography and infarc size: relationship of regional wall motion and thickening to the extent of myocardial infarction in the dog," *Circulation*, vol. 63, p. 739, 1981.

[2] J. Lima, L. Becker, J. Melin, and et al, "Imparied thickening of noischemic myocardium during acute regional ischemia in the dog," *Circulation*, p. 1048, 1985.

[3] E. Zerhouni, D. Parish, W. Rogers, A.Yangand, and E. Shapiro, "Human heart: tagging with MR imaging — a method for noninvasive assessment of myocardial motion," *Radiology*, vol. 169, pp. 59–63, 1988.

[4] M. Guttman and J. Prince, "Image analysis methods for tagged MRI cardiac studies," in *Medical Imaging IV: Image Processing*, Bellingham WA: SPIE, 1990.

[5] E. McVeigh and E. Zerhouni, "Non-invasive measurement of transmural gradients in myocardial strain with magnetic resonance imaging," *Radiology*, vol. 180, no. 3, pp. 677–683, 1991.

[6] J. L. Prince and E. R. McVeigh, "Motion estimation from tagged MR image sequences," *IEEE Transactions on Medical Imaging*, vol. 11, pp. 238–249, June 1992.

[7] M. A. Gennert and S. Negahdaripour, "Relaxing the brightness constancy assumption in computing optical flow," Tech. Rep. Memo No. 975, MIT Artificial Intelligence Laboratory, June 1987.

[8] B. K. P. Horn and B. G. Schunck, "Determining optical flow," *Artificial Intelligence*, vol. 17, pp. 185–203, 1981.

[9] S. M. Song and R. M. Leahy, "Computation of 3-D velocity fields from 3-D cine CT images of a human heart," *IEEE Transactions on Medical Imaging*, vol. 10, pp. 295–306, September 1991.

[10] C.-C. Kuo, B. Levy, and B. Musicus, "A local relaxation method for solving elliptic PDEs on mesh-connected arrays," *SIAM J. Sci. Stat. Comput.*, vol. 8, no. 4, pp. 550–573, 1987.

[11] T. S. Denney Jr. and J. L. Prince, "Optimal brightness functions for optical flow estimation of deformable motion," *IEEE Transactions on Image Processing*, vol. 3, pp. 178–191, March 1994.

DETECTION OF STELLATE DISTORTIONS IN MAMMOGRAMS USING SCALE SPACE OPERATORS

Nico KARSSEMEIJER
University Hospital Nijmegen, Department of Radiology, The Netherlands

INTRODUCTION

Pattern recognition methods are being developed to detect subtle abnormalities in mammographic images. With the introduction of digital mammography this will allow the use of computer assisted screening for breast cancer. Until now, much of the research in mammographic image analysis has been focused on detection of microcalcification clusters. This may have led to the impression that this feature is particularly difficult to detect for the human eye and that missing microcalcifications is a major cause of screening errors. However, retrospective studies analyzing the type of carcinoma's missed in breast cancer screening programs reveal that this is not the case [1][2]. In case radiological signs can be found on previous screening mammograms of patients with breast cancer, these are most often classified as vague densities or masses. Therefore, computer programs for detection of masses are likely to become more important in breast cancer screening than programs for detection of microcalcifications. A recent investigation indeed shows that the radiologic performance can be significantly improved using computer aided detection tumors [3].
Most malignant densities have an irregular appearance and frequently they are surrounded by a radiating pattern of linear spicules. Sometimes the density is very faint and when it is embedded in the parenchymal tissue it may be very hard to perceive. In those cases the stellate pattern of spicules is the most important sign. The aim of this investigation is detection of such stellate patterns without relying on the presence of a central mass. The approach is based on statistical analysis of a map of pixel orientations. The idea is that if an increase of pixels pointing to a given region is found this region may be suspicious, especially if, viewed from the test region, such an increase is found in many directions. It is noted that no attempt is being made to explicitely identify spicules. In this respect the approach is similar to the texture based approach

Y. Bizais et al. (eds.), Information Processing in Medical Imaging, 335–346.

suggested by Kegelmeyer [3], and different from approaches based on using the Hough transform for spicule detection and accumulation of evidence for spicules to detect tumors [4]. The latter approach relies on a separate stage for spicule detection and tends to be less succesful in case the spicules are very faint. Clearly, this is undesirable because these are the cases in which computer assisted reading will probably be the most helpful.

An important feature of the method is the way in which an orientation of the image intensity map is determined at each pixel. A new method is proposed for this purpose, based on application of second order operators. If a line-like structure is present at a given site the method provides an estimate of the orientation of this structure, whereas in other cases the image noise will generate a random orientation. Using scale space theory it will be shown how accurate estimates of line orientation can be obtained at a given scale from the output of only three directional second order Gaussian derivative operators, differing by $\pi/3$ in orientation.

These 'line-based' orientation estimates are used to construct two operators which respond to radial patterns of straight lines. Combination of the output of these operators in a Bayes classifier leads to a very sensitive method for detection of stellate patterns. The method is applied to detect stellate lesions and architectural distortions in mammograms from the MIAS database, which was made available for public use by the Mammographic Image Analysis Society in the UK [5].

A MULTISCALE LINE-BASED ORIENTATION MAP

Line orientation is often determined by using gradient operators. An important reason for this is the fact that two operators are sufficient to esimate edge orientation, while the result is relatively insensitive for changes of linewidth. This makes the approach computationally attractive. If the image function is denoted by $I(x, y)$ the gradient of I in a given direction ϕ can be written as

$$\frac{\partial I}{\partial r} = \frac{\partial I}{\partial x}\frac{\partial x}{\partial r} + \frac{\partial I}{\partial y}\frac{\partial y}{\partial r} = I_x \cos(\phi) + I_y \sin(\phi) \tag{1}$$

By maximization of this first order directional derivative over ϕ if follows that edge orientation can be computed by

$$\theta(x, y) = tan^{-1}\left(\frac{I_y}{I_x}\right) \tag{2}$$

Estimation of I_x and I_y can be performed by applying, for instance, the Sobel operator. For more accurate results a convolution with Gaussian derivatives should be performed [6,7]. For continuous I it can be shown that this leads to homogeneous, isotropic and scale invariant results.
A drawback of the use of first order operators for estimation of line orientation lies in the fact that no valid results are obtained at the central part of the line. Moreover, the method is not sensitive for lines of one pixel in width. These problems can be avoided by using second order operators. Compass gradients have been suggested for this purpose, which roughly approximate second order directional derivatives $\partial^2 I/\partial r^2$ for lines of one pixel in width at a fixed set of angles [8]. Using this approach, line orientation can be estimated by selecting the filter with the highest output. Clearly, the accuracy of the estimates will depend on the number of angles at which the filters are being applied. Generalization to lines of different width leads to a multi-scale representation. An example of this approach using Gabor filters can be found in [9].
From a computational viewpoint the use of a large filter bench is not very attractive. Especially the use of the same neighborhood operator at many different angles seems inefficient. It will be shown here by application of scale space theory that an accurate and more efficient method is possible. More precisely, at a given level of spatial scale σ, convolution of an image with only three filter kernels $W_\sigma(\theta_n)$ is sufficient to make an accurate operator for determining line orientation, with $W_\sigma(\theta_n)$ the second order directional derivatives of the Gaussian kernel $G(r,\sigma)$ in the directions $\theta = n\pi/3$, $(n = 0, 1, 2)$, and

$$G(r,\sigma) = \frac{1}{2\pi\sigma^2}\exp\left(-\frac{r^2}{2\sigma^2}\right). \tag{3}$$

The basis for this approach is a relation derived by Koenderink [6] which shows that for an arbitrary direction $W_\sigma(\theta)$ can be expressed as

$$\begin{aligned} W_\sigma(\theta) &= \frac{1}{3}(1+2\cos(2\theta))\,W_\sigma(0) \\ &\quad +\frac{1}{3}(1-\cos(2\theta)+\sqrt{3}\sin(2\theta))\,W_\sigma(\pi/3) \\ &\quad +\frac{1}{3}(1-\cos(2\theta)-\sqrt{3}\sin(2\theta))\,W_\sigma(2\pi/3) \end{aligned} \tag{4}$$

The three independent line operators W_σ form a non-orthogonal basis (figure 1). Using this relation we can calculate the filter orientation with

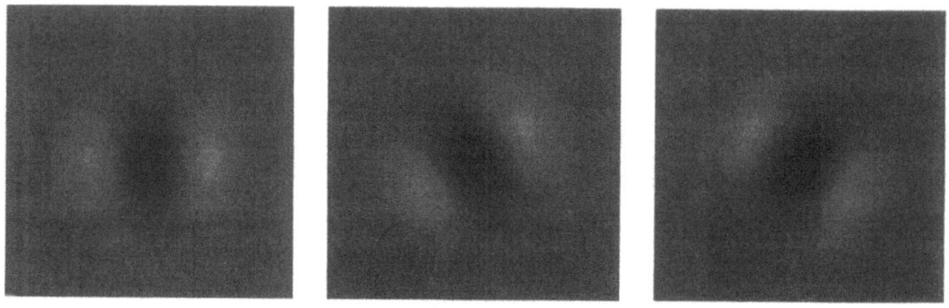

Figure 1: Three directional second order Gaussian derivatives used for estimation of line orientation

maximum output by solving $dW_\sigma(\theta)/d\theta = 0$. This leads to

$$\theta_{min,max} = \frac{1}{2}\left[\arctan\left(\sqrt{3}\frac{W_\sigma(\pi/3) - W_\sigma(2\pi/3)}{W_\sigma(\pi/3) + W_\sigma(2\pi/3) - 2W_\sigma(0)}\right) \pm k\pi\right] \quad (5)$$

where θ gives the orientation perpendicular to the line itself. Thus, for a given site in the image two extrema are found at each scale. By using (4) the output of the filter at these orientations can be calculated, and the orientation with maximum output can be determined. At this point, however, care must be taken because lines with positive contrast will give a strong negative output whereas lines with negative contrast give a positive output. Moreover, each scale will generate its own optimal orientation. Here, to condense the multiscale filter output into one orientation per pixel, the orientation with the maximum absolute value is taken, at the scale at which this value is maximum. In this way the filter orientation is fitted for line like structures of both positive and negative contrast. The intensity $W_\sigma(\theta_{max})$ at the selected scale is stored for each pixel, to allow selection of pixels in a given range of intensities at a later stage of processing.

FEATURES FOR DETECTION OF STELLATE PATTERNS

Two new features have been defined for detection of stellate patterns of straight lines. These features are derived from the map of pixel orientations ϑ_i determined as described above, with $\vartheta_i = \theta_i + \pi/2$. As these orientations reflect the line structure of the image, embedded in a background of random orientations, stellate patterns may be detected in this map by an increase of the number of pixels directed towards a central area.

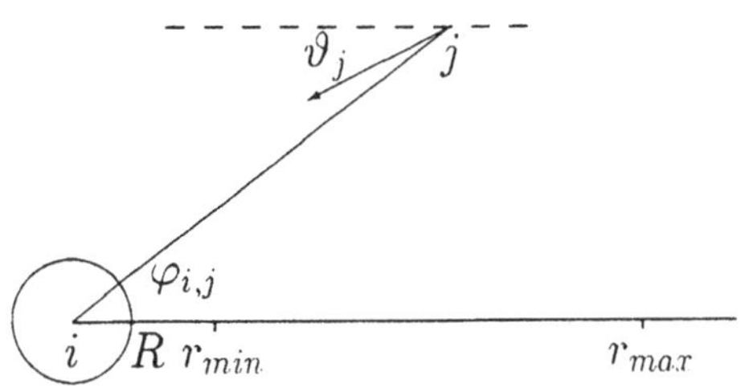

Figure 2: Pixels j located at a distance $r_{i,j}$ between r_{min} and r_{max} and directed towards a disk of radius R centered at the test site i are counted.

The features are defined to quantify such an increase. At a given site i they are calculated from the orientations of pixels in a neighborhood $\mathcal{N}_i$ representing pixels j with distances $r_{i,j} \in [r_{min}, r_{max}]$ from i. Not all pixels in this neighborhood are evaluated. Using a selection criterion on the intensity of the filter output $W_\sigma(\theta_{max})$ a subset of pixels $\mathcal{S}$ is determined representing potential sites of interest. Sites with negative contrast and sites where the intensity of the line feature is very low are excluded. To calculate the features, all pixels in the neighborhood of the test site i that belong to $\mathcal{S}$ and that are directed towards a small area around i are counted as a function of the direction φ in which they are located (figure 2). Dividing the space around i in K bins of different directions the number of neighboring pixels $n_{i,k}$ in bin k that are oriented towards i is calculated by

$$n_{i,k} = \sum_{j \in \mathcal{N}_{i,k} \cap \mathcal{S}} h(\vartheta_j, \varphi_{i,j}, r_{i,j}) \tag{6}$$

with the set $\mathcal{N}_{i,k}$ denoting the neighboring pixels in direction φ_k, and

$$h(\vartheta_j, \varphi_{i,j}, r_{i,j}) = \begin{cases} 1 & \text{for} \quad |\varphi_{i,j} - \vartheta_j| < \frac{R}{r_{i,j}} \\ 0 & \text{otherwise} \end{cases} \tag{7}$$

The parameter R is used to set the accuracy with which pixels need to be directed to i to be counted.

The first feature which is defined is the total number of pixels with directions pointing to the center n_i, which can be calculated by summation of $n_{i,k}$ over k. In order normalize this feature, the mean value of n_i and its variance are estimated under the assumption that the pixel orientation map is a uniformly distributed random noise pattern. The mean probability p that a pixel in this random map is pointing to the test site i

is

$$p = \frac{2}{\pi N_i} \sum_{j \in \mathcal{N}_i \cap \mathcal{S}} \left(\frac{R}{r_{i,j}}\right) \tag{8}$$

with N_i the total number of pixels in $\mathcal{N}_i \cap \mathcal{S}$. The normalized feature then is defined by

$$f_{1,i} = \frac{n_i - pN_i}{\sqrt{N_i\, p(1-p)}} \tag{9}$$

Because of the normalization the sensitivity of the feature and its range do not change systematically when the neighborhood or target size R are changed. This enables changing these parameters adaptively and avoids problems at the edge of the image.

If an increase of the number of pixels oriented towards a region is found in a few directions only, it isn't very likely that the site being evaluated belongs to the center of a stellate pattern. On the other hand, if evidence for spicules is found in many directions this should increase the likelihood of a stellate structure being present. To represent this property, a second operator is constructed. In each direction bin k the mean probability of finding $n_{i,k}$ pixels oriented to the center out of a total of $N_{i,k}$ is calculated by applying (9) for each bin separately. Using binomial statistics it is determined how many times $n_{i,k}$ is larger than the median value calculated for random orientations. Denoting this number by n_+ and with K' the number of bins, the second feature is defined by

$$f_{2,i} = \frac{n_+ - K'/2}{\sqrt{K'/4}} \tag{10}$$

where $K'/2$ is the expected value of n_+ when no signal is present. To avoid boundary effects, only bins in which a minimum number of contributing sites is present are considered. Therefore, near the breast edge the actual number of bins K' that contribute is smaller than K. The standard deviation of random fluctuations in the denominator normalizes the expression.

To demonstrate the performance of the two operators an artificial stellate pattern was generated, which was hidden by adding i.i.d. Gaussian noise. Figure 3a shows this pattern generated at a signal to noise ratio of 1.0. In 3b and 3c it is shown that both operators give a high output at the centre of the test pattern. Calculation of the features was performed by applying the directional derivative filters at scales $\sigma = 1, 2$, and 4, where the width of the spicules was about 4. The test pattern in figure 3d was

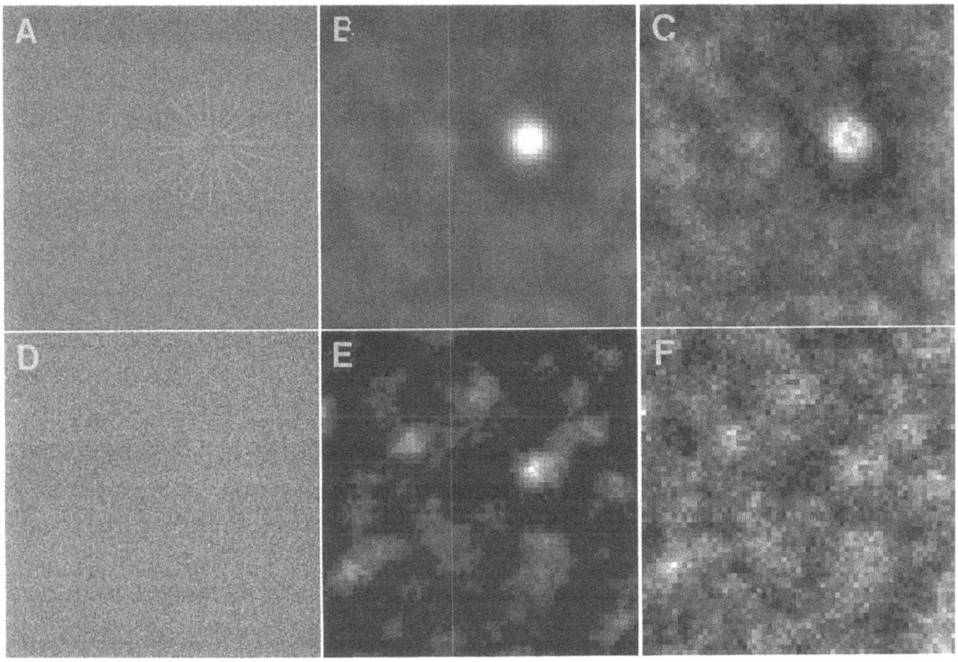

Figure 3: Figure (a) shows a test pattern embedded in Gaussian noise at a signal to noise ratio of 1.0. The features f_1 and f_2 constructed to detect the pattern are shown in figures (b) and (c). Both features have a strong peak at the center of h he stellate pattern. Figures (d),(e) and (f) show the same sequence for a pattern generated at a signal to noise ratio of 0.25. Still a peak is found in f_1 but the signal is lost in f_2.

generated at a signal to noise ratio of 0.25. Figure 3e shows that even in this very noisy background the feature f_1 still gives shows an increase at the position of the stellate signal.

APPLICATION TO MAMMOGRAMS

Stellate lesions in mammograms have a much more complex appearance than the artificial images of figure 3. To deal with this a few refinements of the feature calculation were introduced, and some preprocessing of the image data was performed. Some problems were due to the fact that calculation of the Gaussian derivatives was performed in the Fourier domain. Although this method is convenient and accurate, it easily leads to artifacts near the matrix boundary. Mammograms were preprocessed by the following operations. First the breast image was shifted away from the matrix boundary, and an automatic procedure was applied to segment

the breast area from the pectoral muscle in the oblique views. Next, the region around the breast was replaced with a smoothed average of the neighboring breast tissue. In addition, in the region near the breast skin line the fall of due to a decrease of the tissue thickness is corrected for, allowing embedding of the breast tissue in a more or less homogeneous background. By way of example figures 4a and 4b show a mammogram before and after preprocessing.
A refinement of the feature calculation was introduced to deal with the fact that spiculated patterns vary a lot in size. In mammograms malignant stellate patterns often have a central mass. If this mass is small, the spiculated area normally is closer to the centre than in case of a larger mass. Therefore, if a central mass is likely to be present at a given site, its estimated size can be used to set the size of the neighborhood to be evaluated for spicules. This idea was used by testing for the possible presence of a mass by application the the Laplace operator ∇_σ^2 at a number of different spatial scales. If the output of this operator exceeded a certain threshold T_L the neighborhood size parameter r_{max} was scaled with σ_{max}, the scale level with the highest output. To avoid artifacts near the matrix boundary, the Laplace output at sites i was taken to be invalid if the distance between i and the matrix boundary was less than 2σ. Figure 4c shows the maximum of the Laplace operator over a number of scales ranging from $\sigma = 1.6$ to 8 mm.
It appeared that the presence of distinct, clear structures in the mammograms like blood vessels, skin folds or sharp outlines in the glandular tissue detoriated the performance of the detection algorithm. This was caused by the fact that at strong boundaries between regions the second order directional derivatives give ripples in parallel to the boundary. This leads to structure in the orientation map which is not related to the presence of lines in the range of chosen widths. To reduce this effect, at a coarse scale regions with a high gradient magnitude were determined, and within these regions pixels with orientations ϑ_i perpendicular to the gradient orientation (thus in parallel to the boundary) were removed from the subset $\mathcal{S}$ of contributing sites. Gradient magnitude and orientation were calculated by using first order Gaussian derivatives at a relatively large value of σ for differentiation.
The method was applied to mammograms taken from the MIAS database [5]. All malignant stellate lesions (9) and architectural distortions (10) were selected. By adding the first 31 normals from the database (003 to 046) a test set of 50 images was obtained. Calculation of the features

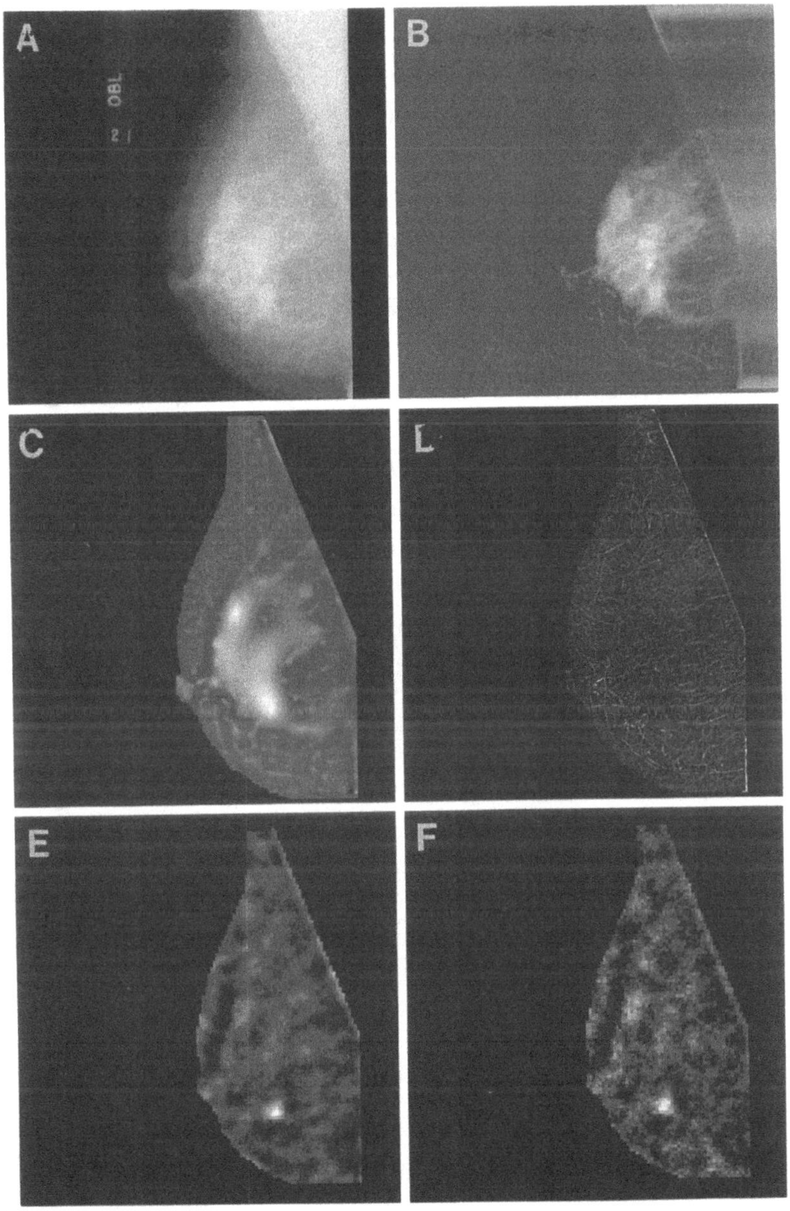

Figure 4: Image mdb120 of the MIAS database, showing a malign stellate distortion (a). In (b) the mammogram is shown after preprocessing, and (c) shows the output of the Laplace operator maximized over a range of spatial scales. In (d) the intensity of the line map is shown. The orientation map of all pixels with a positive output on the line intensity map is used to calculate the feature images f_1 and f_2 shown in figures (e) and (f), where the Laplace output is used to adaptively set the size of the neighborhood. Both features have a strong peak at the position of the tumor. The cancer is detected even at a specificity level of 0.02 false positives per images.

was performed on 1k×1k images. Images from the MIAS database were reduced to this size by subsampling the original 50 micron/pixel data, merging 4 × 4 blocks of 8 bits pixels into one 12 bits pixel by addition. Subsequently, adaptive noise equalisation [10] was applied to obtain a uniform noise level over the image.

Calculation of stellate features was performed for sites inside the tissue area sampled at 1.6 mm intervals. Each of these sites was classified individually, using Bayesian decision theory to combine the two features. To reduce noise the feature images were slightly smoothed. The joint probability density functions of the two features, for pixels labeled as normal or abnormal, were estimated from a set of 14 training images using a non-parametric approach. The training images were digitized mammograms from the Nijmegen breast cancer center, all showing a malignant stellate lesion. They were not used in the test set. For each mammogram showing a malignant stellate pattern the true labeling was entered manually under supervision of an expert radiologist. Only pixels inside the central region of a lesion were labeled as abnormal, not the spicule pattern itself. To measure the performance of the classification method the number of false positives was determined as a function of the detection sensitivity, where the latter was varied by changing the prior probability of a pixel to belong to a stellate lesion. To reduce influence of noise, regions labeled as abnormal that were smaller than 500 pixels were removed. This minimum corresponds to the size of a circlular area of 0.5 cm in diameter. The remaining regions were compared with the annotated true masks. If the overlap between a region labeled abnormal and the true lesion was more than 50% of either the true or detected area the lesion was regarded as detected, otherwise a false positive was counted.

RESULTS AND DISCUSSION

Parameter settings were chosen by optimizing the output of the algorithm on the training set. For estimation of line orientation the directional second order derivatives were applied at three spatial scales $\sigma = 0.1,\ 0.17,\ 0.29$ mm. The gradient operator used to remove ripples at strong gradients was computed at $\sigma = 1$ mm. Pixels for which the gradient magnitude was larger than $T_g = 50$ and for which the difference between line and gradient orientation was smaller than $\pi/6$ were not considered for calculation of f_1 and f_2. To determine the neighborhood size the Laplace operator was applied at $\sigma = 1.6,\ 2.4,\ 3.6,\ 5.4$, and 8.1 mm. Fixed values values $r_{min} = 4$ mm and $r_{max} = 16$ mm were used when the

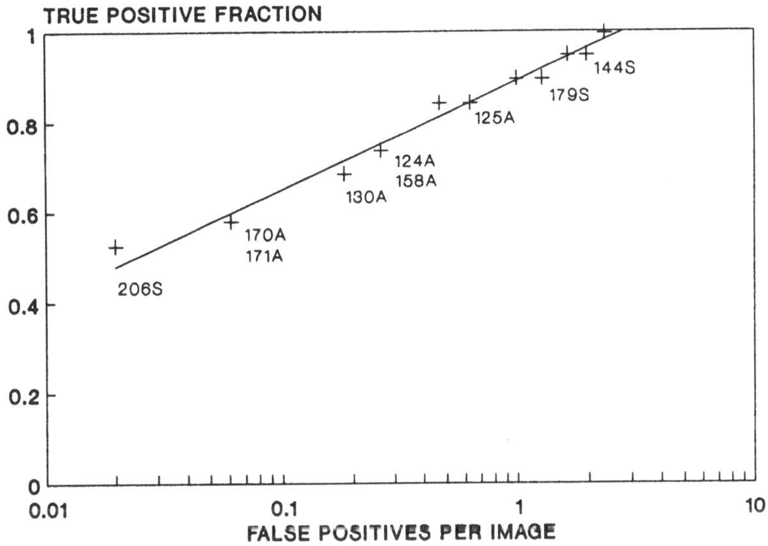

Figure 5: FROC curve computed for 50 images of the MIAS database, including 19 malignant stellate patterns and 31 normals. The spiculated lesions (S) and architectural distortions (A) which are lost at a given level of sensitivity are denoted under the curve.

Laplace operator had a low output. For sites at which the output was higher than $T_L = 20$ the parameter r_{max} was set to $3\sigma_{max}$. Determination of pixels oriented to the center was performed by using $R = 2$ mm. For calculation of f_2 the number of direction bins K was set to 16. As an example, figure 4d shows the intensity of the linemap and 4e and 4f represent the two features calculated for the image of 4a. A fairly obvious case was selected to guarantee visibility of the tumor in the reproduction. The cancer was detected even at a sensitivity level of 0.02 FP/image.

Overall results are shown in figure 5. Numbers under the curve denote malignant cases which are lost at a particular level of the sensitivity. On average about 50% of the tissue pixels contributed to calculation of the features. For these pixels the intensity of the line operator exceeded a small threshold set to select lines with positive contrast. Experiments with an increased value of the threshold to ignore pixels at which the line operator output was low did not improve the results.

The results show that on the basis of a line based pixel orientation map many subtle spiculated lesions and architectural distortions can be detected at a high degree of specificity. In the test set spiculated lesions are more easily detected than architectural distortons. Lesions which are lost first when the sensitivity is reduced are those without many spicules or

those in which spicules are very short. For instance, the lesion in image mdb144 which has the appearance of a small irregular mass is missed first. To recognize such lesions the focus should be on detection of the central mass as well. All lesions were detected at a false positive rate less than 2.4 FP/image.
The computations are based on a new method for estimation of line orientation described in section 2. Given the fact that many of the stellate paterns in the test set are quite faint and noisy, the method appears to be very robust. Attempts to improve results by using a threshold on the line operator output were not successful. This indicates that even at sites where the presence of a line seems rather unlikely the estimated orientations still have a positive contribution on average.

REFERENCES

1. J.A.M. van Dijck, L.M. Verbeek, J.H.C.L. Hendriks, R. Holland. "The current detectability of breast cancer in a mammographic screening program", *Cancer*, Vol 72, pp 1933-1938, 1993

2. C.J. Savage, A.G. Gale, E.F. Pawley, A.R.M. Wilson. "To err is human, to compute divine?", in *Digital Mammography*, Gale, Astley, Dance, Cairns eds, pp 405-414, Elsevier, Amsterdam, 1994

3. W.P. Kegelmeyer Jr., J.M. Pruneda, P.D. Bourland, A. Hillis, M.W. Riggs, M.L. Nipper, "Computer-aided mammographic screening for spiculated lesions", *Radiology*, Vol 191, pp 331-337, 1994

4. S.L. Ng, W.F. Bischof, "Automated detection and classification of breast tumors", *Computers and Biological Research*, Vol 25, pp 218-237, 1992

5. J. Suckling, J. Parker, D.R. Dance, S. Astley, I. Hutt, C.R.M. Boggis, I. Ricketts, E. Stamatakis, N. Cerneaz, S.L. Kok, P. Taylor, D. Betal, J. Savage, "The mammographic image analysis society digital mammogram database", in *Digital Mammography*, Gale, Astley, Dance, Cairns eds, pp 375-378, Elsevier, Amsterdam, 1994

6. J.J. Koenderink, A.J. van Doorn, "Generic neighborhood operators", *IEEE PAMI*, Vol 14, pp 597-605, 1992

7. L.M.J. Florack, B.M. ter Haar Romeny, J.J. Koenderink, M.A. Viergever, "Scale and the differential structure of images", *Image and Vision Computing*, 10, pp 376-388, 1992

8. A.K. Jain, *Fundamentals of digital image processing*, Prentice Hall Int, London, 1989

9. G.D. Daugman, "Complete discrete 2-D Gabor transforms by neural networks for image analysis and compression", *IEEE Trans. ASSP* , Vol 36, pp 1169-1179, 1988

10. N. Karssemeijer, "Adaptive noise equalization and image analysis in mammography", in "Information Processing in Medical Imaging", Barret, Gmitro eds, pp 472-486, Springer-Verlag, Berlin, 1993

GEOMETRICAL INTERPRETATION OF MOTION ARTEFACT PATTERNS IN CT IMAGES

Heinrich WALLSCHLÄGER
Siemens AG, Med, Erlangen, Germany

Motion artefacts in CT images result in deformations of the contour of a moving object, in artefacts near the object boundary and in long streaks across the whole image. Related displacements of CT-values must be accounted for in quantitative CT. Typical motions resulting in artefacts are swallowing, respiration, peristaltic, cardiac motion and inferred secondary motions (of contrast medium, of air in the bowel etc.).

This report will help to understand the genesis of artefact patterns of moving objects by simple geometric considerations. We start with the observation that the shape of an object is primarily determined by its boundary. In computed tomography this image boundary results from the superposition of all undershoots near object boundaries in the convoluted projections during backprojection. The locus of an undershoot in a single convoluted projection is a straight line tangential to the instantaneous object boundary. In the following we study the motion of this tangent on the boundary of a moving or resting object. This leads to a qualitative understanding of the artefact patterns of moving objects.

For the sake of simplicity all calculations are done for 180° parallel projections.

We found in particular:

1. The motion distorted form of the object boundary is the envelope of all tangents to the moving object.
2. There are two different types of artefact patterns: longe range or streak artefacts and proximity artefacts.
3. Streak artefacts occur if the tangents to the object in the first projection and in the last projection are different. They consist of a bright and a dark part and they run over the full images.
4. Dark/bright proximity artefacts result if there are more/less undershoots at the boundary of the moving object compared with the stationary object. This happens where the motion distorted object contour has a greater/smaller curvature than the nondistorted contour.

Y. Bizais et al. (eds.), Information Processing in Medical Imaging, 347–348.

5. There are two kinds of proximity artefacts differing in genesis and appearance. Proximity artefacts of the first kind occur where the curvature of the distorted object contour changes continously. Proximity artefacts of the second kind occur where the curvature of the distorted object contour changes discontinously.

6. Proximity artefacts of first and second kind in the same image can be separated by their different shape. Similar artefact shapes in one image correspond to the same kind of artefact.

7. Artefact patterns of simple objects have mirror or rotation symmetries but similar symmetries in two images need not result from similar motions. The foldiness of a rotation symmetry is not related to the periodicity of the underlying motion.

Artefact reduction techniques (such as interleaving of redundant pairs of projections or the use of overscans) will mainly reduce the amplitude of the streak artefacts. The amplitude of proximity artefacts and the distortion of the object boundary will not be reduced to a greater extent.

The application of this model to 360° fanbeam projections results in two additional observations: First, the generation of streak artefacts also depends on inconsistencies between first and last projections (and not on inconsistencies between opposite projections). Second, there are more types of motion leading to streak artefacts in fanbeam geometry.

With this procedure the motion distorted boundary of a moving object and the related artefact patterns can qualitatively be predicted without any data simulation and subsequent reconstruction.

INTER- AND INTRA-SLICE INTENSITY CORRECTION IN MR IMAGES

Alex P. ZIJDENBOS[1], Benoit M. DAWANT[1], Richard A. MARGOLIN[2]

[1]Dept. of Electrical and Computer Engineering, [2]Dept. of Psychiatry Vanderbilt University, Nashville, TN 37235, U.S.A.

1. INTRODUCTION

Inter- and intra-slice intensity variations in MRI data affect automatic and semi-automatic methods designed for the segmentation of the images. This increases the amount of required user-interaction and limits the usefulness of segmentation techniques as reliable quantitative analysis methods. In this paper we propose two correction methods; one for intra- and the other for inter-slice intensity variations. We also demonstrate the efficacy of these methods for the segmentation of white matter lesions (WML) in 2-D spin echo images, the separation of brain and CSF in data acquired with a fast 3-D sequence, and the elimination of visual artifacts in reformatted images.

2. METHODS

The intra-slice intensity correction method we propose herein is an extension of a method we have presented earlier [1]. It is based on a multiplicative model of the intensity variation artifact and the method consists in estimating an intensity correction surface from the data. As was the case in our earlier work, we rely on thin-plate splines to estimate this surface but we have automated is computation. Previously, we required the user to carefully select reference points which were used for the surface computation. We now rely on a presegmentation step which provides the algorithm with the reference points without requiring their labeling in each individual slice. The inter-slice intensity intensity correction we propose is based on the following multiplicative model: $f_i(\omega)=\alpha_i f_{i-1}(\omega)$, with $f_i(\omega)$ the intensity value of tissue class ω in slice i. This model requires the estimation of the parameter α for each pair of adjacent slices. Rather than computing a single value with, for example, the ratio of

Y. Bizais et al. (eds.), Information Processing in Medical Imaging, 349–350.

average intensity values in adjacent slices, the parameter α is estimated from local correction factors. These correction factors are computed by first identifying patches of identical tissues in adjacent slices and then computing intensity ratios on these patches, point by point. The global parameter α is finally computed from the distribution of local correction factors. This approach permits the calculation of α even when the type of brain structures visible in adjacent slices changes significantly as is the case in the presence of tumors and abnormalities.

3. EXPERIMENTS AND RESULTS

The effect of both the inter- and intra-slice intensity algorithms on segmentation results has been investigated in a study aimed at quantifying WML in MR images. Two operators used a segmentation method we have previously developed [2] to classify 6 brain volumes obtained from patients with a clinical diagnosis of Alzheimer's disease and/or multi-infarct dementia. These analyses were performed with and without the intensity corection algorithms, repeated over time, and compared against repeated manual measurements of WML in terms of the intra- and interoperator variability. Results demonstrate that both inter- and intra-rater variability in the measurement were reduced when the correction algorithms were applied. We also present results which demonstrate the usefulness of the algorithms to separate brain tissue from CSF in high resolution 3-D images as well as its utility to eliminate alternate light and dark lines visible in reformatted MR volumes.

4. REFERENCES

[1] B.M. Dawant, A.P. Zijdenbos, and R.A. Margolin. Correction of intensity variations in MR images for computer-aided tissue classification. *IEEE Transactions on Medical Imaging*, 12(4):770-781, Dec. 1993.

[2] A.P. Zijdenbos, B.M. Dawant, R.A. Margolin, and A.C. Palmer. Morphometric analysis of white matter lesions in MR images: method and validation. *IEEE Transactions on Medical Imaging*, 13(4):716-724, Dec. 1994.

HOW CAN WE DEFINE NORMALITY IN A MEDICAL IMAGE?

Alexander S. HOUSTON, Paul M. KEMP, Murdo A. MACLEOD
Department of Nuclear Medicine, R.N.H. Haslar, Gosport, U. K.

1. INTRODUCTION

An attempt is made to define normality in a medical image using HMPAO rCBF SPECT images as an example. For this purpose, six methods of creating a "normal atlas" are compared and the parameters of each atlas optimised.

2. MATERIALS AND METHODS

A "normal atlas" was created using a mean and several principal component (PC) images from a database of 50 normal rCBF images following image co-registration and normalisation [1]. The mean and k PCs were obtained using (a) the voxels of the whole image; (b) the voxels of the cortex. In each case, the images were first normalised to (i) cerebellar counts; and (ii) total counts in the voxels being analysed. For any subsequent image, a "nearest normal equivalent" image was obtained from the atlas formed using each method. The significance (in normal SDs) of deficits found for voxels in the cortical region was calculated. An image was defined as abnormal if it contained at least one deficit with N connected voxels above a threshold of x SDs. A similar analysis was performed on count density values in eight cortical regions. Cerebellar and totalised values were again used for normalisation. In this case, an image was defined as abnormal if it contained at least one deficient region above a threshold of x SDs.

In order to determine the optimal values of k and N, 200 patients with dementia were compared with a further 40 normal controls. Using a range of values of x as decision thresholds, ROC curves were obtained for different values of k, for all methods, and N, for voxel methods, and the statistic d_a [2] determined in each case.

3. RESULTS

For methods involving cerebellar normalisation, a value of $k = 0$ generally gave the best discrimination. For voxel methods, the optimal value of N was around

Y. Bizais et al. (eds.), Information Processing in Medical Imaging, 351–352.

10 or 11. These methods gave much better discrimination (optimal at d_a = 1.91 for both methods with $k = 0$, $N = 11$) than the regional method (optimal at d_a = 1.32 for $k = 0$). For normalisation to totalised values, the optimal value of k for voxel methods was around 4 or 5 with the optimal value of N again around 10 or 11. Voxel methods again gave better discrimination (optimal at d_a = 1.96 for method (a) with $k = 5$, $N = 11$) than the regional method (optimal at d_a = 1.39 for $k = 2$).

4. CONCLUSIONS

It is concluded that a "normal atlas" is best designed using a voxel method following normalisation to total counts with, in this example, $k = 4$ or 5 and N = 10 or 11.

5. REFERENCES

[1] A.S. Houston, P.M. Kemp, M.A. Macleod. "A method for assessing the significance of abnormalities in HMPAO brain SPECT images", *J Nucl Med*, 35(2), pp 239-244, 1994

[2] J.A. Swets, R.M. Pickett. *Evaluation of diagnostic systems: methods from signal detection theory*, p 29, Academic Press, New York, 1982

NOISE CHARACTERIZATION AND OBJECTIVE IMAGE-QUALITY ASSESSMENT OF SPECT IMAGING

Edward J. SOARES, Harrison H. BARRETT and Craig K. ABBEY
Department of Radiology, University of Arizona, Tucson, AZ

1. ABSTRACT

In this paper, we investigate the noise character of SPECT images as propagated through the attenuation correction process, and objectively assess image quality. The correction techniques examined include the methods of Chang, Bellini, Inouye, and Tretiak & Metz, as well as a new Metz method. Noise character is determined from the image covariance matrix, while image quality is measured by the performance of the ideal, non-prewhitening (NPW) and region-of-interest (ROI) model observers. We demonstrate the noise character for a uniformly emitting and attenuating disk object. In addition, image-quality assessment is performed for the tasks of detecting a 10% contrast cold disk signal, at center and edge locations, on a uniformly emitting and attenuating circular background.

2. RESULTS

Figure 1 shows a plot of noise variance versus radial distance from the center of the disk for the five methods. The magnitude of the variance for the Chang method is proportional to the square of the Chang correction factor. The increase in variance of the Tretiak & Metz method is due to the exponential weighting during backprojection. The behavior of the variance for the Bellini, Inouye, and new Metz methods can be attributed to how each method weights the conjugate projection view Fourier coefficients when estimating the unattenuated data.

Table 1 lists SNRs derived using the three model observers for each of the five correction methods for the center and edge detection tasks. The results demonstrate

Y. Bizais et al. (eds.), Information Processing in Medical Imaging, 353–354.

that model-observer performance is remarkably invariant across attenuation correction method for the center task. However the Inouye and new Metz methods demonstrated higher observer performance for the edge task. For this task, the performance of the Bellini and Chang methods was hindered due to the fact that opposing projection views were averaged prior to correction. The Tretiak & Metz method yielded the lowest performance, most likely as a result of the notched-ramp filtering.

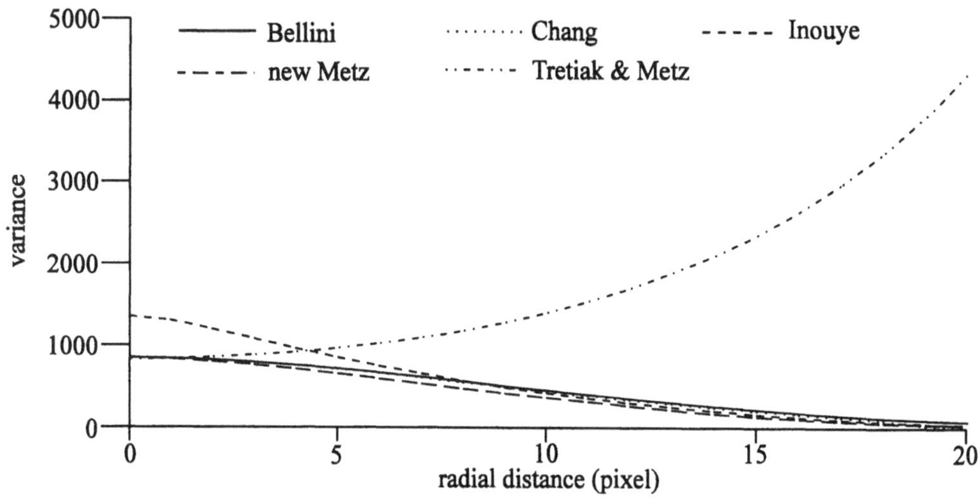

Figure 1. A plot of noise variance vs. radial distance for the five correction methods.

	SNR_{ideal}	SNR_{NPW}	SNR_{ROI}
Bellini	3.12/4.83	2.47/3.79	1.92/2.96
Chang	3.12/4.83	2.49/3.78	1.93/2.94
Inouye	3.12/5.48	2.47/4.06	1.91/3.14
new Metz	3.12/5.39	2.47/4.03	1.92/3.02
Tret.& Metz	3.12/4.08	2.47/1.79	1.92/1.32

Table 1. SNRs derived using the three model observers for center and edge detection tasks for each of the five correction methods.The data are listed in each cell as SNR center/SNR edge respectively.

EXPLORING THE RELIABILITY OF BAYESIAN RECONSTRUCTIONS

Kenneth M. HANSON and Gregory S. CUNNINGHAM
Los Alamos National Laboratory, MS P940
Los Alamos, New Mexico 87545

1. INTRODUCTION

Because of uncertainties in measurements and/or because of a lack of sufficient data to define an unambiguous answer (in the absence of regularizing priors), there often is no unique answer to many real data-analysis problems. The Bayesian approach to modeling allows one to combine measurement data with prior knowledge about models of reality to draw inferences about the validity of those models. Because it provides the degree of certainty of every possible solution, Bayesian analysis can be used to ascertain the reliability of the solution.

Skilling et al.[1] suggested the technique of displaying a sequence of distinct solutions drawn from the posterior probability distribution as a video loop, to give one a general feeling for the uncertainty in a Bayesian solution. We propose a new technique to directly probe the reliability of the MAP solution. The brevity of this summary does not permit adequate development of the background material or the particulars of our new scheme. The reader is referred to [2-5] for details.

2. BAYESIAN MECHANICS

Bayesian analysis provides the means to study the reliability of a parameterized model through the posterior probability, which characterizes the degree of certainty of any possible combination of parameter values. The parameters are typically chosen to maximize the posterior, yielding the maximum a posteriori (MAP) solution. The standard approach [6] to estimating the uncertainties in the parameters, as well as their correlations, entails calculation of the reciprocal of the curvature matrix of φ, the negative logarithm of the posterior. When dealing with many (10^4-10^6) parameters, this approach is impractical and does not provide one with an instinctual result.

We suggest an approach for probing the reliability of a MAP solution that makes use of an analogy between φ and a physical potential. Then, just as in physics, the gradient of φ wrt a parameter is analogous to a force acting on that parameter. The condition for the MAP solution, that the derivative of φ wrt all parameters is zero, states that the net force on each parameter is zero. When a variable is perturbed slightly from the MAP solution, the force pulls it back toward the MAP solution. The phrase "force of the data" takes on real meaning in this context. The reliability of the MAP solution is directly related to its

Y. Bizais et al. (eds.), Information Processing in Medical Imaging, 355–356.

rigidity, which is indicated by the rate at which the restoring force increases as one perturbs parameters from their MAP values. Correlations between parameters may be explored by applying a force on some parameters and then allowing all parameters to readjust to minimize φ. The correlations between those parameters pushed on and the rest is demonstrated by how much the latter parameters change. Ideally, these correlations should be experienced through direct interaction with a rapidly-responding dynamical Bayesian system.

3. USE WITH DEFORMABLE MODELS

Our approach is particularly intuitive when the reconstructed object is defined in terms of its geometric shape. The prior on the geometry [3] is defined in terms of the default shape together with a prescription for the probability of other possible shapes, reasonably taken to be $\exp(-\beta W)$, where W is the energy required to deform the geometry from the default shape into a new shape. The parameter β regulates one's strength of conviction in the default geometry.

We have demonstrated the usefulness of our novel Bayesian tool for reliability exploration in the context of tomographic reconstruction of a geometrically-defined object model. The problem involved very limited projection data, from two to four views. We have shown that when one pushes or pulls on the boundary of a reconstructed object, the restoring force is nonlinearly related to the displacement, has a different characteristic when pushed inward than when pulled outward, and varies with position along the boundary. It would be very difficult to treat all of these effects in the standard approach to the estimation of parameter uncertainty that is based on the covariance matrix. The reader is referred to Refs. 4 and 5 for further results.

REFERENCES

[1] J. Skilling, D.R.T. Robinson, and S.F. Gull. Probabilistic displays. In Jr. W.T. Grandy and L.H. Schick, eds, *Maximum Entropy and Bayesian Methods*, pp. 365-368. Kluwer Academic, 1991.

[2] K. M. Hanson. Introduction to Bayesian image analysis. *Proc. SPIE,* vol. 1898, pp. 716-731, 1993.

[3] K.M. Hanson. Bayesian reconstruction based on flexible prior models. *J. Opt. Soc. Amer. A10*, pp. 997-1004, 1993.

[4] K.M. Hanson and G. S. Cunningham. The hard truth. To be published in J. Skilling, ed., *Maximum Entropy and Bayesian Methods*, pp. 399-406. Kluwer Academic, 1993

[5] K. M. Hanson and G. S. Cunningham. Exploring the reliability of Bayesian reconstructions. To be published in *Proc. SPIE,* vol. 2434, 1995.

[6] P. R. Bevington. *Data Reduction and Error Analysis for the Physical Sciences.* McGraw-Hill, New York, 1969

NONLINEAR FOURIER ANALYSIS OF fMRI TIME SERIES

Nicholas LANGE

National Institutes of Health, Bethesda, MD, USA

1. INTRODUCTION

In human brain mapping by functional magnetic resonance imaging (fMRI), significant activations as small as a single voxel can be critical. The goal of this short paper is to improve activation detection by extending a current method [1] to allow the hemodynamic response to neuronal activity induced by controlled stimuli to differ voxel-by-voxel.

2. METHOD

The approach is to work with the temporal discrete Fourier transforms of the image intensities at each voxel. One can then restrict the analysis to a small number of frequencies corresponding to the forcing frequency of the controlled stimulus and its major harmonics. A nonlinear parametric model that relates signal $Y(t,u)$ at time t and location u to a controlled stimulus $x(t)$ is

$$\begin{cases} Y(t,u) = X_\lambda(t,\theta_u)\beta_u + \varepsilon(t,u) \\ X_\lambda(t,\theta_u) = \int \lambda(s,\theta_u)x(t-s)ds. \end{cases} \qquad \text{(Eq. 1)}$$

In (1), $X_\lambda(t,\theta_u)$ is the convolution of a known global stimulus $x(t)$ with an unknown local function $\lambda(t,\theta_u)$. At brain location u, the local regression coefficient β_u carries the magnitude of the dependence of signal Y on stimulus X, and the function λ, through θ_u, carries the temporal delay and dispersion of that signal. The magnitude parameter enters into (1) linearly but the delay and dispersion parameter enters nonlinearly. Hence their joint estimation involves iteration, in this case an iteratively reweighted least-squares algorithm. The model is fit in the frequency domain to simplify the convolutions and for easy accommodation of voxel-specific temporal and spatial autocorrelations.

3. APPLICATION

A coronal slice through calcarine cortex is analyzed (64×64 voxel images and a echo-planar GE sequence at 4.0 Tesla, TR 3 s, TE 25 ms, acquisition time 41ms). A single subject was imaged in darkness for 30 s (10 images), followed by visual stimulation for 30 s by flashing checkerboards reversing at 16 Hz, darkness, checkerboards, etc., to yield three, 20–image cycles and a total of 60

Y. Bizais et al. (eds.), Information Processing in Medical Imaging, 357–358.

images. The previous method [1] assumed that a single, global hemodynamic response function delay was constant over the entire brain. The results of the two approaches are compared in the following figure:

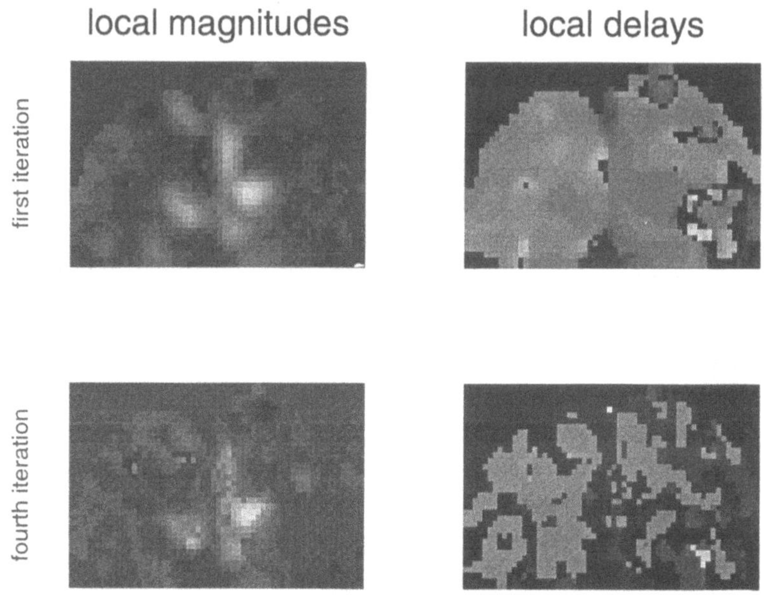

This figure displays t-fields of local magnitudes (right) and local delays (left) at the first (top) and fourth (bottom) iterations of the algorithm. Local delays are assumed to be Poisson functions, as in [1]. Reweighted least-squares updates were obtained for many more iterations yet appeared essentially unchanged after the fourth iteration. Note that the estimated magnitudes indicate activation in similar yet less diffuse regions than those reported previously [1, cf. Figure 3]. Note also the considerable degree of heterogeneity and hemispheric asymmetry among the estimated delays and dispersions.

4. CONCLUSION

These results suggest that a nonlinear model that employs locally varying delay functions reveals more specific focal activations than those obtained under a constant global delay assumption. It is reasonable to conjecture that this finding applies generally to higher-level cognitive fMRI experiments that require very locused functional-anatomical localization of responses to more complex stimuli.

5. REFERENCE

[1] K.J. Friston, P. Jezzard, R. Turner. "Analysis of fMRI time series", *Human Brain Mapping*, 1, 153–171, 1994

PRECOMPUTED ORTHOGONAL CURVES FOR DEFORMABLE TEMPLATE MATCHING

HEMANT D. TAGARE
Dept. of Diag. Radiology, Dept. of Comp. Sci.,
Yale University, CT 06520.

1. Deformable Template Matching

In medical image analysis, prior knowledge of the shape of an organ is often available as a standard template along with inner and outer limits of deformation of the template. Orthogonal curves can be extended from every point of the template to the inner and outer limits by (1) conformally mapping an annulus first between the template and the outer limit, and then between the template and the inner limit; and (2) using the image of radial rays of the annulus as the orthogonal curves (fig.1) [1]. Since the template shape is known a priori, the orthogonal curves can be precomputed.

The orthogonal curves provide a systematic way of deforming the template by simply moving each point of the template along its orthogonal curve. Internal and external energies can be associated with the deformed template. Internal energies keep the deformation smooth and close to the template, while external energies pull the deformation close to regions of high image gradient. At optimal deformation, the template has the least sum of internal and external energies. The optimal deformation of the template is obtained by dynamic programming (details are in [1]).

We have successfully used precomputed orthogonal curves and the dynamic programming algorithm for deformable template matching in a number of cases [1][2]. Figure 2 shows a typical template with its precomputed orthogonal curves. Figure 3 shows the optimal deformation found by the dynamic programming algorithm when the template is placed over a wrist CT image.

Y. Bizais et al. (eds.), Information Processing in Medical Imaging, 359–360.

Simulations with additive noise indicate that the algorithm performs robustly with respect to noise as well as initial placement of the template.

2. Conclusion

Precompute orthogonal curves can be used along with dynamic programming to optimally deform a template.

References

1. Tagare H. D., "Deformable 2-D Template Matching Using Orthogonal Curves," Technical Report, Div. of Imag. Sci., Diag. Radiology, Yale University, 1994.
2. Tagare H. D., Elder K., et. al., "Location and Geometric Description of Carpal Bones in CT Images," Annals of Biomedical Engineering, vol. 21, 715-726, 1993.

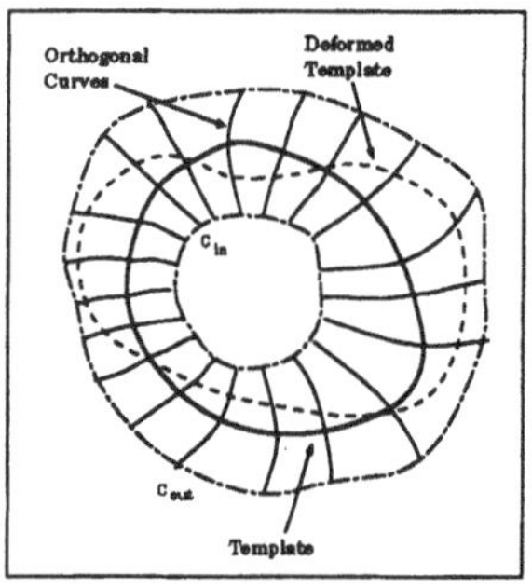

Figure 1. Smooth Orthogonal Curves.

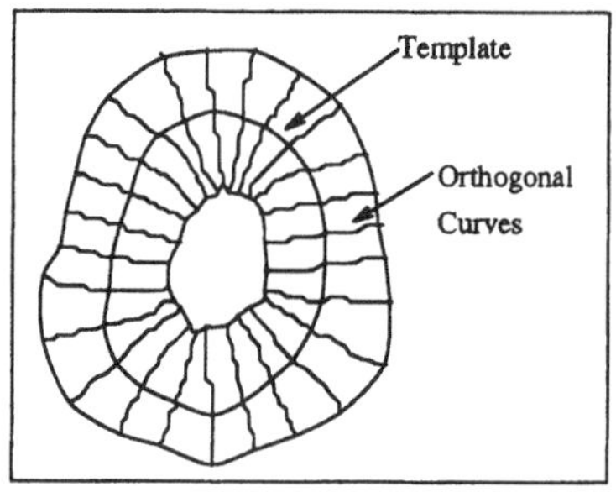

Figure 2. Template for Wrist CT

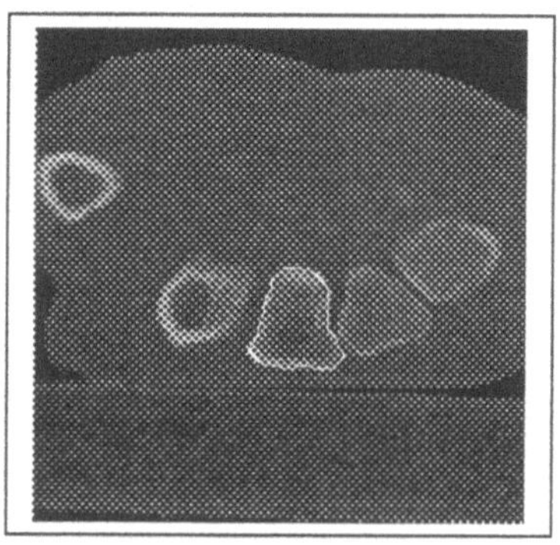

Figure 3. Optimal Deformation for the Wrist Template

A MODEL-BASED APPROACH TO 3D FREEHAND ULTRASOUND IMAGING

Michael Syn and Richard Prager
Dept. of Engineering, Cambridge University, U.K.

1. ABSTRACT

An iteratively refined shape recovery procedure for use in 3D freehand ultrasound imaging is presented. This procedure utilises models of both 3D shape and shape variation, by using free vibration modes derived from finite element analysis of the initial shape model, and updating the model using principal component analysis (PCA). The difficulty in applying common PCA schemes, which require prior segmentation of training examples, is thus avoided. This is very important, since manual segmentation in 3D ultrasound is impractical for a reasonably sized training set, particularly if it has to be performed for each new class of shapes. A scale and rotation invariant shape registration procedure is also presented. Such a procedure allows the quantitative comparison of segmented shapes from ultrasound with other modalities for the purposes of validation. It also allows us to address the issue of automated specification of homologous landmarks, by registering a densely specified example which has just been segmented, with the current model. The issues of model construction and volume measurement are also discussed.
Keywords: shape models, vibration modes, principal components, automated registration, model construction, volume measurement

2. INTRODUCTION

Full copies of the paper originally submitted can be obtained by contacting the authors **(mhs@eng.cam.ac.uk)** or using anonymous ftp **(svr-ftp.eng.cam.ac.uk)**

Most efforts in 3D ultrasound imaging have concentrated on the use of volume images which are rendered directly. These are of limited use in general because of noise artifacts obscuring details of interest. The nature of noise processes and the physics of image formation in this modality, make automated segmentation procedures such as those commonly used with MR images inadequate. These problems suggest that, in order to move on to examing useful clinical applications for 3D ultrasound, a more practical approach would recognise the fact that the ultrasonographer must have already visualised the structure of interest when performing a 3D scan, and that a synergistic combination of the

Y. Bizais et al. (eds.), Information Processing in Medical Imaging, 361–362.

ability of a trained clinician to segment and visualise the 3D structure, with a computer's ability to quantify structure can lead to more immediate dividends in clinical application.

3. SHAPE RECOVERY PROCEDURE

The initial shape model of an organ or structure can be arrived at in a number of ways. The initial estimate of the organ's shape variation due to inter-patient variability, is derived by considering the free vibration modes of a finite element model of the organ. This bootstrapped model is used in constrained segmentation of the volume image, using a GUI which allows specification of corresponding points in the volume image and shape model. Our model of mean shape and shape variation is now updated, eventually converging to the principal components of the class variation of normal organs, under the assumption of stationarity. Quantitative analysis (length, volume etc) is performed on the recovered shape. The relative invariance of the lower vibration modes to changes in model shape leads to a class of invariant features which can be used in shape registration of two models with arbitrary meshing, scale and orientation. This allows the recovered shape from the above to improve local fit by considering image evidence directly. This newly resampled mesh can then be automatically registered with the current model to update current estimates of mean shape and shape variation.

MEASUREMENT OF TAG DEFORMATIONS WITH DP SNAKES AND SNAKE GRIDS

AMIR AMINI*, RUPERT CURWEN†, AND JOHN GORE*
*Yale University**, *New Haven, CT*
CMA Associates†, *Schenectady, NY*

1. DYNAMIC PROGRAMMING SNAKES

MRI is an excellent imaging technique for measuring non-rigid tissue motion and deformation. Conventional MR or CT imaging however can not be used to infer the actual trajectories of individual tissue elements, limiting the accuracy of measurements of motion or deformation. To overcome such limitations, tagging methods and phase-contrast measurements of motion have been developed. In this paper, we apply B-spline snakes in 1D (curves) as well as 2D (grids) to MR tagging data [1, 2, 4]. B-spline curves are suitable for representing a variety of industrial and anatomical shapes. The advantages of B-splines are: (1) they are continuous parametric curves which can represent open or closed curves, (2) they are completely specified by few control points, and (3) individual movement of control points will only affect their shape locally. To localize a radial tag line, we minimize a snake external energy term which arises from normalized correlations of a template window of simulated tag profiles with tag lines in different orientations [2]. The simulation for a single profile is described in [3]. Dynamic Programming (DP) can be used to globally optimize the open snake curve in the space of B-spline control points. DP can also be used to optimize closed B-spline curves.

2. DEFORMABLE SNAKE GRIDS

Deformable snake grids are a sequence of spatially ordered B-splines, constructed by having the horizontal and vertical grid lines share control points [2]. By moving a spline control point, the corresponding vertical and horizontal snakes deform. To localize and track SPAMM tag lines, we optimize

Y. Bizais et al. (eds.), Information Processing in Medical Imaging, 363–364.

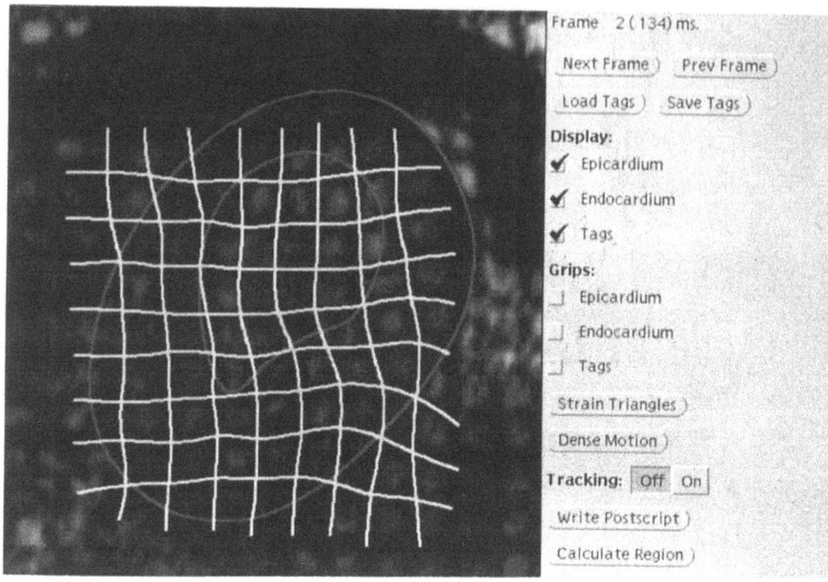

Figure 1. Localizing tag lines in SPAMM images. An intermediate frame in a sequence of tagged images is shown. The LV contours in the figure are hand-drawn B-spline curves.

the following energy function

$$\mathcal{E}(p_{12}, \cdots, p_{M,N-1}) = \lambda_1 \sum_k \int I(\alpha_k(u))du + \lambda_2 \sum_{ij} SSD(v_{ij}) \qquad (1)$$

where v_{ij} are intersection points of horizontal and vertical snakes, p_{ij} are spline control points which are initially placed at tag intersections, and λ_i are weights. The SSD function determines the sum-of-squared-differences of pixels in a window around point v_{ij} in the current frame with a window around the corresponding grid intersection in the next frame. We minimize (1) by gradient descent, using only the snake points which lie on the tissue.

References

1. A. A. Amini and et al. Energy-minimizing deformable grids for tracking tagged MR cardiac images. In *Proc. of Computers in Cardiology*, pages 651–654, October 1992.
2. A. A. Amini and et al. MR physics-based snake tracking and dense deformations from tagged MR cardiac images (oral presentation). In *AAAI Symposium on Applications of Computer Vision to Medical Image Processing*, Stanford University, Stanford, California, March 1994.
3. M. Guttman, J. Prince, and E. McVeigh. Tag and contour detection in tagged MR images of the left ventricle. *IEEE-TMI*, 13(1):74–88, 1994.
4. A. Young, D. Kraitchman, and L. Axel. Deformable models for tagged MR images: Reconstruction of two- and three-dimensional heart wall motion. In *Proceedings of IEEE Workshop on Biomedical Image Analysis*, pages 317–323, Seattle, WA, June 1994.

STIMULATED CORES: APPLICATIONS IN MEDICAL IMAGING

Daniel S. FRITSCH, David EBERLY, and Stephen M. PIZER
Medical Image Display and Analysis Group, University of North Carolina, Chapel Hill, NC, USA

1. INTRODUCTION

A variety of medical image analysis tasks require robust representations of image objects, including tasks such as segmentation, registration, recognition, and measurement (e.g., area, volume, length, etc.). In previous work we have shown that to measure geometric properties of image objects independent of object size (a property we call *zoom invariance*), one must make those geometric measurements at spatial scales proportional to local object width, which varies along the length of an object. A multiscale method of object representation we call *cores* allows for the selection of places in scale space where object middles are optimally manifested. Cores are computed directly from the greyscale image intensities by first deriving a measurement called *medialness,* at each position in the image and at a range of scales, and then finding *scale space ridges* in the medialness function. Operationally, medialness is defined as the degree to which a given position behaves like an object middle when measured at a particular scale. Medialness is derived from the linkage of boundariness measurements across image space, where in order to provide for zoom invariance the linkage distance is proportional to the scale (aperture width) at which the boundariness measurements are made. High values of medialness indicate positions and scales that are the most consistent with the presence of an object middle of width proportional to scale. For width-varying objects, medialness values are relatively high along a track running spatially down the middle of the object and in scale proportional to local object width at each middle point. This track of especially high medialness is the core, defined as a ridge in medialness.

The Core has been shown by us to have the following properties [1]:

(1) It is derived directly from the image intensities — typically via convolution of the image with scaled, radially-symmetric *medialness kernels* (see [2] for a discussion of various options for computing medialness) — without requiring the preliminary extraction of a boundary locus.

(2) It is invariant to rotation, translation, zoom and linear variation of intensity.

Y. Bizais et al. (eds.), Information Processing in Medical Imaging, 365–367.

(3) Because it is derived from boundary information measured with an aperture size proportional to local object width, it captures figural shape in a way that is insensitive to small scale (e.g., spatially uncorrelated) noise, blurring of a scale small compared to the width of the object, small scale local deformation, and object boundary properties that are small compared to the figural width.

(4) We have shown experimentally that it accords with the behavior of the human visual system in extracting and representing image objects [3].

2. STIMULATED CORES

Because of the computational burden associated with computing medialness at every image location and at a range of scales, we have recently developed a more efficient method called *stimulated cores* that begins from an approximate position and scale (i.e., from knowledge that an object is "about here" and "about this large") and which generates medialness values only as needed — first to locate an initial core point and then to traverse the core in scale space. This algorithm, based on the mathematics of scale space ridges developed by Eberly [4], derives its efficiency by restricting the computation of medialness to a small subvolume of scale space and allows for the extraction of a core for an object in a 2D image in a few seconds. This increased computational efficiency, coupled with the invariance and insensitivity properties of cores, gives rise to core-based solutions to a number of medical applications that require robust, automatic means of object representation.

3. APPLICATIONS

Cores have been extracted in images from a variety of modalities including CT, MRI, ultrasound, and projection x-rays (e.g., portal images and angiograms). The core is usable for all object-based image operations including segmentation (i.e., object definition), shape measurement (volume, width, length, width rate of change and curvature, medial curvature, etc.), object-based registration [5,6,7], and object recognition. Because of its built-in insensitivities to small scale image variations, the core carries out these operations in an especially stable manner. (We present examples of the use of the core in these tasks in the poster session).

4. ACKNOWLEDGMENTS

This research was supported in part by the National Library of Medicine (NLM) under the HPCC grant LM05508-1, by an NLM/National Cancer Institute Medical Informatics Research Training grant 5 T15 LM07071, and by a

National Institutes of Health program project grant P01 CA47982. The authors gratefully acknowledge the contributions of Bryan Morse, Edward Chaney, Matthew McAuliffe, Andrew Thall, Alan Liu and Aziz Boxwala to this work.

5. REFERENCES

[1] B.S. Morse, S.M. Pizer, and A. Liu. "Multiscale medial analysis of medical images", in *Information Processing in Medical Imaging*, Barrett, Gmitro eds, pp 112-131, Springer-Verlag, Berlin, 1993

[2] D. Eberly, S.M. Pizer, and D.S. Fritsch. "Objects, boundaries and medical image analysis", *UNC Computer Science Dept. Technical Report #TR94-069*. Submitted to *ICCV*, 1995

[3] C.A. Burbeck & S.M. Pizer. "Object representation by cores: Identifying and representing primitive spatial regions", *UNC Computer Science Dept. Technical Report #TR94-048b*. Accepted for publication in *Vision Research*, 1994

[4] D. Eberly, R.B. Gardner, B.S. Morse, S.M. Pizer, and C. Scharlach. "Ridges for image analysis", *J. Math. Imaging and Vision*, vol 4, pp 351-371, 1994

[5] D.S. Fritsch, S.M. Pizer, E.L. Chaney, and A. Liu. "Cores for image registration", *Proc. SPIE Medical Imaging '94: Image Processing*, pp 128-142, 1994

[6] D.S. Fritsch, S.M. Pizer, B.S. Morse, D. Eberly, and A. Liu. "The multiscale medial axis and its applications in image registration", *Pattern Recognition Letters*, vol 15, pp 445-452, 1994

[7] A. Liu, S.M. Pizer, D. Eberly, B.S. Morse, J. Rosenman, E.L. Chaney, E. Bullitt, and V. Carrasco, "Volume registration using the 3D core", *Visualization in Biomedical Computing*, pp 217-226, 1994

EFFICIENT SHAPE-BASED RETRIEVAL IN MEDICAL IMAGE DATABASES USING KD-TREES

GLYNN. P. ROBINSON AND HEMANT. D. TAGARE
Yale University, Dept. Diag. Radiol. and Comp. Science, USA

1. EFFICIENT, NON-RIGID CURVE MAPPING

Geometric properties of anatomical organs are hard to describe precisely in words; it is much easier to use visual examples. Consequently, interaction with medical image databases is improved if visual examples are added to queries. Moreover, the contents of the database have to be indexed for efficient retrieval. Here we report a similarity measure which compares shapes of curves. The similarity measure is used with KD-trees [1] for efficient retrieval of similarly shaped curves.

We use the measure for image retrieval in a database of cardiac MR images which has the visible chambers outlined and stored [3] During retrieval by shape similarity, an outlined curve in an example image is selected and all images containing a curve with similar shape are retrieved. More precisely, if C_{ex} is the example curve, $E(C_{ex}, C)$ is a measure of dissimilarity of the shapes of C_{ex} and C, and D is a user defined threshold, then all curves C satisfying $E(C_{ex}, C) \leq D$ are retrieved. This query can be used in conjunction with other similarity measures to create more complex queries.

The dissimilarity measure is based on the following observations. First, most organs are imaged from only a few view points. Hence, the same organ in different subjects appears similarly oriented in its image and shape comparison is required to be orientation sensitive. Second, some non-rigid slack is required while comparing organs of different subjects to compensate for normal variation in organ development. However, the slack cannot be too large or it will compensate for large changes of shape instead.

The measure $E(C_{ex}, C)$ is created by finding a function $\phi : C_{ex} \rightarrow C$ such that for every point $x \in C_{ex}$ the tangent vector $T_{ex}(x)$ is as similar as possible to the tangent vector $T(\phi(x))$ at the point $\phi(x) \in C$. Tangent

Y. Bizais et al. (eds.), Information Processing in Medical Imaging, 369–370.

vectors make the shape comparison orientation sensitive. The function ϕ mimics a "non-rigid slack". The amount of slack is constrained. Once the appropriate ϕ is found, the integrated square difference in orientations of $T_{ex}(x)$ and $T(\phi(x))$ are used to compute $E()$. We use dynamic programming to find the correct ϕ. In spite of the intermediate dynamic programming step, KD-trees can still be used to index curves. Details can be found in Robinson *et al* [2]

2. EXAMPLES FROM THE CARDIAC DATABASE

Figure 1 shows the result of retrieval using an example query from the image database. The top row shows the original images, and the bottom row the curves corresponding to the left-ventricle for these images. The curve shown in column a) is the example curve used in the query. The curves shown in columns b) - d) were the three curves retrieved as being the "closest" to the query curve. Figure 2 shows the mapping between the query curve of figure 1 and the curve in column d) of figure 1. The non-rigid nature of the mappings is evident in this figure.

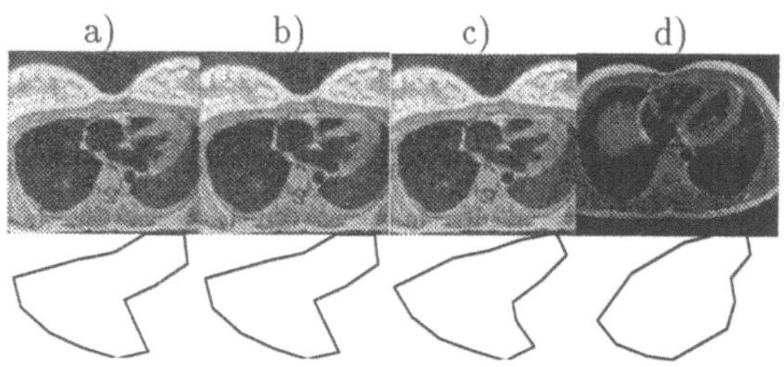

Figure 1. An example query from the database

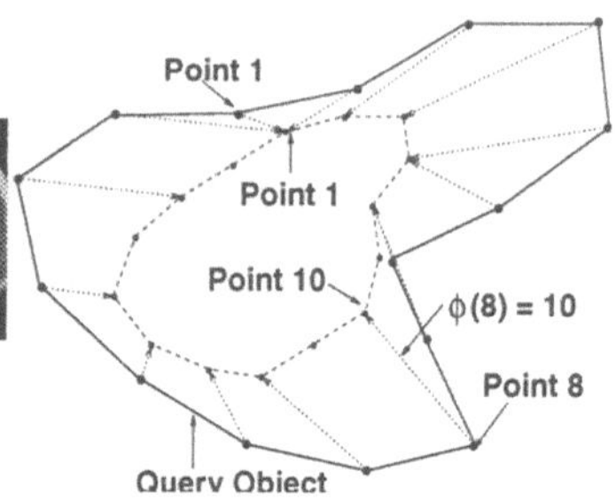

Figure 2. Example mapping

3. CONCLUSIONS

In spite of a non-rigid slack, it is still possible to index curves using a KD-tree for retrieval by shape similarity.

References

1. J.L. Bentley. .Multidimensional binary search trees used for associative searching. *Communications of the ACM*, 18,9:599–617, 1975.
2. G.P. Robinson and H. Tagare. Efficient medical database queries using shape properties and KD tree storage. Yale Uni. Dept. of Diag. Radiol. Report 94-1, 1994.
3. H.D. Tagare, F. Vos, C.C. Jaffe, and J.S. Duncan. Arrangement: A spatial relation comparing part embeddings and its use in medical image comparisons. In H.H. Barrett and A.F. Gmitro, editors, *IPMI XIII*, pages 133–148. Springer-Verlag, 1993.

MEDICAL IMAGE INTERPRETATION USING ACTIVE SHAPE MODELS : RECENT ADVANCES

T.F.Cootes, A.Hill, C.J.Taylor

Department of Medical Biophysics
University of Manchester.
email: bim@wiau.mb.man.ac.uk

1 INTRODUCTION

Many structures of interest in medical images can change in shape, either because they are naturally flexible or due to differences between individuals. We can interpret these images more reliably if we model this variability explicitly. Flexible models have proved extremely useful in tackling such problems. One approach is the Active Shape Model (ASM) framework [1]. Our poster describes three modifications to this framework which give considerable increases in speed, robustness and quality of fit compared to earlier methods.

1.1 Multi-Resolution Search Strategy

Given a rough starting position we can find the pose and shape parameters which best match an instance of an ASM to an image. This is done using a coarse–to–fine search strategy on a gaussian image pyramid. We train grey–level models for shape model landmark point at each level of the pyramid. During search we first locate the object in the coarsest image, using the grey–level models trained at that level, then refine the result by searching at increasingly fine levels until the original image is reached [2]. This method is considerably more robust and accurate than the original single resolution approach and typically reduces computational cost by about 80%.

1.2 Incremental Increase of Variability Allowed During Search

The amount of shape variation we can detect depends on the resolution of the image. Thus as the search progresses from coarse to fine we increase the number of shape modes which we allow to vary. This gives a significant improvement in the quality of the final fit, and also reduces the total computational load.

Y. Bizais et al. (eds.), Information Processing in Medical Imaging, 371–372.

1.3 Directional Weighting Scheme

In the original ASM search scheme the pose and shape parameters were updated at each iteration by estimating values which minimise the sum of squares of distances between the model points and the target points. By minimising the distances weighted along particular directions (such as normals) we can allow the model points the freedom to slide along found image edges [3]. This can give a further improvement in performance, allowing more rapid convergence to a better result.

2 DISCUSSION AND CONCLUSIONS

The methods described above are some facets of a broad programme of research aimed at improving the performance and applicability of shape model based systems. Other aspects include how physically based models can be combined with statistical shape models [4] and how non–linear shape models can give a better representation of the variability of certain classes of shapes [5]. All the methods described can be extended to 3D.

The techniques reviewed above are complementary and should significantly enhance our ability to interpret images of variable objects.

REFERENCES

[1] T.F.Cootes, A.Hill, C.J.Taylor, J.Haslam. "The Use of Active Shape Models for Locating Structures in Medical Images", *Image and Vision Computing Vol.12, No.6 July*, pp 355–366, 1994

[2] T.F.Cootes , C.J.Taylor, A.Lanitis, "Active Shape Models : Evaluation of a Multi–Resolution Method for Improving Image Search", in *Proc. British Machine Vision Conference,* (Ed. E.Hancock) BMVA Press, pp 327–338, 1994

[3] T.F.Cootes, "Directional Weighting For Active Shape Model Search", *Internal Report, Wolfson Image Analysis Unit, Manchester University,* 1994

[4] T.F.Cootes , C.J.Taylor, "Combining point distribution models with shape models based on finite element analysis", in *Proc. British Machine Vision Conference,* (Ed. E.Hancock) BMVA Press, pp 419–428, 1994

[5] P.D.Sozou, T.F.Cootes, C.J.Taylor, "A non–linear generalisation of point distribution models using polynomial regression", in *Proc. British Machine Vision Conference,* (Ed. E.Hancock) BMVA Press, pp 397–406, 1994

3D BRAIN STRUCTURES EXTRACTION USING FULLY AUTOMATED MRI SEGMENTATION

Laurent VERARD[1], P. ALLAIN[2], J.M. TRAVERE[2], D. BLOYET[3]
URA 1829 CNRS - GIP CYCERON, Caen, France[1]
CEA/DSV/DRIPP - GIP CYCERON, Caen, France[2]
URA 1526 CNRS - GREYC - ISMRA, Caen, France[3]

1. INTRODUCTION

By 3D Magnetic Resonance Imaging (MRI), brain anatomy can now be investigated at high resolution, using procedures referred to as "segmentation" [1][2]. The method proposed in this paper shows good quality and consistent results for the fully automated gross segmentation (encephalon, brain and hemispheres) of the human brain from 3D MRI (T_1-weighted-SPGR sequence with a 1.5 T magnetic field). Our work is an important criterion for the purpose of clinical application.

2. METHOD

The pre-processing step consists of a grey scale histogram analysis which allows automatic thresholds selection for a scene analysis and conditional dilatations. After a grey scale thresholding, which also reduces parasitic connections with the surrounding structures, a scene analysis in the inter-hemispheric plan is used to cut automatically true anatomical links. Then, a binary erosion is applied twice in order to insure the breaking of some parasitic links. During the labelling process, if the segmentation choice is N, the largest N objects are preserved and the region to be segmented is selected automatically [3]. Thus, the conditional dilatation recaptures the lost voxels during the erosion, through the use of grey scales given by the calculated histogram and a 3D connectivity algorithm, until no new voxel is added to the seed. While N is different from 0, the process loops to the

Y. Bizais et al. (eds.), Information Processing in Medical Imaging, 373–374.

region selection with a decremented N. As soon as N=0, the process stops and gives the result.

3. RESULTS AND DISCUSSION

Up to now, for a set a 20 normal patients considered who do not present with a pathology which itself alters the observed grey level scaling, the process applied never failed. The originality of this work is the 3D visualization of the basal and mid-sagittal sulci and allows an automated detection of gyri on lateral faces [4]. These segmentations are useful for a registration on Positron Emission Tomography (PET) scans [5] and also as an *a priori* information in PET reconstruction algorithms. This work has been performed under the auspices of the Centre Pôle "Traitement et Analyse d'Images"-Image Processing Centre.

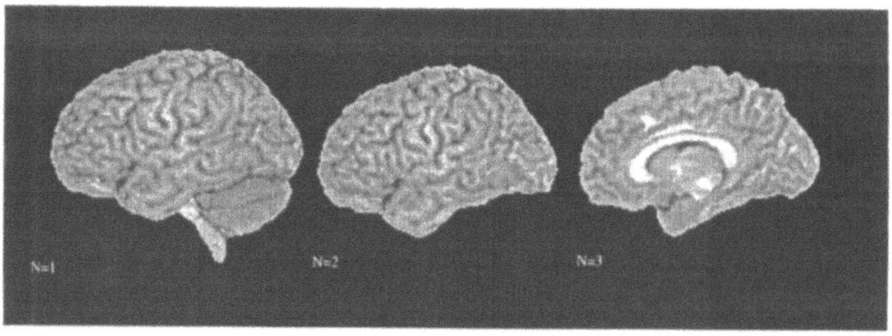

REFERENCES

[1] K. H. Höhne and W. A. Hanson, "Interactive 3D segmentation of MRI and CT volumes using morphological operations", *Journ. Comp. Assist. Tomog.*, vol. 16, No. 2, pp. 285-294, Mar./Apr. 1992.

[2] M. E. Brummer, R. M. Mersereau, R. L. Eisner and R. R. J. Lewine, "Automatic detection of brain contours in MRI data sets", *IEEE Trans. Med. Imag.*, vol. 12, No. 2, pp. 153-166, June 1993.

[3] J. Talairach, G. Szikla, P. Tournoux, A. Prossalentis, M. Bordas-Ferrer, L. Covello, M. Jacob, E. Mempel, "Atlas d'anatomie stéréotaxique du télencéphale : études anatomo-radiologiques", Ed. Masson & Cie, Paris, 1967.

[4] M. Desvignes, H. Fawal, M. Revenu, D. Bloyet, J.M. Travere, P. Allain, J.C. Baron, "Reconnaissance du sillon latéral du cortex sur des images tridimensionnelles", proposition de communication au *congrès AFCET RFIA*, Paris, 1994.

[5] P. Allain, J.M. Travere and al., "Accurate PET positionning with reference to MRI and neuroanatomical data bases", *Brain PET'93 Akita Proceedings,* Japan, pp. 401-408, 1993.

CLASSIFICATION IN SCALE-SPACE: APPLICATIONS TO TEXTURE ANALYSIS

Andrew P. BRADLEY, Paul T. JACKWAY, Brian C. LOVELL.
Dept. of Electrical and Computer Engineering, University of Queensland, Brisbane 4072, AUSTRALIA

To fully analyse the structure of a signal it is necessary to relate information from a number of different scales. A method of combining this information, proposed by Witkin [5], is to treat scale as a *continuous* variable rather than a parameter. Signal features measured at different scales can then be related if they lie on the same feature path in "scale-space". In this paper we have applied this principle to the investigation of measures of image texture at varying scales. We propose a method to incorporate the information from all these scales in a meaningful way, hence significantly reducing the dimensionality of the data, whilst maintaining as much information as possible. Specifically we analyse the texture of Papanicolaou stained cervical cell nuclei using features derived from a Grey Level Co-occurrence Matrix (GLCM) [2, 3].

The conventional method of texture analysis using the GLCM is to treat features, extracted at different distances (scales), as independent features. Selecting a small subset of scales (often a single scale) that gives the highest discriminatory power. We propose to treat these measurements of texture at each scale as sample points of a *continuous* function through scale-space. This function can then be reconstructed by interpolating the sample points of the functions extracted from the GLCM. This allows a whole new range of classical mathematical techniques for comparing functions to be used in the analysis and classification of textures.

We define the Grey Level Co-occurrence Function (GLCF), $g_{\sigma\theta}(\alpha,\beta)$, to estimate the probability that a pair of grey levels $[\alpha,\beta]$ will be found at a displacement $[\sigma\cos\theta, \sigma\sin\theta]$ apart. Scale dependent texture features $T(\sigma)$ are then extracted by applying some functional ψ to the GLCF. If this functional integrates over α,β and θ, then the texture feature of interest can be expressed solely as a function of scale. We write:

$$T(\sigma) = \psi\left[g_{\sigma,\theta}(\alpha,\beta)\right]. \qquad (1)$$

The scale dependent functionals used in this work are continuous versions of those commonly used for discrete textural feature measures [2, 4], for example, Energy:

$$En(\sigma) = \int_\theta \iint_D g^2_{\sigma,\theta}(\alpha,\beta)\, d\alpha\, d\beta\, d\theta \qquad (2)$$

We also used: Inverse Difference Moment (Local Homogeneity); Entropy; Correlation; Inertia; Cluster Shade; and Cluster Prominence.

The data set used was as detailed by Walker *et al* [4], and consisted of 61 segmented cell nuclei, there being 30 normal and 31 abnormal cells.

Y. Bizais et al. (eds.), Information Processing in Medical Imaging, 375–376.

The 7 features extracted from the GLCM were measured at distances of 1 to 15, at odd intervals only. Polynomial models were then fitted to each of these scale dependent texture features. We then compared the classification accuracy of selecting N parameters from these models and selecting N distances (single scales) from all those available. All features were normalised to [0,1] and then selected using a Sequential Forward Search with the Bhattacharyya distance measure for determining feature separation [1]. Classification accuracy was measured using leave-one-out random sub-sampling with K-Nearest Neighbours (K=1).

No. Features	1	2	3	4	5
Best Scales	90%	92%	88%	87%	88%
Best Model	83%	90%	94%	94%	94%

Table 1: Classification accuracies for the best N features

Table 1 shows the classification accuracy using only N of the 56 distance measures[1] and using N of the 21 parameters from the models of the texture functionals[2]. The highest classification accuracy of 94% was obtained using the proposed technique, and was found to be for 3, 4 or 5 parameters. The difference between 92% and 94% accuracy on a data set of 61 examples is only 1 case, but, the proposed technique significantly reduced the dimensionality of the original data (from 56 to 21), with no loss in classification performance.

This technique can also be applied to other scale dependent functionals, such as, features extracted from a morphological or a gradient watershed scale-space. This project is supported by the Cooperative Research Centre for Sensor Signal and Information Processing. The authors are grateful to P. Shield and G. Perkins of the Royal Women's Hospital in Brisbane, for providing the slides and classifying the cell images used in this study.

References

[1] Devijver,P. A., Kittler, J. (1982) *Pattern Recognition: A statistical approach.* Prentice Hall, London.

[2] Haralick, R. M., Shanmugham, K. S., Dinstein, I. (1973) "Texture features for image classification", *IEEE Trans Syst Man Cybern,* **Vol. 3 no. 6**, pp. 610-621.

[3] Julesz, B. (1962) "Visual Pattern Discrimination", *IRE Transactions on Information Theory,* **Vol. 8**, pp. 84–92.

[4] Walker, R., Jackway, P., Lovell, B., Longstaff, D. (1994) "Classification of cervical cell nuclei using morphological segmentation and textural feature extraction", *Proceedings of the 2nd Australian and New Zealand Conference on Intelligent Information Systems,* pp. 297–301.

[5] Witkin, A. P. (1983) "Scale-space filtering", *Proceedings of the International Joint Conference on Artificial Intelligence.* Kaufman, Palo Alto, pp. 1019–1022.

[1] 7 features each at 8 distances.

[2] 7 models each with, on average, 3 parameters.

SOME ASPECTS OF DEMPSTER-SHAFER EVIDENCE THEORY FOR CLASSIFICATION OF MULTI-MODALITY MEDICAL IMAGE

Isabelle BLOCH
Département Images, ENST,
46 rue Barrault, 75634 Paris Cedex 13, France, bloch@ima.enst.fr

Among the numerical data fusion methods, probabilistic fusion and Bayesian inference are the most popular. But they suffer from strong drawbacks. Alternatives to this approach are fuzzy sets theory, possibility theory, and Dempster-Shafer evidence theory [4]. We will focus here on the Dempster-Shafer approach (DS). Looking at medical applications, it appears that DS has been applied mainly for topics derived from artificial intelligence, and thus concerns medical diagnosis. Evidences issued from the image signal (of "iconic" type rather than propositional) are seldom considered. To our knowledge, only a few papers report on image fusion by DS in medical imaging [3], [2], [5]. However, DS offers a number of advantages which could be of major interest for medical image fusion and deserve to be detailed. We aim here at contributing to this task. The originality of our work, described more in detail in [1] lies in underlining how some features of Dempster-Shafer decision theory may be further exploited in medical imaging. It is illustrated on a classification problem from multi-echo MRI (the registration is supposed to be done).

DS has several advantages since it allows to represent both imprecision and uncertainty, partial and global ignorance, and is able to fit a large variety of situations (for instance on Figure 1, the two images give different information about the classes). The modelization by DS consists in defining mass functions on the power set of the class set D, from which credibility and plausibility functions are derived. A combination of masses issued from the images to be fused is then performed, from which a decision is taken according to several possible rules.

The key features of our approach are the following. First, an adapted modelization is proposed in the DS framework, in particular by assigning masses to compound hypotheses representing the unability of an image to discriminate between two classes on one hand, and partial volume effect on the other hand. Such a modelization constitutes already an improvement over other applications of DS. Second, even a rough definition of mass functions turns out to be sufficient and robust. Figure 2 shows results with 3 classes (brain, CSF and pathology). Experiences with more

Y. Bizais et al. (eds.), Information Processing in Medical Imaging, 377–378.

classes have been performed and provide also good results, for instance for discriminating grey matter and white matter. Third, reliability of an image concerning each class can be introduced by weighting accordingly the corresponding masses, and its influence on the combination and on the conflict has been tested. At last, decision has been taken according to two rules: a classical one, where a decision is always taken in favour of a simple hypothesis, and a second one, where it can also be decided in favour of a compound hypothesis. This second rule fits the reality, by highlighting in particular regions with partial volume effect, and is adapted to the physician's way of reasoning (see Figure 2).

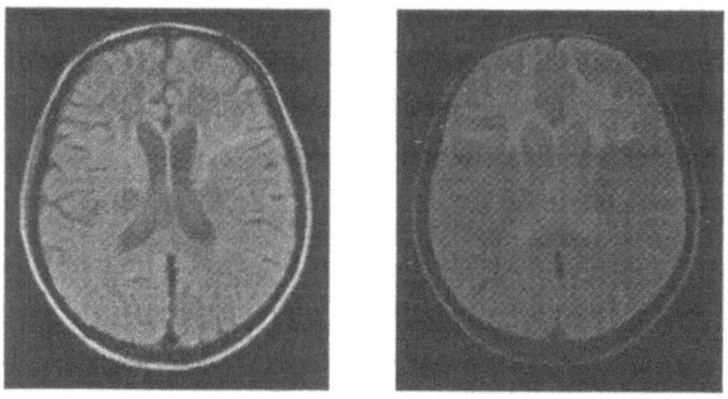

Figure 1: Initial images: dual echo brain MRI (one slice). The first image shows a good discrimination between brain, ventricles and CSF but the pathology (ALD) is not discriminated. On the contrary, the second image shows ALD very clearly (the white area), but the ventricles have almost the same grey levels than grey matter and their contours are imprecise.

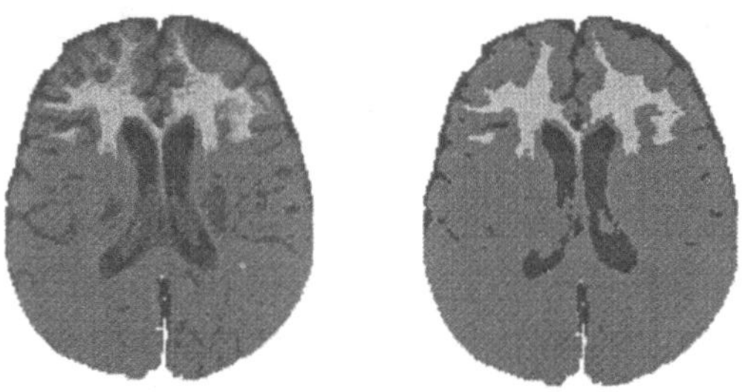

Figure 2: Decision image by taking the maximum of belief over all hypotheses but D (left), and decision image by taking the maximum of belief over all simple hypotheses (right): white = ALD, grey = brain, black = CSF. Intermediate grey levels on the left image correspond to compound hypotheses (brain $\cup$ ALD and brain $\cup$ CSF), representing ambiguous regions (in particular partial volume effect).

References

[1] I. Bloch. *Some Aspects of Dempster-Shafer Evidence Theory for Classification of Multi-Modality Medical Images*, Technical Report, Ecole Nationale Supérieure des Télécommunications, 1995.

[2] S. Y. Chen, W. C. Lin, and C. T. Chen. "Evidential Reasoning based on Dempster-Shafer Theory and its Application to Medical Image Analysis", in *SPIE, volume 2032*, pp. 35–46, 1993.

[3] R. H. Lee and R. Leahy. "Multi-Spectral Classification of MR Images Using Sensor Fusion Approaches", in *SPIE Medical Imaging IV: Image Processing, volume 1233*, pp. 149–157, 1990.

[4] G. Shafer. *A Mathematical Theory of Evidence*. Princeton University Press, 1976.

[5] D. Y. Suh, R. M. Mersereau, R. L. Eisner, and R. I. Pettigrew. "Automatic Boundary Detection on Cardiac Magnetic Resonance Image Sequences for Four Dimensional Visualization of the Left Ventricle", in *First Conference on Visualization in Biomedical Computing*, pp. 149–156, Atlanta GE, 1990.

MORPHOMETRY AND IDENTIFICATION OF BRAIN SULCI ON THREE-DIMENSIONAL MR IMAGES

Nicolas ROYACKKERS[1], Houssam FAWAL[1], Michel DESVIGNES[1], Marinette REVENU[1], Jean-Marcel TRAVERE[2]
GREYC-ISMRA (CNRS, URA 1526) 6, Bd Maréchal Juin 14050 Caen, France[1]
CYCERON (CEA/DSV) Bd Becquerel BP 5229 14074 Caen, France[2]

1. INTRODUCTION

Positron emission tomography (PET) is widely used for the study of human cerebral activity. As PET images do not reflect brain anatomy of patients, functional areas identified in such examinations cannot be localized precisely. Thus, a matching between PET and anatomical data from other sources is necessary to make the most of PET images. An approach to this problem is the direct recognition of cortical sulci on 3D magnetic resonance images (MRI) in order to build an accurate parcellation of brain for the localization of functional areas found in PET examinations.

2. AN INDIVIDUAL MODEL OF CORTICAL TOPOGRAPHY

The idea is to extract a statistical model of cortical topography from a database composed of MR images where voxels belonging to some major sulci have been identified. The computed model should then be used for sulci identification on other brain images.

The main point is that a single model cannot be built once for all: extracted data must be individually adapted to each MR image under examination. As a matter of fact, sulci features depend on the shape and size of the different brains contained in the database, which are very variable [2]. Therefore, those characteristics cannot be statistically merged in order to be used for sulci identification in another image. To tackle this issue, a correction mechanism is proposed. The coordinates of the points belonging to sulci contained in the database are converted into the coordinate system of the image where recognition has to take place, by means of Talairach's proportional grid [3]. Next, a set of features is extracted so as to represent the cortical topography. Thus, data are homogeneous because they are computed in the same anatomical

Y. Bizais et al. (eds.), Information Processing in Medical Imaging, 379–380.

coordinate system. Moreover, they take into account the brain characteristics of each patient.
To build a model of the cortical topography of the brain under examination, geometrical features are computed for each sulcus of the database (localization, direction, length, depth and straightness). For each couple of sulci appearing in the same database image, geometrical and topological relations are also determined (connectivity, relative orientation and localization). Statistics on each extracted feature are finally computed.

3. IDENTIFICATION OF SIX MAJOR SULCI

Precentral, central, postcentral, lateral, superior frontal and superior temporal sulci in both hemispheres are currently recognized. To identify these sulci in an MR examination, a processing is applied on the MR image so as to obtain the superficial part of the whole sulci as a set of 3D digital curves. Next, some of these curves are identified as parts of the sought sulci. This recognition is performed by means of the topography model computed from the database described in §2. A detailed description of our identification method can be found in [1], in the case of lateral sulcus. Finally, a recurrent voxel following is performed from the previously identified parts of the sulci in order to find missing voxels in deeper and deeper layers of the brain. The result is a thin and continuous 3D surface for each one of the recognized sulci.

4. RESULTS AND PERSPECTIVES

The superficial identification process has proved to be robust and efficient, since more than 90% of the sulci are correctly recognized on 13 MR images. The 3D identification step is still under improvement. Though its results need to be validated by neuroanatomists, they seem to be very promising.

Other improvements are also planned, such as the development of a unique identification protocol for all sulci, the extension of the sulci model in order to take into account ramified structures, or the building of a more declarative representation of processes so as to improve identification flexibility.

[1] M. Desvignes, H. Fawal, M. Revenu, D. Bloyet, P. Allain, J.M. Travere, J.C. Baron. "Reconnaissance du sillon latéral sur des images RMN 3D", *RFIA*, Paris, pp 685-690, 1994
[2] N. Royackkers, H. Fawal, M. Desvignes, M. Revenu, J.M. Travere. "Feature extraction for cortical sulci identification", to appear in 9th *SCIA*, Uppsala, 1995
[3] J. Talairach, P. Tournoux. *Co-planar stereotaxic atlas of the human brain*, Georg ThiemeVerlag, Stuttgart, 1988

PERFORMANCE EVALUATION OF IMAGE SEGMENTATION TECHNIQUE

Tianhu LEI, Wilfred SEWCHAND
University of Maryland School of Medicine
Baltimore, Maryland 21201, USA

1. INTRODUCTION

References [1-4] developed unsupervised stochastic model-based image segmentation techniques, where image is modeled by Finite Normal Mixture (FNM). Tasks of unsupervised image segmentation techniques based on FNM model are divided into three steps: 1) detecting the number of image regions, 2) estimating the parameters of FNM model, and 3) classifying the pixels into the different image regions. These three problems are solved by using a) information theoretic criteria to detect the number of image regions, b) CM and EM algorithms to estimate the model parameters, and c) Bayesian classifier to segment the observed image. This type of techniques has been applied to the various medical images and the satisfied segmenation results have been obtained. However, what are the detection, estimation, and classification performances of this type of techniques? Or, in other words, how good are they? [1-4] do not give the answers. This paper addresses these problems.

2. DETECTION PERFORMANCE

Let K_0 and K_1 denote the correct and the assumed (by the information theoretic criterion) number of the image regions, respectively, our study [5] shows that probabilities of the over-detection and the under-detection of the number of image regions are

$$P_{ov} = (\gamma + \gamma^{-1})^{-2m-1} \sum_{l=0}^{m} C_{2m-l}^{m}[\gamma(1+\gamma^2)^l e_0]$$
$$P_{ud} = (\gamma + \gamma^{-1})^{-2m-1} \sum_{l=0}^{m} C_{2m-l}^{m}[\gamma(1+\gamma^2)^l + \gamma^{-1}(1+\gamma^{-2})^l(1-e_1)]$$

where $\gamma^2 = \frac{K_1}{K_0}\sqrt{\frac{1+SNR_0}{1+SNR_1}}$ and e_0, e_1 are functions of K_0, K_1, σ_0, and σ_1.

3. ESTIMATION PERFORMANCE

Our study [5] shows that CM and EM algorithms produce asymptotically unbiased ML estimates of model parameters in the case of no-overlap and

Y. Bizais et al. (eds.), Information Processing in Medical Imaging, 381–382.

derives Cramer-Rao bounds of variances of these parameter estimates.

$$CRB_{\hat{\pi}_k} = J^{-1}\pi_k(1-\pi_k)$$
$$CRB_{\hat{\mu}_k} = \sigma_k^2/J\hat{\pi}_k$$
$$CRB_{\hat{\sigma}_k^2} = (2\sigma_k^4/J\hat{\pi}_k)(1-\frac{1}{J\hat{\pi}_k})$$

where $\hat{\pi}_k$, $\hat{\mu}_k$, and $\hat{\sigma}_k^2$ are the estimates of weight, mean, and variance.

4. CLASSIFICATION PERFORMANCE

Image segmentation is performed by pixel classification, which is carried out by Bayesian classifier. Our study [5] shows that the probability of misclassification is

$$P_{mis}(\bullet|k_0) = \hat{\pi}_{k_0} \sum_{k=1,k\neq k_0}^{K_0} (\Phi(\frac{x_k'' - \hat{\mu}_{k_0}}{\hat{\sigma}_{k_0}}) - \Phi(\frac{x_k' - \hat{\mu}_{k_0}}{\hat{\sigma}_{k_0}}))$$

where $G_k: \{x_k' < x < x_k''\}$ and $\Phi(y)$ is CDF of standard Gaussian.

5. RESULTS

Due to the page limitation, we don't have space to include numerical tables and images. A brief summary of our results is given below. The results obtained by applying this performance evaluation method to the simulated images show that even SNR of the image decreases to 10.2 db, a) the numbers of image regions suggested by detection criterion are still correct; b) all parameter estimates (weight, mean, and variance) are in Cramer-Rao intervals (when $SNR > 14.5$ db); c) misclassification probability is less than 5%. These results demonstrate that the detection procedure is robust, the paramter estimates are accurate, and the segmentation errors are small.

6. REFERENCES

1. Lei, T. and Sewchand, W., "Statistical approach to X-ray CT imaging and its applications in image analysis - Part 2: A new stochastic model-based image segmentation technique for X-ray CT image," *IEEE Trans. on Medical Imaging*, Vol. MI-11, No. 1, Mar. 1992.
2. Zhang, J. and Modestino, J.W., "A model fitting approach to cluster validaty with application to stochastic model based image segmentation," *IEEE Trans. on Patt. Anal. and Machine Intell.*, Vol. PAMI-12, Oct. 1990.
3. Liang, Z., Jaszczak, R.J., and Coleman, R.E., "Parameter estimation of finite mixture using the EM algorithm and information criteria with application to medical image processing," *IEEE TNS*, Vol. 39, 1992.
4. Liang, Z., "Tissue classification and segmentation of MR images," *IEEE EMB Magazine*, Vol. 12, 1993.
5. Lei, T., "Performance evaluations of segmentation technique", (unpublished note), University of Maryland at Baltimore, 1991.

THIN NETS EXTRACTION IN MEDICAL IMAGES

N. Armande[1], O. Monga[2] and P. Montesinos[1]
LGI2P, Parc Scientifique G. Besse Nimes, France [1]
INRIA, BP. 105, 78153 Le Chesnay Cedex, France [2]

Abstract

We describe a new approach for extracting *thin nets* in grey-level images. The key point of our approach is to model thin nets as the crest lines of the image surface. These lines are defined using first, second and third derivatives of the image. We apply this approach to the extraction of blood vessels in medical images.

1 Introduction

In many images, Thin Nets (TN) correspond to important features [3]. For instance, in medical images, TN are attached to blood vessels. TN are formed by the points where the grey level is locally extremum in a given direction. This direction is the normal to the curve traced by the TN at this point. Here, we propose a new method to detect TN using differential geometry. An important point of our approach is its ability to identify TN by crest lines on the surface defined by the image [5][4]. We have tested this method on medical images in which the goal is to extract blood vessels. The quality of the results obtained clearly shows the flexibility and the pertinence of our approach.

2 Extraction of the TN

Let us consider the surface $\vec{S}(x,y)$ associated to the grey level intensity of a 2D image $I(x,y)$ described by the equation $\vec{S}(x,y) = (x, y, I(x,y))^t$. The first and second Fundamental Forms F_1 and F_2 of this surface, can be used to compute the principal curvatures k_{max}, k_{min} and principal directions $\vec{t_{max}}$, $\vec{t_{min}}$ at each point $P(x,y)$ of the surface $\vec{S}(x,y)$ [1]. TN can be characterized as the zero-crossings of the directional derivative of the maximum principal curvature in the corresponding principal direction defined by $e_m = \nabla \vec{k_{max}} \cdot \vec{t_{max}}$. We select the zero-crossings of the e_m such that the maximum curvature k_{max} is greater than a given positive

Y. Bizais et al. (eds.), Information Processing in Medical Imaging, 383–384.

threshold. This thresholding stage yields only the points where the maximum curvature is locally maximum. The expression of e_m contains up to the third partial derivatives of the 2D image. We estimate these partial derivatives using recursive filters approximating the Gaussian filter and its derivatives [2] [3]. We have extracted blood vessels in medical images using this method as illustrated in Figure 1.

3 Conclusion

We have presented a new approach for extracting blood vessels in medical images using differential properties of the image surface. This work shows that even third order differential features can be robustly extracted from grey level images.

References

[1] M.P. Do Carmo, *Differential Geometry of Curves and Surfaces*, pp 153-175, Prentice Hall, 1976.

[2] R. Deriche, "Recursively implementing the Gaussian and its derivatives", *INRIA Research Report*, no 1893, 1993.

[3] O. Monga, N. Armande and P. Montesinos, "Thin Nets and Crest Lines : Application to Satellite Data and Medical Images", *INRIA Research Report*, 1995.

[4] O. Monga, R. Lengagne and R. Deriche, "Crest lines extraction in volumic 3D medical images : a multiscale approach", In *International Conference on Pattern Recognition (ICPR)*, Jerusalem, Israel, October 1994.

[5] J. Ponce and M. Brady, "Towards a surface primal sketch", In *Proceedings, IJCAI*, 1985.

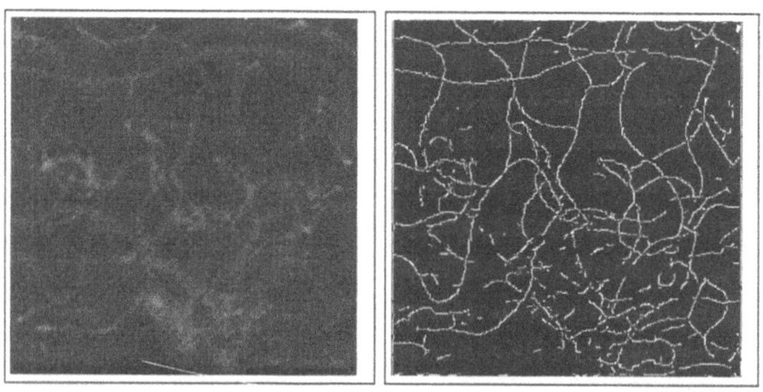

Figure 1: Left : Original image of blood vessels. Right : TN extraction ($\sigma = 1$).

A HYBRID ALGORITHM FOR THE SEGMENTATION OF 2D/3D IMAGES

Stelios ORPHANOUDAKIS, George TZIRITAS, Kostas HARIS
Institute of Computer Science, FORTH and
Department of Computer Science, University of Crete
P.O. Box 1385, 71110 Heraklion, Crete, Greece

The hybrid segmentation algorithm described in this communication assumes a piece-wise image model and is based on the integration of edge and region based techniques. First, adaptive smoothing is applied to the input image followed by gradient estimation, the detection of watersheds in the gradient magnitude image, and hierarchical region merging to get around the problem of oversegmentation which is characteristic of the watershed algorithm.

Edge-preserving adaptive smoothing of the input image is achieved by the statistical local classification method described in [1]. Then, the Canny gradient operator [2] is applied to the smoothed image and a preliminary segmentation is obtained by detecting the watersheds of the gradient magnitude image [3]. Oversegmentation, a known drawback of the watershed algorithm, is limited by the first stage of noise reduction and by thresholding the gradient magnitude in order to retain regions with a relatively high contrast. However, there still remain neighboring regions which can be merged to obtain a more accurate segmentation. For this purpose, a hierarchical region merging process has been implemented which merges the most similar pair of neighboring regions at each iteration. A brief description of this process follows.

Having obtained a good initial segmentation of the input image into K segments, a cost function F is defined over the space of the K-partition. Assuming that the optimal $(K-1)$-partition with respect to F is produced from the optimal K-partition by merging a pair of neighboring regions, good suboptimal solutions to a difficult combinatorial optimization problem are found by a stepwise optimization approach [4]. Partitions are represented by the Region Adjacency Graph (RAG) [5]. The *Most Similar Neighbor Graph (MSNG)* is introduced to allow consideration of the fact that it is not necessary to store all edges of the RAG so that the memory and computational requirements of the algorithm may be reduced.

Y. Bizais et al. (eds.), Information Processing in Medical Imaging, 385–386.

The hybrid segmentation algorithm described above has been applied to 2D MR images with very good results as shown in Figure 1. The total execution time for a 256x256x8bit image is approximately 10sec on a SUN4. The algorithm has been extended in a straightforward manner for the segmentation of 3D images and can be easily adapted for use in an interactive medical image processing environment by allowing the user to control the stopping point of the iterative hierarchical merging process.

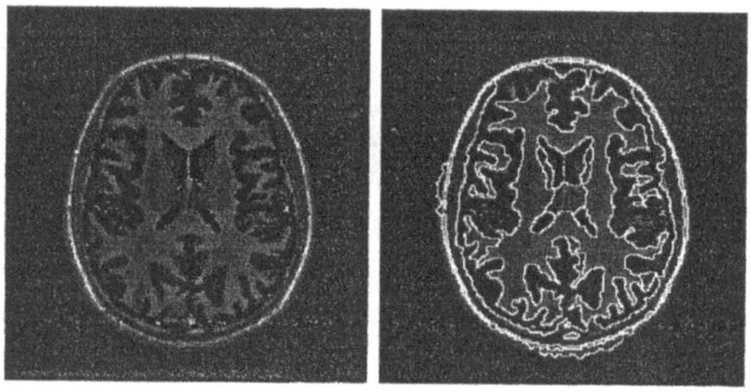

Figure 1: Original MR image (left) and segmented image overlaid on original (right)

Acknowledgements

This work has been funded in part under project COVIRA of the CEC AIM programme.

References

[1] K. Haris, G. Tziritas, and S. Orphanoudakis. Smoothing 2-d or 3-d images using local classification. In *Proceedings of EUSIPCO'94*, Ediburg, September 1994.

[2] J. Canny. A computational approach to edge detection. *IEEE Trans. on Pattern Anal. and Mach. Intell.*, 8(6):679–698, November 1986.

[3] L. Vincent and P. Soille. Watersheds in digital spaces: An efficient algorithm based on immersion simulations. *IEEE Trans. on Pattern Anal. and Mach. Intell.*, 13(6):583–598, June 1991.

[4] J. Beaulieu and M. Goldberg. Hierarchy in picture segmentation: A stepwise optimization approach. *IEEE Trans. on Pattern Anal. and Mach. Intell.*, 11(2):150–163, February 1989.

[5] X. Wu. Adaptive split-and-merge segmentation based on piecewise least-square approximation. *IEEE Trans. on Pattern Anal. and Mach. Intell.*, 15(8):808–815, August 1993.

IMAGE REGISTRATION METHOD FOR INTER-PLANE INTERPOLATION OF CARDIAC PET FDG IMAGES

Kang-Ping LIN[1], Sung-Cheng HUANG[2], Heinrich R. SCHELBERT[2], Michael E. PHELPS[2]

Dept. of Electrical Engineering, Chung-Yuan University, Chung-Li, Taiwan[1]

Dept. of Molecular and Medical Pharmacology, UCLA School of Medicine, Los Angeles, USA[2]

Positron emission tomographic images that provide three-dimensional data usually have a larger distance between consecutive planes (6.675 mm) than between adjacent pixels within a plane (2.6 mm). Multiple image planes are usually resliced to long axial orientation for cardiac study based on linear interpolation. However, if inter-plane distance is much larger than inter-pixel distance, most linear interpolation methods can not provide accurate estimation of pixel values between planes. In this study, we develop an interpolation method that is based on elastic transformation to improve the accuracy of image reslicing in cardiac PET FDG images.

The strategy of the present interpolation method is to estimate a transformation that provides local moving vectors for each pixel in one image plane onto an adjacent image plane. The magnitude of the estimated transformation can be adjusted by a linear factor from zero to one to indicate the distance of the intermediate plane between the two original image planes. The estimation of the transformation between two adjacent planes is achieved by matching each local area of the images at different locations and then to combine all local matching information to achieve a global matching. The approach is iterative, the global matching performance will follow the local matching results as the iteration progresses. The iterative process is stopped as the sum of square error oscillations within 5% for the last five iterations.

The present method was evaluated by using computer simulated images of two cylinders tilted 30 degrees with respect to the image planes. The new interpolation method was used to generate between-plane images, with coarse compared to the true images and to the ones obtained by linear interpolation. Comparing to linear interpolation, the present elastic transformation method had 3%, 9%, 18%, and 26% improvements when the original images planes were separated by 3 mm, 6 mm, 9 mm, and 13 mm respectively. The new method was also applied to PET measured FDG cardiac images. The image of

Y. Bizais et al. (eds.), Information Processing in Medical Imaging, 387–388.

a middle plane was used as a reference. The new method was used to generate an interpolated image from the two measured images of the adjacent planes. The interpolated images was then compared pixel-by-pixel with the reference image. The present method gives 5% mean errors (maximum error is 11%) in pixel estimation as compared to 16% mean errors (maximum error is 29%) for the linear interpolation method. The method does not require any image pre-processing, or any expected landmarks. The method can also be applied to other 3D tomographic images and for interpolations used in image rendering.

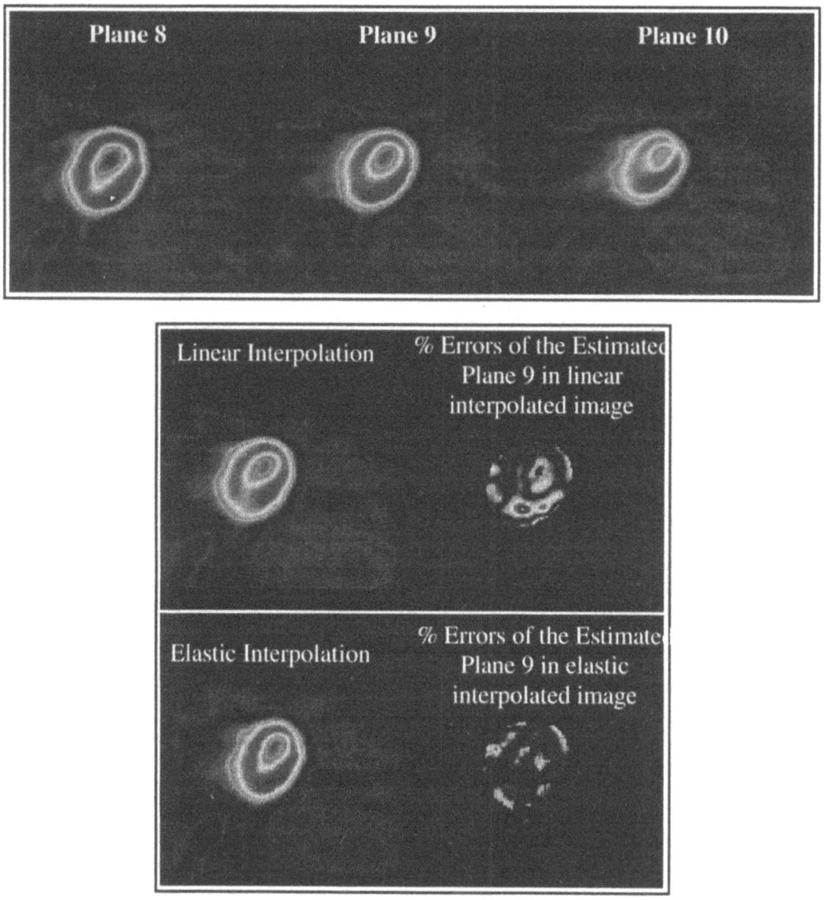

Figure 1: The results of comparing the linear interpolation method and the present elastic interpolation method for a normal cardiac PET FDG images. The plane-9 image has been interpolated based on the two adjacent image planes, plane-8 and plane-10, and has also been compared to the true plane-9 image. The maximum percent error is 29% by linear interpolation with respect to 4% by the elastic interpolation at the same location around the left ventricular myocardium.

AUTOMATICAL REGISTRATION OF BRAIN VOLUME DATASETS

Frithjof Kruggel[1], Peter Bartenstein[2]
Max-Planck-Institute for Cognitive Research, Leipzig, Germany[1]
Clinic for Nuclear Medicine, Klinikum rechts der Isar, Munich, Germany[2]

We describe an automatical two-step procedure for the intra- and intermodal registration of brain volume datasets. It combines the advantages of hull- and volume-based approaches, allows non-rigid matching and the detection of regions with suspected changes in time-series experiments.

The *first step* is based on an *alignment of the hulls* in both modalities by a variant of the well-known „chamfer-matching" approach. The hull of the source dataset (which is approximated by any number of points on the hull) is aligned to the hull of the reference dataset by a 3D rigid transformation M. We need to find the best set of coefficients in the matrix M, for which overall squared distance between the surfaces is minimized. This optimization process is performed by a combination of a genetic algorithm for global search and a local-gradient based method for faster convergence [3]. The source dataset is transformed by this optimized matrix and interpolated to give a volume dataset that is surface-aligned to the reference.

SPECT-MRI registration tasks often show convergence problems due to „soft transitions" between active and inactive regions in the functional datasets. We overcome this problem by introducing *fuzzy borders*. Instead of using only the border voxels we include a definable transition region in the optimization process, in which the contribution of each voxel is intensity-weighted.

To correct anisotropy effects in both modalities and to allow non-rigid matching, we use a *second step*, which is *volume-based* [1]. The source dataset is cut into small volumes, usually of 2x2x2 voxel size. We move these subvolumes around in a certain (local) region and let them find their best matching partner in the reference dataset. The matching criterion is the total sum of the squared intensity differences in the subvolumes. We thus get a set of displacement vectors for the subvolumes. To overcome heavy local distortions, a global filtering of the displacement vectors by a Kohonen network is performed. This iterative process runs until the global matching error is minimized. A postponed warping step yields a registered source dataset suitable for quantitative comparison with the reference.

If we leave out this filtering/warping step the displacement vectors denote regions of highest local mismatch. In MRI-MRI time-series examinations this

Y. Bizais et al. (eds.), Information Processing in Medical Imaging, 389–390.

mismatch stems either from non-linear movements (f.ex. in the brain stem) or from *changes due to an underlying disease process*. We apply the direction and length of these vectors to track resorption processes in intracranial hemorrhages or to monitor tumor sizes under therapy.

This two-step procedure has been implemented in the C++ language in the context of a multi-modal image editor for the neurologic clinic [2]. We have collected experiences with this algorithm during a 5 month evaluation period, in which we have run MRI-MRI, PET-MRI and SPECT-MRI registrations in about 50 different datasets. For a typical series of 32 axial slices we find computation times in the range of 10-20 minutes using a conventional SGI Indy workstation. The computation time of the „fuzzy border" version depends mostly on the size of the transition region, typical times are in the range of 100-150 minutes. A parallel formulation of the procedure seems reasonable and simple to implement.

[1] Y. Kosugi, M. Sase, H. Kuwatani, N. Kinoshita, T. Momose, J. Nishikawa, T. Watanabe „Neural network mapping for nonlinear stereotactic normalization of brain MR images", *Journal of Computer Assisted Tomography* 17, pp 455-460, 1993

[2] F. Kruggel, A. Horsch, G. Mittelhäußer, M. Schnabel. „Image processing in the neurologic sciences", *IEEE Workshop on Biomedical Image Analysis*, Seattle June 1994, pp 214-223, 1994

[3] L. H. Staib, X. Lee „Intermodality 3D medical image registration with global search", *IEEE Workshop on Biomedical Image Analysis*, Seattle June 1994, pp 224-233, 1994

THREE-DIMENSIONAL SPECT MYOCARDIAL REST - STRESS SUBTRACTION IMAGES AFTER AUTOMATED REGISTRATION AND NORMALIZATION

Catherine PERAULT, Hubert WAMPACH, Jean-Claude LIEHN
Department of Nuclear Medicine, Institut Jean Godinot, Reims, France

A method is proposed for the automated 3D comparison of Single Photon Emission Computed Tomographic (SPECT) stress and rest myocardial images. Firstly, the two voxel sets are geometrically registered by optimizing the correlation coefficient between the data over a myocardial volume of interest, using a steepest descent mixed with a grid search (Fig. 1).

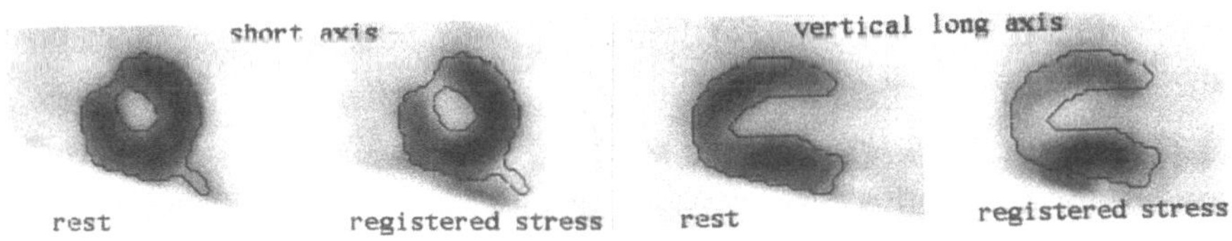

Fig. 1. Geometric registration : registration accuracy can be checked by using superimposed isocontours drawn from the reference rest image.

A two parameter grey level normalization is subsequently performed by optimizing a robust criterion, i.e. the number of stochastic sign changes (ssc) in the subtraction images [2], over the same volume of interest using a multi-scale grid followed by a simplex algorithm. Two parameters are clearly needed (Fig. 2). A robust method is also needed because normalization is performed between images which are different in case of myocardial ischemia.

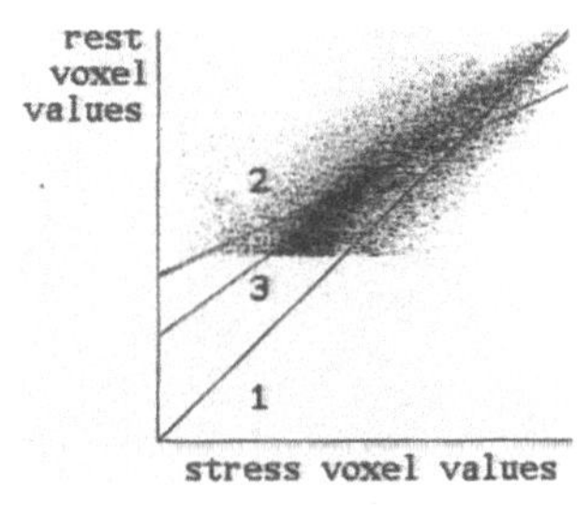

Fig. 2. Grey level normalization : correlation scatter plot between rest and stress myocardial values. Lines 1, 2 and 3 have been drawn with the parameter values obtained by normalizing to the maximum of each image, by linear regression and by using the ssc criterion respectively.

Subtraction volumes are then obtained from the differences between the rest and stress registered and normalized radial maxima surfaces (Fig. 3).

Y. Bizais et al. (eds.), Information Processing in Medical Imaging, 391–392.

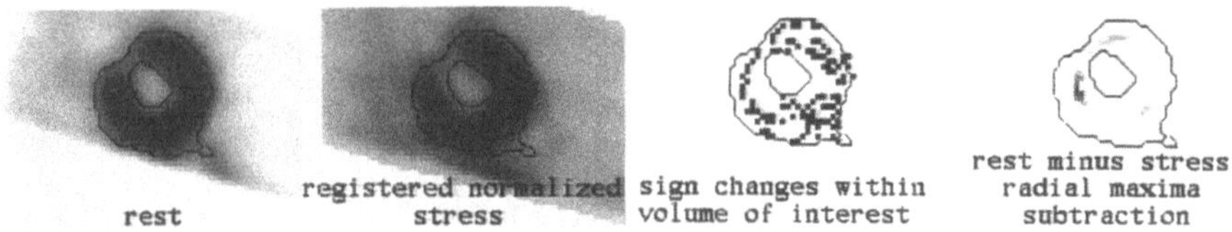

Fig. 3. Subtraction image after geometric registration and grey level normalization : registration and normalization accuracy can be checked by looking at the binary image of sign changes, where a random distribution of black and white voxels (sign change and no sign change) is expected everywhere except in the areas demonstrating differences, i.e. in the septal wall (around 9 o'clock) in this case.

Registered images and subtraction images are then visualized using a shape preserving 3D display[1], which shows the actual shape of the left ventricle myocardium together with the location, extent and shape of the areas demonstrating differences, as well as the magnitude of those differences between rest and stress data. This display needs a color scale. However, 3D black and white activity images are shown in Fig 4, without any depth shading. In conclusion, this method provides an objective operator-independent tool for the comparison of stress and rest myocardial images.

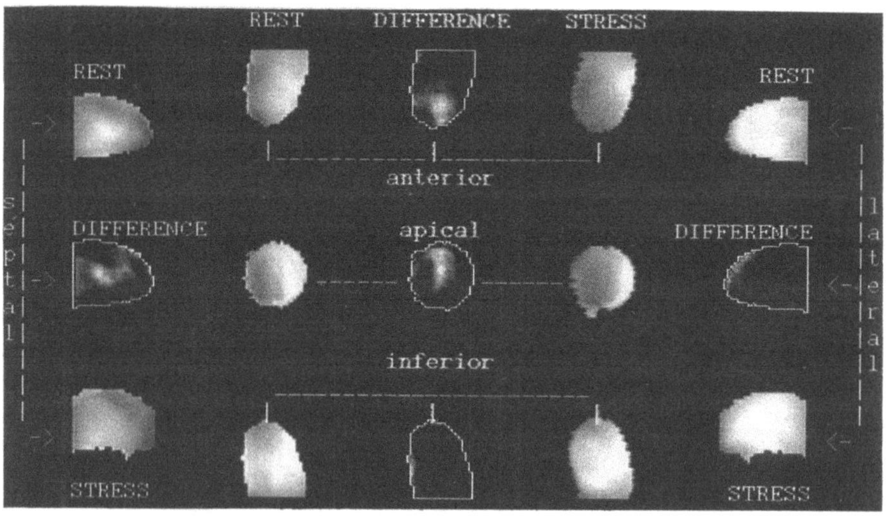

Fig. 4. 3D display of rest, registered normalized stress and rest minus stress subtraction images from five angles. Antero-septo-apical areas show a stress defect which is reversible at rest. These areas are clearly exhibited on the subtraction image.

References :

[1] A. Loboguerrero, C. Pérault, C. Gibold, J. Ouzan, T. Pron, A. Bouchard, A. Lepailleur, J.C. Liehn. "Shape preserving three-dimensional display of myocardial scintigraphic data", *Nucl Med Communications*, 15, pp 417-421, 1994

[2] A. Venot, J.C. Liehn, J.F. Lebruchec, J.C. Roucayrol. "Automated comparison of scintigraphic images", *J Nucl Med*, 27, pp 1337-1342, 1986

DESCRIPTION OF INHALED AEROSOL DEPOSITION IN THE LUNG USING A HEMISPHERICAL TRANSFORM OF MULTIMODALITY IMAGING DATA

John S. FLEMING[1], Peter HALSON[1], Joy CONWAY[1], Elizabeth MOORE[1], Michael A. NASSIM[2], Adel H. HASHISH[3], Adrian G. BAILEY[3], Stephen T. HOLGATE[1], Theodore B. MARTONEN[4]

Medical Physics and University Medicine, Southampton General Hospital, U.K.[1]

Clinical Pharmacology, Fisons plc, Loughborough, U.K.[2]

Department of Electrical Engineering, Southampton University, U.K.[3]

U.S. Environmental Protection Agency, Research Triangle Park, U.S.A.[4]

1. INTRODUCTION

Measurement of pulmonary deposition of aerosol is important in drug inhalation therapy and in studies on health effects of atmospheric particulate pollutants. Planar radionuclide imaging has been used extensively for this purpose and more recently three dimensional assessments of deposition distribution have been obtained using single photon emission computed tomography (SPECT). The use of aligned anatomical images enables both improved accuracy of quantification and localisation of deposition to be achieved. This paper describes methods of analysing spatial depostion patterns using data from multi-modality imaging.

2. METHODS

SPECT imaging of the 3D distribution of a radiolabelled aerosol in the lung was carried out using a dual head gamma camera on twelve normal subjects. On a separate occasion MR images of the thorax were obtained. The two sets of images were aligned based on the alignment of skin surface point positions. Two SPECT studies were carried out on each subject in order to compare the deposition patterns of two different nebulisers which produced particles of different size. The SPECT data were analysed using a hemispherical transform using definition of the lung outline and the position of the 'centre' of the lung from aligned MR data. Each voxel in the lung space was assigned polar angles θ and ϕ relative to the lung centre as the origin (Figure 1). The line between the lung centre and the origin was then extrapolated to the lung periphery and the fractional radial distance (r) of the voxel from the origin along this line

Y. Bizais et al. (eds.), Information Processing in Medical Imaging, 393–394.

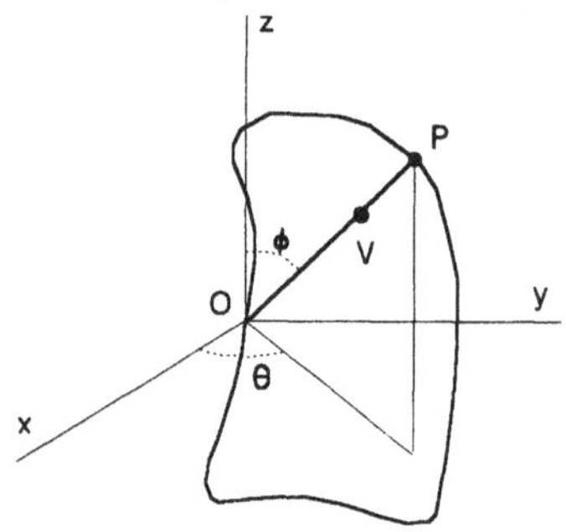

Figure 1. Diagram illustrating the formation of the hemispherical transformation

calculated. This allowed reconstruction of the SPECT data as a hemisphere. This standard transformation enabled data on each aerosol to be averaged and compared. The one dimensional variation of deposition with radial distance was also assessed and analysed quantitatively to give a 3D central to peripheral concentration ratio. A conceptual model of the three dimensional morphology of the airways allowed further transformation of the data to give deposition by airways generation. The results of the above analysis were compared to those from planar image analysis.

3.0 RESULTS

The three dimensional analysis of SPECT deposition data using the hemispherical transform yielded considerably more qualitative information on the deposition patterns of the two aerosols than was available from planar imaging. The 2D central to peripheral concentration ratios were 2.61 and 2.03 for the coarse and fine aerosols respectively, the corresponding 3D ratios being 2.19 and 0.93. When calculated as actual airway concentration ratios between conducting and alveolated airways the values were 5.27 for the coarser aerosol and 2.43 for the fine.

4.0 CONCLUSIONS

Multimodality imaging and the subsequent analysis based on the hemispherical transform considerably enhanced the differences between the distribution patterns of the two aerosols in the lung as compared to planar imaging. It provided new data on aerosol deposition which should be of value in drug inhalation therapy.

CHARACTERIZATION OF THE RADIAL CYCLIC MOTION OF THE LEFT VENTRICLE AS RESTORED FROM ECHOCARDIOGRAPHIC DATA

Lars FLOREBY, Göran SALOMONSSON
Signal processing group, Lund University, Sweden

1. INTRODUCTION

In a short-axis echocardiogram of the left ventricle, an essentially contracting/expanding cyclic motion can be observed. Continuous evaluation of such a moving image, e.g. during surgical operation, can be an awkward task, which may also be fatiguing if it is done during long time. Thus, there is a need for automated and objective evaluation of the echocardiographic image.

2. ESTIMATION OF RADIAL MOTION

Our method for motion estimation was based on the idea of *optical flows* introduced by Horn and Schunck [1]. When Mailloux et al. [2] tested this idea on echocardiograms, the resulting velocity field showed to be essentially radial towards some point inside the ventricle. Thus, we restricted the method to restoration of a *scalar* (radial) velocity field computed on a polar grid covering a *circular region of interest* (ROI) in the image. In continuous time and space, the functional minimized by such a (time-varying) field is given by

$$J_{of,r}[v] = \iint \left(\left(\frac{\partial I}{\partial r} v + \frac{\partial I}{\partial t} \right)^2 + \alpha^2 \left(\frac{\partial v}{\partial r} \right)^2 + \beta^2 \left(\frac{\partial v}{\partial \phi} \right)^2 \right) dr d\phi. \qquad \text{(Eq. 1)}$$

where I is image intensity, v is the radial velocity while the parameters α and β control the smoothness of the resulting field. The discretized solution is found by the iteration

$$v^{(n+1)} = \frac{2(\alpha^2 \overline{v}_r^{(n)} + \beta^2 \overline{v}_\phi^{(n)}) - I_r I_t}{I_r^2 + 2\alpha^2 + 2\beta^2} \qquad \text{(Eq. 2)}$$

where $\overline{v}_r$ and $\overline{v}_\phi$ are *local averages* of v.

Y. Bizais et al. (eds.), Information Processing in Medical Imaging, 395–396.

3. CHARACTERIZATION OF CYCLIC MOTION

In order to reduce the amount of data obtained from the motion estimation, the "momentum" of eight different sectors of the heart wall were computed as sums of the radial velocities weighted by the intensity values of the corresponding image pixels. By carrying out this procedure for each frame, a collection of time series describing the ventricular motion could be aquired without estimating the actual location of the wall. In our example, three echocardiographic sequences were used as described by Table 1.

Sequence Nr	Frames	Heart rate (fr^{-1})	Complete cycles
#1	42	1/24	1
#2	96	1/24	3
#3	96	1/19	4

Table 1: Contents of three echocardiographic sample sequences from one patient.

In Fig. 1, we see the coefficients of a least squares sinusoidal fit applied to each cycle of the momentum time series described above. Such a model has two parameters, equivalent to an amplitude and a phase delay in relation to the ECG peak. From the plots, it can be concluded that the radial motion in most cases is "minus sine" ($A \approx 0$, $B < 0$), which corresponds to a contracting motion following immediately after the ECG excitation.

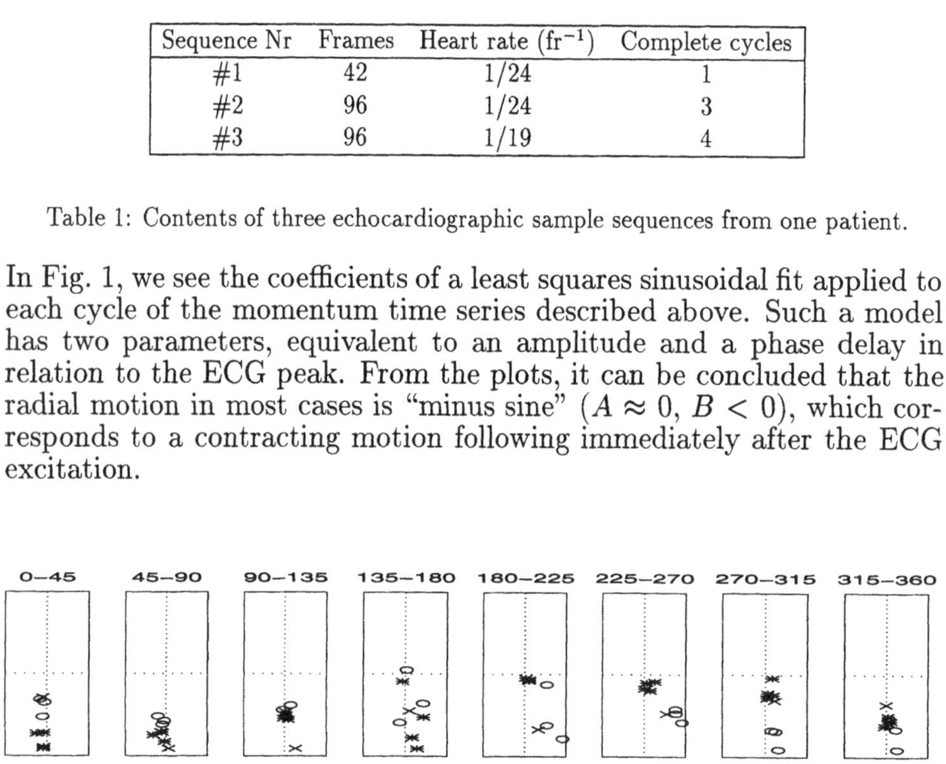

Figure 1: Coefficients of the least squares fit $A \cdot \cos(2\pi n/N) + B \cdot \sin(2\pi n/N)$ to radial momentum cycles. Sequence #1 – $\times$, #2 – o, #3 – $*$ (horizontal – A, vertical – B).

REFERENCES

[1] B. K. P. Horn & B. G. Schunck, "Determining optical flow", *Artificial Intelligence*, Vol 17, pp 185–203, 1981.

[2] G. E. Mailloux, A. Bleau, M. Bertrand & R. Petitclerc, "Computer analysis of heart motion from two-dimensional echocardiograms", *IEEE Trans on Biomed Eng*, Vol 34 No 5, pp 356–364, 1987.

A MULTIRESOLUTION APPROACH TO FLOW FEATURE FROM PHASE CONTRAST MRA

Abhir H. BHALERAO[1], Paul E. SUMMERS[2]
Medical Physics, St. Thomas' Hospital, London, UK[1]
Radiological Sciences, UMDS, Guy's Hospital, London, UK[2]
a.bhalerao@umds.ac.uk p.summers@umds.ac.uk

Phase contrast magnetic resonance angiography provides a uniquely rich information content in a relatively short imaging time. In effect, not one, but four image data sets are obtained - a T1 weighted anatomical image, and three images of orthogonal velocity components. The information is only crudely communicated in MIP views which are based on the square root of the sum of squares of the velocity component images. Similarly, surface rendered views on the vessels represent only a thresholding of the net velocity (figure 1(a)). In particular, directionality and magnitude of velocity are not conveyed to the viewer. To incorporate the function information into an image which also portrays the spatial arrangement of vessels is a difficult task with the large volumes of data which may be present.

This work considers the automatic extraction of a concise symbolic representation of vasculature (e.g. [3]). The segmentation algorithm employs a multiresolution 3D curve extraction method based on an image model where the centre lines of blood vessels are modelled as interconnected, piece-wise linear, 3D curves [2]. Parameters of the model, and hence the segmentation, are obtained by first expanding the image data into an oct-tree representation. Curves segments are identified across a number of voxel resolutions(figure 1(b)), which represent the current best fit of the data to the model. Probable curves are then inferred through these 'trace' points by means of an adjacency graph, which form the basis of the symbolic description (figures 1(c)(d)). The method has been demonstrated on real MRA data to be both robust and accurate, and its computational cost is low (equivalent to less than a $3 \times 3 \times 3$ filter kernel) [1].

This model-based approach makes direct use of the vector magnitude and direction information to form velocity averages over a succession of scales. The resulting oct-tree representation in turn is well suited to further extensions and manipulations for display and processing of the data. Moreover, the data volume is reduced in accordance with the sparsity of the vascular network making subsequent manipulations more efficient than is possible when handling the full data volume. Thus far, the output segmentation has been sucessfully employed to anisotropically filter the original data, having first derived estimates of vessel calibre along the extracted curves. This has proved effective for noise supression in 'limited' MIP displays. Experimentation is currently in progress to determine

Y. Bizais et al. (eds.), Information Processing in Medical Imaging, 397–398.

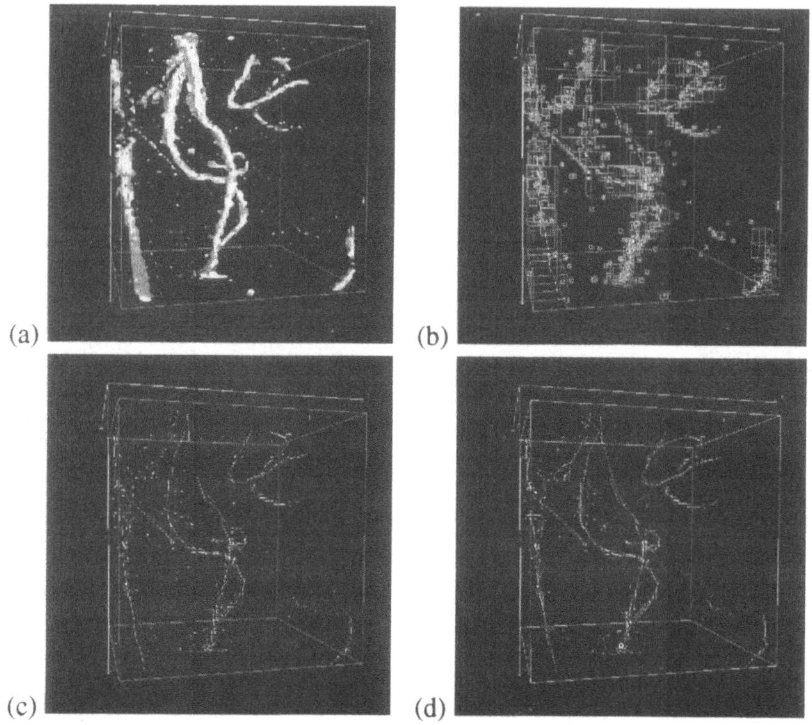

Figure 1: Results on one 64^3 part of head data set

relative pressure gradients from time series PCA data, which are subsequently visualised as colour overlays on a surface rendered view.

REFERENCES

[1] A. Bhalerao and P. Summers. "Visualisation and Segmentation of 3D Phase Contrast Angiography Images". Technical report, Department of Medical Physics, St. Thomas' Hospital, London SE1 7EH, July 1994.

[2] A. Calway and R. Wilson. "Curve Extraction In Images Using the Multiresolution Fourier Transform". In *Proc. IEEE Conf. Acoust., Speech, and Signal Processing*, pages 2129–2132. IEEE, 1990.

[3] G. Gerig, Th. Koller, G. Székely, Ch. Brechbühler and O. Kübler. "Symbolic Description of 3-D Structures Applied to Cerebral Vessel Tree Obtained from MR Angiography Volume Data". In H. H. Barret and A. F. Gmitro (Eds.), editors, *Lecture Notes in Computer Science, NO. 687, Information Processing In Medical Imaging IPMI'93*, pages 94–111. Springer Verlag, 1993.

SIMULATION OF MRI CLUSTER PLOTS

Andrew SIMMONS [1], Simon R. ARRIDGE [2], Gareth J. BARKER [3]
Department of Neurology, Institute of Psychiatry, London, UK [1]
Computer Science Department, University College London, London, UK [2]
NMR Research Group, Institute of Neurology, London, UK [3]

1. INTRODUCTION

Segmentation of MRI data has become increasingly widespread, with methods based upon cluster analysis techniques [1] being the most popular, yet little attention has been paid to how the choice of images affects the quality and ease of segmentation. We have developed a realistic simulation of MRI cluster plots which allows an informed choice of pulse sequence, acquisition parameters and data pre-processing for cluster analysis to be made.

2. METHODS AND RESULTS

The simulation uses a multi-slice multi-compartmental anthropomorphic software model of anatomy with each compartment characterised by a number of MRI parameters such as mean proton density (PD), T_1, and T_2. The simulation is controlled by additional parameters such as slice thickness, field strength, TR and TE. Using the anthropomorphic model and the various parameters, simulated MR images are created which are then used to construct a cluster plot.

Image Contrast has been modelled for the full range of standard pulse sequences. The choice of pulse sequence and acquisition parameters strongly influences image contrast which is reflected by the separation of the centre of image clusters. It is therefore possible to greatly alter cluster separation.

Signal to Noise Ratio (SNR) is varied with a wide range of acquisition parameters and is one component affecting the compactness of clusters. It has important implications for the possibility of cluster overlap.

Field Strength : The variation of T_1 with field strength has been modelled [2]. Although this has an affect on image contrast, the associated variation in SNR is more important. Higher field strengths give significant advantages in this respect, so cluster analysis results at high field strenths may not be reproducible

Y. Bizais et al. (eds.), Information Processing in Medical Imaging, 399–400.

in reasonable imaging times at lower field strengths.

Tissue Heterogeneity is modelled using data on the measured mean and standard deviation of tissue T_1, T_2 and PD. Tissue heterogeneity affects cluster compactness and leads to approximately elliptical clusters with orientation dependent upon the pulse sequence(s) and timing parameters used.

Correlation Between T_1, T_2 and PD is simulated using conditional probability distribution functions. Correlation affects the shape and orientation of clusters, in a similar way to tissue heterogeneity, as well as their position.

The Partial Volume Effect is modelled by combining multiple model slices to produce a single simulated image slice. Partial volume voxels are represented by broad bands of pixels between cluster centres.

Image Non-Uniformity is modelled by multiplying simulated image intensity values by normalised scans of uniform oil phantoms for different RF coils. Image non-uniformity leads to clusters being 'stretched'.

Image Pre-Processing : The simulation includes the ability to apply an array of different pre-processing techniques [3] such as non-uniformity correction or anisotropic smoothing.

3. CONCLUSIONS

The simulation provides a valuable tool for optimising the quality of intensity-based MRI segmentation, whether this is done by classical supervised or unsupervised classifiers, artificial neural networks or other means.

[1] M.W. Vannier, R.L. Butterfield, D. Jordan, W.A. Murphy, R.G. Levitt, M. Gado. "Multispectral analysis of magnetic resonance images", *Radiology*, 154, pp 221-224, 1985

[2] H.W. Fischer, P.A. Rinck, Y. van Haverbeke, R.N. Muller. "Nuclear relaxation of human brain grey matter and white matter", *Mag Res Med*, 16, pp 317-334, 1990

[3] A. Simmons, S.R. Arridge, G.J. Barker, A.J. Cluckie, P.S. Tofts. "Improvements to the quality of MRI cluster analysis", *Mag Res Im*, 12, pp 1191-1204, 1994

PRECISION AND ACCURACY OF A HIGH DIMENSIONAL TRANSFORMATION AND SEGMENTATION OF MR IMAGES OF THE HIPPOCAMPUS

John W. HALLER, Gary E. CHRISTENSEN, Sarang C. JOSHI, Mokhtar GADO, Michael I. MILLER, and Michael W. VANNIER

Washington University School of Medicine
Mallinckrodt Institute of Radiology
and Department of Psychiatry
4940 Children's Place, St. Louis, MO 63110 USA

1. GOALS

The goals of the present study were to determine the precision (repeatability) and accuracy (validity) of a high dimensional image transformation method for the segmentation and measurement of magnetic resonance (MR) neuroimages. A related objective was to compare the high dimensional image transformation and segmentation method to manual segmentation of 2-D MR images of the hippocampus.

2. DESIGN

MR images of five subjects (2 normal controls and 3 schizophrenic subjects) were acquired using a 3-D gradient echo sequence on a Siemens 1.5 Tesla scanner (Siemens Medical Systems, Inc., Iselin, New Jersey). Comparable parasagittal slices of the hippocampus were selected from MR images, and area measurements were made from segmentations of each subject's hippocampus. Both manual segmentations and atlas-based segmentations were performed twice.

Atlas-based transformations of a segmented atlas image to individual MR images were carried out in two steps. In the first step, control points were designated in pairs of images (the atlas and subject's image) and the points were used to provide a global transformation of the image. The globally transformed image was then locally transformed using a high-dimensional, fluid

Y. Bizais et al. (eds.), Information Processing in Medical Imaging, 401–402.

transformation [1,2]. High dimensional transformations of a 2-D "digital atlas" to individual MR brain images were used to segment the hippocampus [2, 3]. Thus, the expertly segmented atlas-template is precisely transformed to the individual's MR brain image, resulting in an automatic segmentation of the latter. Precision and accuracy of manual and atlas-based segmentations of hippocampal areas were determined.

3. RESULTS

The results demonstrate superior repeatability of MR image area measurements using the atlas transformation method. The intra-class correlation (ICC) for the repeatability of the high dimensional transformation and segmentation was 0.998. The ICC for manual segmentation by a single rater (intra-rater ICC) was 0.995. The repeatability between the methods (inter-method ICC) was 0.955.

4. CONCLUSIONS

The results demonstrate that hippocampal morphometry is feasible by a fluid transformation of a digital brain atlas to an individual's MR image. The results demonstrate the high repeatability of manual segmentation and segmentations provided by high dimensional transformations of an atlas template. The validity of high dimensional transformations was demonstrated by a high correlation with manual segmentations. We conclude the quantitative morphometry of brain substructures (e.g. hippocampus) is feasible by use of a high dimensional transformation of a digital brain atlas to an individual MR image. The atlas transformation method automates the search for neuromorphological correlates of schizophrenia by a new mathematically robust method with *unprecedented sensitivity to small local and regional differences.*

[1] G. E. Christensen , R.D. Rabbitt, M. I. Miller. "3D brain mapping using a deformable neuroanatomy," *Phys. Med. Biol.* 39, pp 609-618, 1994

[2] G. E. Christensen , R.D. Rabbitt, M. I. Miller, S. C. Joshi, U. Grenander, T.A. Coogan, D.C. Van Essen, "Topological properties of smooth anatomic maps," In Proceedings of IPMI95, Ile de Berder, France, June 26-30, 1995

[3] J. Haller G. E. Christensen, M. I. Miller, S.C. Joshi, M. Gado , J. Csernansky, M. Vannier "A Comparison of Automated and Manual Segmentation of Hippocampus MR images," Proc. SPIE in press, 1995.

[4] M. I. Miller , G.E. Christensen , Y. Amit, , U. Grenander, "Mathematical textbook of deformable neuroanatomies," *Proc. Nat. Acad. Sci.* 90, pp 11944-11948, 1993

QUANTITATIVE MEASUREMENT OF BLOOD VESSELS BASED ON CONE-BEAM CT IMAGES

Noboru NIKI[1], Yoshiki KAWATA[1], and Tatsuo KUMAZAKI[2]
Department Information Science, University of Tokushima, Japan[1]
Department of Radiology Nippon Medical School, Japan[2]

1. INTRODUCTION

We present a prototype system for three-dimensional (3D) image analysis of blood vessels using cone-beam CT. The purpose is to enhance the performance of clinician in assessing anatomical information such as the orientation of blood vessels branches, cross-sectional area, and aneurysm volume and morphology.

2. METHOD

The prototype system consists of two parts, the 3D blood vessels image reconstruction and its analysis. The Digital angiograms collected by an X-ray rotational angiographic system [1] are approximately reconstructed into 320x320x320 voxels (a voxel size 0.5 mm) by the short scan cone-beam filtered backprojection algorithm [2],[3]. The analysis of the 3D blood vessels image is based on a thinning algorithm for a 3D binary image of blood vessels and a procedure for a graph description of the centerlines of blood vessels [4]. Additionally, the curved surfaces of the blood vessels are quantified by computing surface curvatures of the 3D blood vessels image. Using these 3D image processing techniques, the following analysis functions are realized: (a) 3D visualization of the exterior structure and the interior morphology of the blood vessels, (b) extraction of the blood vessels branches orientation, (c) measurement of the cross-sectional area, (d) measurement of the distance between two points along a 3D blood vessel path, (e) evaluation of the aneurysm volume and morphology, (f) representation of the surfaces shapes of the blood vessels.

3. RESULTS

We applied our analysis functions to a patient's abdominal blood vessels with two aneurysms and a stenosis. A complex 3D shape and an internal morphology of the blood vessel were visualized (Fig.1), and quantitative information was obtained.

Y. Bizais et al. (eds.), Information Processing in Medical Imaging, 403–404.

4. CONCLUSION

Results show that the prototype system may be useful for providing 3D qualitative and quantitative information of blood vessels, which are demanded for diagnosis and therapy.

REFERENCES

[1] T. Kumazaki, " Rotational stereo-digital angiography-image processing and clinical usefulness," *Med.Image.Tech.*, Japan, 7, pp.433-439, 1989.

[2] N. Niki, T. Mineyama, H. Satoh, C. Uyama, and T.Kumazaki, " 3-D reconstruction of blood vessels using a rotational angiopraphy system," *Med. and Biol. Eng. and Comput.*, 29, p.649, 1991.

[3] Y. Kawata, N. Niki, H. Satoh, and T. Kumazaki, " Three-dimensional blood vessels reconstruction using a high-speed X-ray rotational projection system," *Trans. IEICE*, Japan, J76-D-II, pp.2133-2142, 1993.

[4] N. Niki, Y. Kawata, and T. Kumazaki, " Three-dimensional image analysis of blood vessels using cone-beam CT," *1994 IEEE Nuclear Science Symp. and medical Imaging Conf.*, Norfolk, in press, 1994.

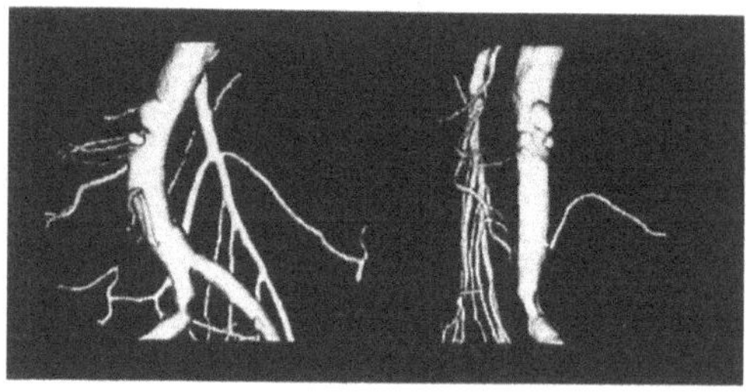

(a)

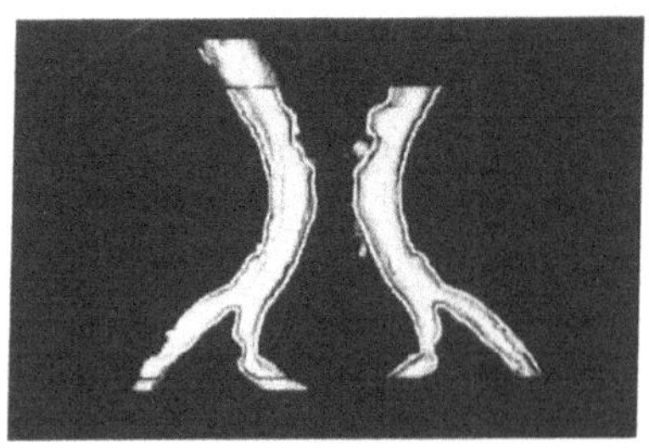

(b)

Fig.1 3D visualization of the exterior structure and the interior morphology of the blood vessels. (a) 3D displays (from two different views) of the concerned blood vessel region extracted from the 3D reconstructed image of the patient's abdomen. (b) 3D longitudinal dissection views of the selected blood vessel image with two aneurysms and a stenosis.

THREE-DIMENSIONAL ELECTRONIC ATLAS OF HUMAN CEREBRAL DEEP STRUCTURES

Wieslaw L. NOWINSKI, Anthony FANG, Shiaofen FANG, Bonnie T. NGUYEN, Raghu RAGHAVAN
Institute of Systems Science, National University of Singapore, Singapore 0511

1. MOTIVATION

Three-dimensional electronic brain atlases overcome limitations of paper brain atlases as well as HyperCard type electronic brain atlases. Nonetheless, existing 3D MRI/CT-based brain atlases have still several shortcomings including coarse anatomical parcellation, low or lack of interactivity particularly for high quality rendering, and/or lack of quantification. For example, the thalamus occupying anatomically and functionally a central position in the brain is treated as a single object. On the other hand, one can delimit more than 150 subdivisions of the human thalamus. Moreover, the constituents of the thalamus and basal ganglia are far from being clearly discernible on MRI images.
In order to fill the existing gap, we have constructed an electronic 3D atlas of human cerebral deep structures. The key requirements of the atlas include detailed parcellation of cerebral deep structures, interactivity, high quality rendering, quantification, registration, and consistency with the other brain atlases and tools developed by our group [3].

2. APPROACH

In order to construct a 3D Talairach-Tournoux (TT) brain atlas [5], we developed a new modelling technique called *zero-potential surface smoothing* [1]. In this paper we propose a more powerful approach for modelling of cerebral structures based on novel geometric modelling combined with volumetric modelling. The geometric model defines the geometry of anatomical structures, and is used for 3D smooth multi-surface display, manipulation, and quantification. The volumetric model labels the space occupied by the cerebral deep structures, and allows to display orthogonal planes, cutting planes and triplanar, and to generate volume rendered images.
We introduce a novel NURBS based modeller which supports smooth branching and provides tools for quantitative analysis, generation of the volumetric model (inside/outside test), and cross-section (contour) re-generation. The initial surface models are reconstructed from contour data, and can later be further manipulated, deformed and/or resliced. The contour data are based on the Schaltenbrand-Wahren (SW) brain atlas [4] which provides a finer parcellation of the thalamic region than the TT atlas does. The data preparation process includes contouring and labelling of the SW coronal

Y. Bizais et al. (eds.), Information Processing in Medical Imaging, 405–406

microseries, 3D anatomical correction of the SW atlas data, geometric modelling, sampling, and mirroring. The volumetric model is generated from the geometric model by sampling the space. The volume sampling rate is controlled by the user.
The 3D atlas is preregistered with the TT atlas similarly to [2]. The MRI-TT registration determines the deformation of the standard atlas geometric model into the personalized atlas geometric model for which, in turn, the personalized atlas volumetric model is generated.

3. IMPLEMENTATION

The atlas system has been implemented in C++ on the Reality Engine Silicon Graphics. Its main components include: neurobrowser, anatomical index, geometric model, volumetric model, contents table, material properties table and text book. *Neurobrowser* displays the hierarchy of anatomical structures. The selected structures are highlighted by *Neurobrowser* and their names are put into *Contents Table*. This table controls *Anatomical Index*, *Geometric Model*, *Volumetric Model*, *Material Properties Table* and *Text Book*. *Anatomical Index* contains the complete index of the SW atlas (about 500 entries) and an extended index of the TT atlas (about 200 entries including the Brodmann areas and gyri). The structures can be queried by name or location. The semi-transparent, multi-surface display of structures is controlled by *Material Properties Table*. This table is set by the user (e.g., the color palette) and *Neurobrowser* (the surface opacity depending on the hierarchy of selected structures). Continuous navigation in the multi-atlas/MRI space [3] is provided.

The approach presented here is suitable for constructing the 3D volumetric TT and SW brain atlases, consistent in all three orthogonal orientations.

REFERENCES

[1] Fang A., Nowinski W. L., Nguyen B. T., Bryan R. N.: "Three-dimensional Talairach-Tournoux brain atlas". *Proc. SPIE Medical Imaging 1995: Image Display*, San Diego, USA, Feb. 1995, SPIE Vol. 2431.

[2] Nowinski W. L., Fang A., Nguyen B. T.: "Schaltenbrand-Wahren/Talairach-Tournoux brain atlas registration". *Proc. SPIE Medical Imaging 1995: Image Display*, San Diego, USA, Feb. 1995, SPIE Vol. 2431.

[3] Nowinski W. L., Fang A., Nguyen B. T., Raghavan R., Bryan R. N., Miller J.: "Talairach-Tournoux/Schaltenbrand-Wahren based electronic brain atlas system". *Lecture Notes in Computer Science*, Vol. 905, 1995, Proc. CVRMed'95, 257-261.

[4] Schaltenbrand G. and Wahren W.: *Atlas of Stereotaxy of the Human Brain*. Georg Thieme Verlag, Stuttgart 1977.

[5] Talairach J. and Tournoux P.: *Co-Planar Stereotactic Atlas of the Human Brain*. Georg Thieme Verlag, Stuttgart 1988.

KNOWLEDGE-BASED INTERPRETATION OF MR BRAIN IMAGES

Milan SONKA[1], **Satish TADIKONDA**[3], **Steve M. COLLINS**[1,2]
Departments of [1]*Electrical and Computer Engineering, and* [2]*Radiology*
The University of Iowa, Iowa City IA 52242, U.S.A., and
[3]*Advanced Imaging Systems, Medical Products Group*
Hewlett Packard Company, Andover, MA 01801, U.S.A.
e-mail: milan-sonka@uiowa.edu

Abstract:
With increased performance and accuracy of CT/MR/PET registration techniques, the functional analysis of the brain requires a larger and larger number of neuroanatomic structures to be accurately identified and labeled (named) in three-dimensional images. The need for robust semi-automated or fully automated brain image interpretation techniques is clear. We have developed a method for fully automated segmentation and labeling of 17 neuroanatomic structures in individual two-dimensional MR images (corpus callosum, fornix, third ventricle, cistern, pineal gland, left/right ventricles, left/right caudate nucleus, left/right internal capsule, left/right putamen, left/right globus pallidus, left/right thalamus).

Our method is based on a *hypothesize and verify* principle and uses a genetic algorithm optimization technique to generate and evaluate image interpretation hypotheses in a self-improving feedback loop. The method begins with a primary segmentation step that divides the original image into a large number of primary regions. The segmentation parameters are chosen to over-segment the image. A primary region adjacency graph is constructed that describes the properties of each primary region as well as its relationships with its neighbors. A genetic algorithm technique is used to generate a population of image interpretation hypotheses that are represented as linear strings of symbols. Each primary region is associated with a position in the string and is assigned a label corresponding to the object to which it is hypothesized to belong. The confidence that a particular hypothesis is correct is evaluated by calculating an objective function that compares the information contained in the image data it-

Y. Bizais et al. (eds.), Information Processing in Medical Imaging, 407–408.

self to a priori knowledge about image regions and objects, their shapes, positions, and relationships. The genetic algorithm serves to retain good hypotheses and eliminate poor hypotheses so that over a series of iterations the process converges to the optimal image interpretation.

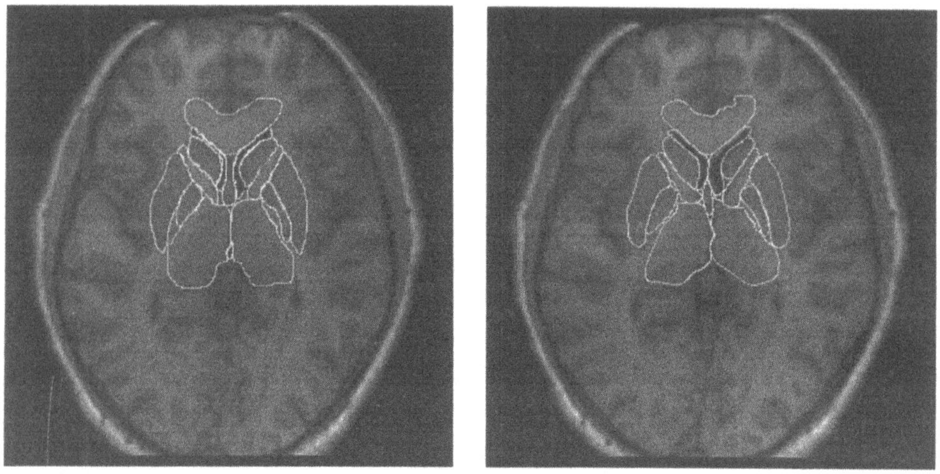

Figure 1: Automated segmentation and labeling of MR brain images from the testing set. Observer-defined borders of the neuroanatomic structures (left) correspond closely with computer-defined borders (right).

The method was trained in 20 individual T1-weighted MR images. Observer-defined contours of neuroanatomic structures were used as a priori knowledge. The method's performance was tested in 8 MR images by comparison to observer-identified independent standards. The genetic image interpretation method correctly interpreted neuroanatomic structures in all images from the testing set. Computer-identified and observer-defined neuroanatomic structure areas correlated very well ($r = 0.99, y = 0.95x - 2.1$). Border positioning errors were small, with a root-mean-square border positioning error of 1.5 ± 0.6 pixel.

Our genetic image interpretation method represents a novel approach to image interpretation. The method has several favorable properties in comparison to conventional methods. These arise from the robustness of the genetic optimization algorithm and its incorporation of segmentation and object interpretation into a single segmentation/interpretation process.

Computational Imaging and Vision

1. B.M. ter Haar Romeny (ed.): *Geometry-Driven Diffusion in Computer Vision.* 1994 ISBN 0-7923-3087-0
2. J. Serra and P. Soille (eds.): *Mathematical Morphology and Its Applications to Image Processing.* 1994 ISBN 0-7923-3093-5
3. Y. Bizais, C. Barillot, and R. Di Paola (eds.): *Information Processing in Medical Imaging.* 1995 ISBN 0-7923-3593-7

Kluwer Academic Publishers – Dordrecht / Boston / London

FSC
www.fsc.org
MIX
Papier | Fördert
gute Waldnutzung
FSC® C083411

Kolibri 360 GmbH
Ferdinand-Jühlke-Straße 7
99095 Erfurt, Deutschland
produktsicherheit@kolibri360.de